Boundary Value Problems
in Linear Viscoelasticity

J. M. Golden   G. A. C. Graham

Boundary Value Problems in Linear Viscoelasticity

With 13 Figures

Springer-Verlag Berlin Heidelberg GmbH

Dr. *John M. Golden*

Roads Division, National Institute for Physical Planning
and Construction Research, St. Martin's House, Waterloo Road,
Dublin 4, Ireland

Professor Dr. *George A. C. Graham*

Department of Mathematics and Statistics, Simon Fraser University,
Burnaby, B.C. Canada V5A 1S6

ISBN 978-3-662-06158-9   ISBN 978-3-662-06156-5 (eBook)
DOI 10.1007/978-3-662-06156-5

Library of Congress Cataloging-in-Publication Data. Golden, J.M. (John M.), 1945 – . Boundary value problems in linear viscoelasticity. Bibliography: p. Includes index. 1. Viscoelasticity. 2. Boundary value problems. I. Graham, G.A.C. (George A.C.), 1939 – . II. Title. QA931.G585 1988 531'.3823 88-4915

© Springer-Verlag Berlin Heidelberg 1988
Originally published by Springer-Verlag Berlin Heidelberg New York in 1988.
Softcover reprint of the hardcover 1st edition 1988

Typesetting: K + V Fotosatz GmbH, 6124 Beerfelden
2153/3150-543210

It is a pleasure to dedicate this book to
Professor *I. N. Sneddon*, O.B.E., F.R.S.
It would not have come into existence had it
not been for his encouragement and advice
on general content.

Preface

The classical theories of Linear Elasticity and Newtonian Fluids, though triumphantly elegant as mathematical structures, do not adequately describe the deformation and flow of most real materials. Attempts to characterize the behaviour of real materials under the action of external forces gave rise to the science of Rheology. Early rheological studies isolated the phenomena now labelled as viscoelastic. Weber (1835, 1841), researching the behaviour of silk threats under load, noted an instantaneous extension, followed by a further extension over a long period of time. On removal of the load, the original length was eventually recovered. He also deduced that the phenomena of stress relaxation and damping of vibrations should occur. Later investigators showed that similar effects may be observed in other materials. The German school referred to these as "Elastische Nachwirkung" or "the elastic aftereffect" while the British school, including Lord Kelvin, spoke of the "viscosity of solids". The universal adoption of the term "Viscoelasticity", intended to convey behaviour combining properties both of a viscous liquid and an elastic solid, is of recent origin, not being used for example by Love (1934), though Alfrey (1948) uses it in the context of polymers.

The earliest attempts at mathematically modelling viscoelastic behaviour were those of Maxwell (1867) (actually in the context of his work on gases; he used this model for calculating the viscosity of a gas) and Meyer (1874). The model proposed by Meyer is generally associated with the names of Kelvin and Voigt who however made their contributions much later. These differential constitutive relations are discussed in section 1.6. Boltzmann (1874) proposed the general hereditary integral form of the constitutive relations which is the basis of most theoretical work on viscoelastic materials of the past three decades, though, historically, it was slow in gaining acceptance against the more specialized and cumbersome differential constitutive relations based on mechanical models, which essentially generalize the work of Maxwell and Meyer. Boltzmann gives the form applicable to isotropic bodies while Volterra (1909) gives the general anisotropic form. Volterra's theory of functionals (1959) is also at the basis of the modern formulation of Viscoelasticity in the general non-linear case. Further discussion of the early history of the subject may be found in the interesting article by Markovitz (1977), and in the book by Love (1934). Until the nineteen fifties, development of the subject was slow. The emergence into common use of a large variety of polymeric materials in the post-war years focussed increasing attention on the topic. References which focus on polymeric materials and their behaviour include Eirich (1956), Staverman and Schwarzl (1956), Ferry (1970) and (more recently) Doi and Edwards (1986). The application of the theory to metals has been surveyed by Zener (1948), while Arutyunyan (1952) and Bazant (1975) survey literature that treats concrete as an aging viscoelastic material.

Viscoelastic boundary value problems have been actively researched now for more than thirty years. During the nineteen fifties, attention centred mainly on the Classical Correspondence Principle and problems to which it was applicable (e.g. see Hunter (1960) and Lee (1960)). Extensions of this principle were discovered during the sixties. Also, several problems not covered by the classical form of the principle received attention. These were of two kinds, extending crack problems and contact problems where the load on the indentor is varying or the indentor is moving across the surface. These problems were of interest in the context of polymer fracture, rebound testing of polymers and the phenomenon of hysteretic friction, respectively. Methods were also developed to handle problems involving thermoviscoelastic behaviour where the dependence on temperatue is non-linear. All of this work presupposed that inertial effects could be neglected. Very little work on inertial boundary value problems was published up to the end of the sixties.

Comprehensive surveys of the application of viscoelastic stress analysis to design were prepared by Rogers (1965) and Lee (1966).

During the seventies, the work on non-inertial problems was consolidated. The main purpose of the present volume is to present a coherent, unified development of this topic, in particular of those problem classes which are not covered by the Classical Correspondence Principle. There has also been some progress on inertial problems. Typically however, to make progress on such problems it is necessary either to confine one's attention to the most idealized configurations or to introduce some approximation. Also, the mathematical techniques used have been generally rather sophisticated. We briefly discuss this work in the last chapter, and derive certain results by comparatively elementary methods.

The theory is developed without any serious attempt at mathematical rigour. However, we also avoid the use of merely heuristic arguments which are particularly common in the literature on fracture. The orientation of the book is applied mathematical, though with the ultimate aim of extracting physically interesting results. Certain required techniques and results, notably the statement and solution of the Hilbert problem and the use of Hilbert transforms, are discussed in several mathematical appendices. Short tables of integrals and other relations are included.

In chapter 1, the properties of the viscoelastic functions are explored in some detail. Also the boundary value problems of interest are stated. In chapter 2, the Classical Correspondence Principle and its generalizations are discussed. Then, general techniques, based on these, are developed for solving non-inertial isothermal problems. A method for handling non-isothermal problems is also discussed and in chapter 6 an illustrative example of its application is given. Chapter 3 and 4 are devoted to plane isothermal contact and crack problems, respectively. They utilize the general techniques of chapter 2. The viscoelastic Hertz problem and its application to impact problems are discussed in chapter 5. Finally in chapter 7, inertial problems are considered.

Exercises are scattered throughout the text, one of their main purposes being to allow the statement, without detailed derivation, of fairly standard or straightforward results. The equations occurring in these problems are numbered separately from the ordinary equations. They are distinguished by the letter "p" occurring after the number.

One of the authors (JMG) wishes to acknowledge gratefully two most pleasant periods spent at Simon Fraser University for the academic year 1983/84 and during the Spring of 1986. It was during these periods that most of his contribution to this work was made. He would also like to acknowledge the encouragement of P. O'Keefe, Head of Roads Division, J. Sheedy, Head of Road Construction Section and A. J. Curran, Head of Road Safety Section of the National Institute for Physical Planning and Construction Research over the years, and for their appreciation of the importance in some contexts, of a fundamental approach to applied research problems. The other author (GACG) acknowledges generous use of the facilities of the Dublin Institute for Advanced Studies particularly during his stay there during the academic year 1986/87.

Both authors are very grateful to Marion Jacques and Cindy Lister for their careful typing of the manuscript and to Dorothy Corr and Eva Wills for obtaining bibliographical information.

This project was supported by funds provided by the Natural Sciences and Engineering Research Council of Canada and the Faculty of Science of Simon Fraser University.

Dublin-Burnaby
December, 1987   *The Authors*

Contents

1. Fundamental Relationships

This chapter deals with fundamental definitions, constitutive equations of a viscoelastic medium subject to infinitesimal strain, and the nature and properties of the associated viscoelastic functions. General dynamical equations are written down. Also, the boundary value problems that will be discussed in later chapters are stated in general terms. Familiar concepts from the Theory of Linear Elasticity are introduced in a summary manner. For a fuller discussion of these, we refer to standard references (Love (1934), Sokolnikoff (1956), Green and Zerna (1968), Gurtin (1972)). Coleman and Noll (1961) have shown that the theory described here may be considered to be a limit, for infinitesimal deformations, of the general (non-linear) theory of materials with memory.

1.1 Stress and Strain

The mechanical system under consideration is a continuum, subject to boundary forces which are in general time-dependent and also body forces, though the latter are usually neglected. These external forces generate stresses throughout the medium. Consider a typical point r within the medium, where r denotes the position vector with Cartesian co-ordinates (x, y, z) or (x_1, x_2, x_3)[1]. The stresses acting in the vicinity of r at time t are completely described by the stress tensor $\sigma(r, t)$ with Cartesian components $\sigma_{ij}(r, t)$, $i, j = 1, 2, 3$. This quantity is related to more fundamental concepts by the following observations. Lets ds denote any small surface in the medium or on its boundary. Let n be the unit normal on one side of it. Then the matter on this side of ds exerts a force per unit area, or stress, on the matter on the other side, given by the stress vector[2]

$$s_i = \sigma_{ij} n_j .$$ (1.1.1)

In particular, this formula relates stress tensor components on the boundary of the medium to the applied external stresses, and for this reason, will be important later. It follows from (1.1.1) that the force exerted by the medium or by the boundary stresses, on any volume V, is given by the surface integral

---

[1] Throughout the book, the subscripts 1, 2, 3 will denote x, y, z Cartesian components, respectively, of vectors and tensors.
[2] It will be assumed that the summation convention is in force here and in later sections, except where stated otherwise.

$$F_i = \int_B ds\, \sigma_{ij} n_j \ , \tag{1.1.2}$$

where B is the boundary of V, and the vector components n_j, $j = 1, 2, 3$ represent the outward normal at a given point on B.

The fact that the outward normal is used is a matter of convention. It implies for example that tensile stresses are positive and compressive stresses negative. Also, normal surface pressures into the medium along a co-ordinate axis generate a negative diagonal component of the stress tensor corresponding to that coordinate.

Problem 1.1.1: Show that if a body occupying the half-space $y > 0$ is subject to a surface shear stress along the positive x-axis, then this stress is equal to $-\sigma_{xy}$ on the surface.

The assumption that each small volume is in equilibrium, so that the resultant moment due to body and surface forces must vanish, gives, by means of a standard argument, that the stress tensor is symmetric. Viscoelastic materials with couple stresses, and therefore a non-symmetric stress tensor have been considered by Misicu (1963, 1964) and Eringen (1967). Also the requirement that the resultant of body and surface forces must vanish gives that the equations of motion take the form

$$\varrho\, \frac{\partial^2 u_i}{\partial t^2} = \frac{\partial \sigma_{ij}}{\partial x_j} + b_i \ , \tag{1.1.3}$$

where the $u_i(r, t)$ are the components of the displacement vector $u(r, t)$ giving the displacement of the point r from its equilibrium position, and ϱ is the density of the medium. The vector components b_i in (1.1.3) are the contributions of body forces such as gravity. The space derivatives of displacements will be assumed to be sufficiently small that products of these quantities can be neglected compared to the quantities themselves. This restriction allows us to construct a linear theory. The state of deformation of the medium is then characterized by the strain tensor $\varepsilon(r, t)$ whose components are

$$\varepsilon_{ij}(r, t) = \frac{1}{2}\left(\frac{\partial u_i}{\partial x_j} + \frac{\partial u_j}{\partial x_i}\right) \ , \qquad i, j = 1, 2, 3 \ . \tag{1.1.4}$$

On physical grounds, we expect a close relationship to exist between the components of stress and strain.

Note that the density ϱ in (1.1.3) will depend upon the state of deformation of the material. However, this correction may be ignored in the linear theory, provided that the acceleration term multiplying ϱ is assumed to be of the same order as the space derivatives of the displacements.

It must be remarked that the linear assumption may be especially restrictive for certain viscoelastic media, namely those which have low modulus. Many polymeric materials are in this category. For such materials, large deformations may occur even at relatively low stresses.

The traces of the tensors ε, σ have special significance, since they are scalar quantities, and thus independent of the coordinate system. It is easy to demonstrate that

$$\text{Tr}\{\varepsilon\} = \frac{\partial u_i}{\partial x_i} = \varepsilon \tag{1.1.5}$$

is the volume strain or the change in volume per unit volume at a given point. The quantity

$$p = \tfrac{1}{3}\text{Tr}\{\sigma\} = \tfrac{1}{3}\sigma \tag{1.1.6}$$

is sometimes referred to as the hydrostatic stress, while

$$s = \sigma - pI \; , \tag{1.1.7}$$

where I is the unit tensor, is the deviator stress tensor. This decomposition into hydrostatic and deviator stresses corresponds to the separation of volume and shear stresses.

We observe that strain is dimensionless and that stress has dimensions of force per unit area.

It is of interest to consider the rate of work done by the external and body forces on the medium. Let us denote this quantity by $\dot{E}$. We have

$$\dot{E} = \int_V b_i \dot{u}_i dv + \int_B \sigma_{ij} n_j \dot{u}_i ds \; , \tag{1.1.8}$$

where V is the total volume of the medium and B is its surface. The quantity dv is a volume element. Applying the Divergence Theorem to the second term and carrying out the differentiation gives that

$$\int_B \sigma_{ij} n_j \dot{u}_i ds = \int_V dv(\sigma_{ij,j} \dot{u}_i + \sigma_{ij} \dot{u}_{i,j}) \; , \tag{1.1.9}$$

where the subscript preceded by a comma indicates differentiation with respect to that space variable. It is easy to show that

$$\sigma_{ij} \dot{u}_{i,j} = \sigma_{ij} \dot{\varepsilon}_{ij} \; . \tag{1.1.10}$$

Using (1.1.3), we finally obtain

$$\dot{E} = \int_V (\varrho \dot{u}_i \ddot{u}_i + \sigma_{ij} \dot{\varepsilon}_{ij}) dv \; . \tag{1.1.11}$$

We infer that the rate of increase of mechanical energy per unit volume at a point r at time t is given by

$$e = \frac{d}{dt}\left(\frac{1}{2}\varrho \dot{u}_i \dot{u}_i\right) + \sigma_{ij} \dot{\varepsilon}_{ij} = \dot{k} + \sigma_{ij} \dot{\varepsilon}_{ij} \; , \tag{1.1.12}$$

where $\dot{k}$ is the rate of change of kinetic energy per unit volume, with time.

We will show later that, in a viscoelastic medium, some of this energy supplied by external forces is stored and some is dissipated.

In order to give physical content to the theory, it is necessary to postulate constitutive relationships between the stress and strain tensors, so that complete knowledge of one determines the other. In this manner, we introduce into the theory the physical characteristics of the material under consideration.

1.2 One-dimensional Linear Viscoelasticity

We consider one-dimensional constitutive equations in some detail before moving on to the general case. In the context of boundary value problems, the one-dimensional case is of limited interest. However, if provides a simple framework in which to discuss a number of points that are of general importance. Also, of course, from the viewpoint of experimental material characterization, one-dimensional configurations are of great interest.

A particular one-dimensional problem is one in which every component of the stress and strain tensors is zero except for instance the xy components $\sigma_{12}, \varepsilon_{12}$. This would occur if a stress σ_{12} were imposed on a medium in which displacement takes place only in the y direction, depending on x alone — perhaps due to constraints. Following this model, we take it that the stress and strain depend only on x. Subscripts will be dropped.

The treatment of the material in this section is similar in many respects to that given by F. Williams (1975)

1.2.1 Linear Hereditary Constitutive Laws

The primary requirement which will be imposed here on the constitutive equation is that it be linear. This implies that the stress, for example, must be given as a linear functional of the strain. In general, this will involve an integral or sum over the strain at various space and time points. We impose the condition that the law be local, which is to say that the stress at a certain position depends only on the strain at that position. This amounts to excluding "action at a distance" forces such as might arise for example if the material were susceptible to electromagnetic fields which in turn were dependent on the mechanical state of the material. Furthermore, the Principle of Causality implies that the stress depends only on present and past values of the strain. We write

$$\sigma(x, t) = f(t; x)\varepsilon(x, t) + \int_{-\infty}^{t} dt' g(t, t'; x)\varepsilon(x, t') . \tag{1.2.1}$$

Since the functions $f(t; x)$, $g(t, t'; x)$ characterize the response of the material, their dependence upon x implies that the material is spatially inhomogeneous. For most of the present volume, a homogeneous material will be assumed. However, there are important problem classes for which inhomogeneity cannot be neglected, notably where it arises from space-dependent environmental effects, in particular temperature. This type of inhomogeneity is discussed in Sect. 1.7.

The space dependence of $f(t; x)$, $g(t, t'; x)$ will not effect the considerations of this section and so explicit reference to it will be dropped. An alternative way of writing (1.2.1) is as follows:

$$\sigma(t) = G(t, -\infty)\varepsilon(-\infty) + \int_{-\infty}^{t} dt' G(t, t')\dot{\varepsilon}(t') , \qquad \text{where} \qquad (1.2.2)$$

$$\frac{d}{dt'}G(t, t') = -g(t, t') , \qquad G(t, t) = f(t) . \qquad (1.2.3)$$

If $g(t, t'), f(t)$ are given, it is always possible to choose $G(t, t')$ such that (1.2.3) is satisfied. This is clear if one observes that the first condition determines $G(t, t')$ only up to an arbitrary function of t, which can be chosen so that the second condition is satisfied.

Integrals over the history of strain (or stress) as occur in (1.2.1, 2) are sometimes referred to as hereditary integrals. Materials whose constitutive equations contain such hereditary integrals are described as having memory.

Observe that (1.2.3) implies that $G(t, t')$ must be differentiable with respect to its second argument. This is a smoothness requirement on the function characterizing the dependence of $\sigma(t)$ on the past history of $\dot{\varepsilon}(t)$.

In may cases, $\varepsilon(-\infty)$ will vanish, thus eliminating the first term in (1.2.2). If the strain is zero before a certain time t_1, then (1.2.1) becomes, with the aid of (1.2.3),

$$\sigma(t) = G(t, t)\varepsilon(t) - \int_{t_1}^{t} dt' \left[\frac{d}{dt'}G(t, t') \right] \varepsilon(t') . \qquad (1.2.4)$$

Under these circumstances, (1.2.2) becomes

$$\sigma(t) = G(t, t_1)\varepsilon(t_1) + \int_{t_1}^{t} dt' G(t, t')\dot{\varepsilon}(t') \qquad (1.2.5)$$

as may be deduced by partial integration of (1.2.4). Equations (1.2.4, 5) are equivalent and equally important forms of the same physical statement. Constitutive equations cast in the form of (1.2.5) are perhaps most commonly used in the literature. It has advantages, particularly where $\varepsilon(t)$ is constant over certain time intervals. Equation (1.2.4) however expresses stress as a linear functional of strain in a straightforward manner, essentially generalizing simple matrix multiplication. It has practical advantages and an adaptation of it will be used in much of the detailed work of later chapters.

If the hereditary integral is absent in (1.2.4), the relation reduces to Hooke's Law where the modulus is time-dependent. Note that the dimension of $G(t, t')$ is the same as the moduli in elastic theory, namely force per unit area.

The integral in (1.2.5) is really a special case (where $\varepsilon(t)$ is differentiable) of a Stieltjes integral. Gurtin and Sternberg (1962) base their rigorous formulation of Linear Viscoelasticity on constitutive equations which have this Stieltjes form. We adopt a convenient notation of theirs, and write (1.2.5) as

$$\sigma(t) = [G*d\varepsilon](t) , \qquad (1.2.6)$$

which we term a Stieltjes product.

One can view G in (1.2.6) as a linear operator acting on ε, which may be thought of as a vector in an abstract space. The concrete realization of G, ε are the functions $G(t, t')$, $\varepsilon(t)$, the relationship between them being a generalization of that between a linear operator and matrix elements or vector and vector components.

1.2.2 The Operator Algebra

It is useful to adopt a somewhat more abstract viewpoint on operators of this type. Our treatment, however, falls far short of thoroughgoing rigour.

Consider the class of operators corresponding to all functions $G(t, t')$ which are differentiable with respect to t, t' and which obey the Causality condition

$$G(t, t') = 0 , \quad t < t' . \tag{1.2.7}$$

We impose a further condition that

$$G(t, t) \neq 0 \tag{1.2.8}$$

for all t, except where $G(t, t')$ is identically zero, for all t, t'. The functions $\varepsilon(t)$ corresponding to the vectors ε on which the operators act will be assumed to be zero for $t < t_1$.

Addition and multiplication of operators of this kind can be defined in a straightforward manner. The sum of two operators G_1, G_2 is that operator associated with $G_1(t, t') + G_2(t, t')$. It is easy to show that addition in this sense obeys all the usual rules. Similarly, multiplication of an operator by a constant may be defined in an obvious manner. The zero operator is that corresponding to $G(t, t')$ equal to zero for all t, t'.

Multiplication of operators is defined as the consecutive application of two operators. In other words, for all ε,

$$[(G_2 * dG_1) * d\varepsilon](t) = [G_2 * d(G_1 * d\varepsilon)](t) . \tag{1.2.9}$$

From this definition, it may be shown, with the aid of (1.2.4, 5), and an interchange in the order of integration, that

$$[G_2 * dG_1](t, t') = G_2(t, t) G_1(t, t') - \int_{t'}^{t} dt'' \left[\frac{d}{dt''} G_2(t, t'') \right] G_1(t'', t') \tag{1.2.10a}$$

$$= G_2(t, t') G_1(t', t') + \int_{t'}^{t} dt'' G_2(t, t'') \frac{d}{dt''} G_1(t'', t') . \tag{1.2.10b}$$

Observe that in the second form, the derivative is with respect to the first rather than the second time argument, which is why differentiability with respect to both arguments is required.

Using essentially the same type of manipulation, one can show that this product is associative, i.e.

$$[G_3 * d(G_2 * dG_1)](t, t') = [(G_3 * dG_2) * dG_1](t, t') . \tag{1.2.11}$$

It is easy to show that it is distributive with respect to addition:

$$[G_3 * d(G_1 + G_2)](t, t') = [G_3 * dG_1](t, t') + [G_3 * dG_2](t, t') . \tag{1.2.12}$$

Finally, one can define a unit operator I, of the form

$$I(t, t') = H(t - t') , \tag{1.2.13}$$

where $H(t)$ is the Heaviside step function defined by $(A\,3.1.3\,p)$. The following properties may be demonstrated:

$$[I * d\varepsilon](t) = \varepsilon(t), \quad [I * dG](t, t') = G(t, t') = [G * dI](t, t') . \tag{1.2.14}$$

It will now be shown that if $G_2 * dG_1$ is zero, then either G_1 or G_2 is zero. The proof requires a fundamental property of Volterra integral equations, discussed in Sect. A4.2. If $[G_2 * dG_1](t, t')$ is zero, then by definition, from (1.2.10a):

$$G_2(t, t) G_1(t, t') - \int_{t'}^{t} dt'' \left[\frac{d}{dt''} G_2(t, t'') \right] G_1(t'', t') = 0 . \tag{1.2.15}$$

Let us assume that G_2 is non-zero. It follows from (1.2.8) that $G_2(t, t)$ is non-zero for all t. Equation (1.2.15) is therefore a homogeneous Volterra equation, for $G_1(t, t')$ at fixed t'. Such equations have only the trivial zero solution, so that $G_1(t, t') = 0$. If G_1 is assumed to be non-zero, it can be shown similarly, with the aid of (1.2.10b) that G_2 must be zero.

Finally, a very important property for later use, is that every operator G has a unique inverse J in this class, such that

$$G * dJ = J * dG = I . \tag{1.2.16}$$

The requirement that J be the right inverse is given explicitly by

$$G(t, t) J(t, t') - \int_{t'}^{t} dt'' \left[\frac{d}{dt''} G(t, t'') \right] J(t'', t') = H(t - t') . \tag{1.2.17}$$

This is an inhomogeneous Volterra integral equation of the second kind for $J(t, t')$ at fixed t', which always has a unique non-zero solution. It gives the form of $J(t, t')$ for $t \geq t'$. For $t < t'$ we take it to be zero so that it obeys the Causality condition. Furthermore, from (1.2.17), it follows that for all t

$$G(t, t) J(t, t) = 1 \tag{1.2.18}$$

so that $J(t, t)$ cannot be zero anywhere. Therefore, J is an operator in the class of interest. To show that it is also a left inverse, we consider

$$K = (J * dG - I) * dJ = J * d(G * dJ) - J = 0. \tag{1.2.19}$$

It follows that one of the factors in the product is zero. Since J is not zero, one must have $J * dG = I$.

An immediate consequence of the existence of an inverse is that the constitutive equation (1.2.5) may be written as

$$\varepsilon(t) = [J * d\sigma](t) = J(t, t_1) \sigma(t_1) + \int_{t_1}^{t} dt' J(t, t') \dot{\sigma}(t') \tag{1.2.20}$$

or, in the form of (1.2.4),

$$\varepsilon(t) = J(t,t)\sigma(t) - \int_{t_1}^{t} dt' \left[\frac{d}{dt'} J(t,t') \right] \sigma(t') \tag{1.2.21}$$

where the function $J(t,t')$ is uniquely determined by $G(t,t')$ and vice versa. We refer to $G(t,t')$, $J(t,t')$ as the relaxation and creep functions, for physical reasons which will be discussed later in Sect. 1.4.

1.2.3 Alternative Notation

Equations (1.2.4), and (1.2.21) may be written as

$$\sigma(t) = \int_{t_1}^{t} dt' \mu(t,t')\varepsilon(t') = [\mu * \varepsilon](t)$$

$$\varepsilon(t) = \int_{t_1}^{t} dt' \gamma(t,t')\sigma(t') = [\gamma * \varepsilon](t) , \qquad \text{where} \tag{1.2.22}$$

$$\mu(t,t') = G(t,t)\delta(t-t') - H(t-t')\frac{d}{dt'} G(t,t')$$

$$= -\frac{d}{dt'} [H(t-t')G(t,t')]$$

$$\gamma(t,t') = J(t,t)\delta(t-t') - H(t-t')\frac{d}{dt'} J(t,t') \tag{1.2.23}$$

$$= -\frac{d}{dt'} [H(t-t')J(t,t')] ,$$

and $\delta(t)$ is the singular delta function, discussed in Sect. A3.1. The upper limits of the integrals in (1.2.22) are understood to be t^+, the limiting value from above, so that the full contribution of the delta functions are picked up. This compact notation is very useful for formal development and manipulation. It will be widely used in later chapters – generally in the context of non-aging materials, which are introduced below. It tends to be avoided in more rigorous mathematical treatments of viscoelastic theory, since rigorous use of the delta function involves some cumbersome mathematical theory. Note that $\mu(t,t')$, $\gamma(t,t')$ obey the Causality relations

$$\mu(t,t') = \gamma(t,t') = 0, \qquad t < t' . \tag{1.2.24}$$

Also, from (1.2.22), it can be shown that

$$\int_{t'}^{t} dt'' \mu(t,t'') \gamma(t'',t') = \int_{t'}^{t} dt'' \gamma(t,t'')\mu(t'',t') = \delta(t-t') \tag{1.2.25}$$

with the aid of an interchange in the order of integration. The first of these is just (1.2.17) in differentiated form.

The notation in (1.2.22) is more formally akin to matrix notation in Linear Algebra, which is a reason why it tends to be easier to use in formal manipulations. In fact, it is precisely the matrix notation but with indices taking a continuous range of values. The operator algebra could also have been formally developed in terms of this notation, that is using (1.2.22) rather than (1.2.4, 5), to define the product operation. It is essentially the algebra of triangular matrices with continuous indices, and non-zero diagonal elements.

1.2.4 Non-aging Materials

We now consider the most important special case of (1.2.2), namely where the mechanical properties of the material do not vary with time, at least over the timescale of interest. This means that $G(t, t')$ must be unchanged under any overall time change, giving for any t_0

$$G(t + t_0, t' + t_0) = G(t, t') \quad \text{or} \tag{1.2.26}$$

$$G(t, t') = G(t - t') . \tag{1.2.27}$$

We will sometimes describe this property by saying that the material properties are time-homogeneous. Equations (1.2.4), (1.2.5) become

$$\sigma(t) = G(0)\varepsilon(t) + \int_{t_1}^{t} dt' \dot{G}(t - t')\varepsilon(t') \tag{1.2.28a}$$

$$= G(t - t_1)\varepsilon(t_1) + \int_{t_1}^{t} dt' G(t - t')\dot{\varepsilon}(t') = [G*d\varepsilon](t) , \tag{1.2.28b}$$

the last form being the compact notation introduced in (1.2.6). The dot notation in the first form indicates time differentiation, with respect to t or $(t - t')$. This is different to (1.2.4), with a consequent difference in sign of that term. For future use, we write down these explicitly, when $t_1 = -\infty$:

$$\sigma(t) = G(0)\varepsilon(t) + \int_{-\infty}^{t} dt' \dot{G}(t - t')\varepsilon(t') \tag{1.2.29a}$$

$$= G(\infty)\varepsilon(-\infty) + \int_{-\infty}^{t} dt' G(t - t')\dot{\varepsilon}(t') = [G*d\varepsilon](t) . \tag{1.2.29b}$$

In the latter form, the term $G(\infty)\varepsilon(-\infty)$ will often be zero since $\varepsilon(-\infty)$ generally vanishes.

The inverted constitutive relations (1.2.21) and (1.2.20) have the form

$$\varepsilon(t) = J(0)\sigma(t) + \int_{t_1}^{t} dt' \dot{J}(t - t')\sigma(t') \tag{1.2.30a}$$

$$= J(t - t_1)\sigma(t_1) + \int_{t_1}^{t} dt' J(t - t')\dot{\sigma}(t') = [J*d\sigma](t) , \tag{1.2.30b}$$

or again, if $t_1 = -\infty$:

$$\varepsilon(t) = J(0)\sigma(t) + \int_{-\infty}^{t} dt' \dot{J}(t-t')\sigma(t')$$

$$= J(\infty)\sigma(-\infty) + \int_{-\infty}^{t} dt' J(t-t')\dot{\sigma}(t') = [J*d\sigma](t) \ .$$

(1.2.31)

Finally, the compact notation (1.2.22), involving a singular kernel, becomes

$$\sigma(t) = \int_{-\infty}^{t} dt' \mu(t-t')\varepsilon(t') = [\mu*\varepsilon](t)$$

(1.2.32a)

$$\varepsilon(t) = \int_{-\infty}^{t} \gamma(t-t')\sigma(t') = [\gamma*\sigma](t)$$

(1.2.32b)

where the lower limits may be replaced by t_1 if $\varepsilon(t)$ and $\sigma(t)$ vanish before that time, and where

$$\mu(t) = G(0)\delta(t) + \dot{G}(t)H(t) = \frac{d}{dt}[H(t)G(t)]$$

(1.2.33a)

$$\gamma(t) = J(0)\delta(t) + \dot{J}(t)H(t) = \frac{d}{dt}[H(t)J(t)] \ .$$

(1.2.33b)

Note that the relations (1.2.32) are convolution integrals, as given by (A3.1.17), since $\mu(t)$, $\gamma(t)$ obey the Causality condition:

$$\mu(t) = \gamma(t) = 0, \quad t < 0 \ .$$

(1.2.34)

In (1.2.32) it is understood as before that the upper limit of integration is t^+, the limit from above, so that the full contribution of the delta function in (1.2.33) is included. It follows from (1.2.32) or (1.2.25) that for any t_1, t_2 where $t_2 \geq t_1$:

$$\int_{t_1}^{t_2} dt' \mu(t_2-t')\gamma(t'-t_1) = \int_{t_1}^{t_2} dt' \gamma(t_2-t')\mu(t'-t_1) = \delta(t_2-t_1) \ .$$

(1.2.35)

Taking $t_1 = 0$, $t_2 = t$, we obtain a more compact result from which the general form (1.2.35) can be deduced:

$$\int_{0}^{t} dt' \mu(t-t')\gamma(t') = \int_{0}^{t} dt' \gamma(t-t')\mu(t') = \delta(t) \ .$$

(1.2.36)

Substituting for $\mu(t)$, $\gamma(t)$ from (1.2.33) gives that for $t \geq 0$,

$$G(0)J(0) = 1$$

(1.2.37a)

$$\dot{G}(t)J(0) + G(0)\dot{J}(t) = -\int_{0}^{t} dt' \dot{G}(t-t')\dot{J}(t')$$

(1.2.37b)

$$= -\int_{0}^{t} dt' \dot{J}(t-t')\dot{G}(t') \ .$$

This is a differentiated form of the following relationship:

$$G(0)J(t) + \int_0^t dt' J(t-t') \dot{G}(t') = J(0)G(t) + \int_0^t dt' G(t-t') \dot{J}(t')$$

$$(1.2.38)$$

$$= 1 \quad \text{for} \quad t \geq 0 \;,$$

which follows directly from (1.2.28a, 1.2.30b); see also (1.2.17). This in turn is a differentiated form of the relation

$$\int_0^t dt' J(t-t')G(t') = \int_0^t dt' G(t-t')J(t') = t, \quad t \geq 0 \;. \tag{1.2.39}$$

Consider the leftmost expression in (1.2.38) at large t. Let us assume that $\dot{G}(t)$ tends to zero at large t. It will emerge from arguments given in Sect. 1.4 that this is physically reasonable. This means that large t' values will not contribute in the integral so that we may replace $J(t-t')$ by $J(t)$ and remove it from the integral, giving

$$\lim_{t \to \infty} G(t)J(t) = 1 \;. \tag{1.2.40}$$

If either $G(\infty)$ or $J(\infty)$ is finite, then the other is also and we have

$$G(\infty)J(\infty) = 1 \;. \tag{1.2.41}$$

Problem 1.2.1: If $G(t), J(t)$ are positive decreasing and increasing functions of time, respectively, show that

$$G(t)J(t) \leq 1 \;. \tag{1.2.1 p}$$

Hint: replace $G(t-t')$ by $G(t)$ in (1.2.38).

Further discussion of relationships between $G(t), J(t)$ may be found in Pipkin (1972). Gross (1953) and Ferry (1970) discuss this topic comprehensively, giving many relations, one or two of which have already been given and others which will be discussed in Sect. 1.4 – 6.

These equations express the close connection between the creep and relaxation functions, either of which, it will be recalled, can be determined in principle from knowledge of the other. This point is discussed further, in general terms, in Sect. 1.5. How it may be done in specific cases is discussed in Sect. 1.6.

Henceforth, it will be assumed, unless otherwise specified, that the material under consideration is non-aging.

Problem 1.2.2: Show that in the non-aging case, operator multiplication as defined by (1.2.10) is commutative. In fact, this is probably perceived more easily by using the alternative form, akin to matrix multiplication, derived from (1.2.32). The easiest way to show it is to take into account a result of Sect. 1.5, namely that in the frequency representation, operator multiplication becomes simple multiplication, by virtue of the Faltung theorem (Sect. A3.1).

1.3 Energy Considerations in the One-dimensional Case

It is of interest to seek an expression for the energy stored in the medium at a given time, as a result of a specified history of strain. Linear Elasticity Theory would indicate that we should expect an expression quadratic in the strain. Carrying through the assumptions of Sect. 1.2, we would also wish to include a dependence on the history of strain. Therefore, we write the energy density stored at a given point (excluding kinetic energy) in the form

$$V(t) = E_1 \varepsilon^2(t) + \varepsilon(t) \int\limits_{-\infty}^{t} dt' E_2(t-t')\varepsilon(t')$$

$$+ \int\limits_{-\infty}^{t} dt' \int\limits_{-\infty}^{t} dt'' E_3(t-t',t-t'')\varepsilon(t')\varepsilon(t'') \ ,$$

(1.3.1)

where we have confined our attention to non-aging materials. It will be assumed that the strain vanishes in the distant past. Equation (1.3.1) can then be written more compactly as

$$V(t) = \frac{1}{2} \int\limits_{-\infty}^{t} dt' \int\limits_{-\infty}^{t} dt'' K(t-t',t-t'')\dot{\varepsilon}(t')\dot{\varepsilon}(t'')$$

(1.3.2)

where

$$E_1 = \frac{1}{2} K(0,0) \ , \qquad E_2(t) = \frac{1}{2} \frac{d}{dt}[K(t,0)+K(0,t)] \ ,$$

$$E_3(t,t') = \frac{1}{2} \frac{d}{dt} \frac{d}{dt'} K(t,t') \ .$$

(1.3.3)

Problem 1.3.1: Show that if E_1, $E_2(t)$, $E_3(t,t')$ are given, it is always possible to choose $K(t,t')$ such that (1.3.3) holds.

Note that $K(t,t')$ may be taken to be symmetric in its arguments. Differentiating $V(t)$, we obtain

$$\dot{V}(t) = \dot{\varepsilon}(t) \int\limits_{-\infty}^{t} dt' K(t-t',0)\dot{\varepsilon}(t')$$

$$+ \frac{1}{2} \int\limits_{-\infty}^{t} dt' \int\limits_{-\infty}^{t} dt'' \frac{\partial}{\partial t} K(t-t',t-t'')\dot{\varepsilon}(t')\dot{\varepsilon}(t'') \ .$$

(1.3.4)

From (1.1.12), we have that the rate of increase of mechanical energy density at a point, due to the work done by boundary forces, is $\sigma(t)\dot{\varepsilon}(t)$, neglecting the kinetic energy term. This must be equal to the rate of increase of stored elastic energy plus the rate of energy dissipation, per unit volume. Comparing with (1.3.4), we deduce that

$$\sigma(t) = \int\limits_{-\infty}^{t} dt' K(t-t',0)\dot{\varepsilon}(t') \ ,$$

(1.3.5)

and the quantity

$$\dot{D}(t) = -\frac{1}{2} \int_{-\infty}^{t} dt' \int_{-\infty}^{t} dt'' \frac{\partial}{\partial t} K(t-t', t-t'') \dot{\varepsilon}(t') \dot{\varepsilon}(t'') \tag{1.3.6}$$

is the rate of dissipation of mechanical energy into heat, per unit volume. Comparing (1.3.5) and (1.2.29b), we see that the function $K(t, t')$ is related to the relaxation function $G(t)$ by

$$G(t) = K(t, 0) . \tag{1.3.7}$$

This relationship does not in general determine $K(t, t')$, so that knowledge of constitutive behaviour does not determine the stored energy. This is unsatisfactory since, in principle at least, knowledge of the constitutive relationship combined with the dynamical equations determines the behaviour of the material under any system of forces. At least, this is true if a uniqueness theorem can be proved. This topic is referred to briefly in Sect. 1.8. If it is the case that $K(t, t')$ cannot be deduced from $G(t)$, then this implies that the distribution of stored energy in the material is not determined even if the complete history of stresses and strains are known. This question is discussed in some detail by Christensen (1982), Chap. 3, with references to the literature.

If, however, it is assumed that

$$K(t, t') = F(f(t, t')) , \tag{1.3.8}$$

where $f(t, t')$ is a known, preferably simple function of $t, t' > 0$ with the property that the range of $f(t, t')$ for $t' > 0$ is contained in the range of $f(t, 0)$, then knowledge of $G(t)$ for $t > 0$ will suffice to determine $K(t, t')$. The simplest choice is to take

$$K(t, t') = G(t + t') . \tag{1.3.9}$$

In this case,

$$V(t) = \frac{1}{2} \int_{-\infty}^{t} dt' \int_{-\infty}^{t} dt'' G(2t - t' - t'') \dot{\varepsilon}(t') \dot{\varepsilon}(t'')$$

$$\dot{D}(t) = -\frac{1}{2} \int_{-\infty}^{t} dt' \int_{-\infty}^{t} dt'' \frac{\partial}{\partial t} G(2t - t' - t'') \dot{\varepsilon}(t') \dot{\varepsilon}(t'') . \tag{1.3.10}$$

Observe that (1.3.10) shows clearly that the presence of mechanical energy dissipation is a direct consequence of the dependence of $G(t)$ upon time. If such dependence is not present, the theory reduces to linear elastic theory.

Note that $V(t)$ was chosen to be purely quadratic, without linear terms. It may be checked that the presence of such contributions would lead to terms in the resulting constitutive relationship which are independent of the field quantities. This would correspond to non-zero stress in an undeformed medium, for example. Such phenomena will not be considered.

There is a fundamental requirement that the total stored energy in any volume be non-negative, so that $V(t) \geq 0$. Consider a strain history where a sudden strain is imposed at $t = 0$ so that $\varepsilon(t) = \varepsilon_0 H(t)$. Using (A3.1.11), we deduce from (1.3.2) that

$$K(t, t) \geq 0 \tag{1.3.11}$$

and, assuming that (1.3.9) applies,

$$G(t) \geq 0 \ . \tag{1.3.12}$$

Also, $\dot{D}(t)$ must be non-negative. It follows by the same argument that

$$\frac{dG(t)}{dt} \leq 0 \tag{1.3.13}$$

again, relying on the validity of (1.3.9). Christensen (1972, 1982) formulates a fading memory hypothesis which, combined with (1.3.13), gives the further restriction that

$$\frac{d^2 G(t)}{dt^2} \geq 0 \ . \tag{1.3.14}$$

Detailed discussion of these and related matters may be found in Christensen (1982). Thermodynamic considerations are also treated by Hunter (1983).

1.4 Creep and Relaxation

In this section, we discuss in more detail the physical content of the constitutive relations introduced in Sect. 1.2. Consider the behaviour of a material suddenly subjected to a constant stress σ at time t_1, having been unstrained before then. From (1.2.30) we have that

$$\varepsilon(t) = J(t - t_1)\sigma \ . \tag{1.4.1}$$

In other words, the resultant strain is not instantaneous but develops over time to a final value

$$\varepsilon = J(\infty)\sigma \tag{1.4.2}$$

if $J(\infty)$ is finite. We shall see that in fact $J(\infty)$ is often, though not always finite.

It is observed experimentally, and is anyway intuitively reasonable, that the suddenly imposed stress causes a certain instantaneous strain. Subsequently, the strain increases either to some final value or indefinitely.

This behaviour is termed creep and, as noted previously, $J(t)$ is termed the creep function. If creep ceases eventually, this corresponds to saying that $J(\infty)$ is finite. Equation (1.4.2) suggests that $J(\infty)$ may be regarded as a natural generalization of the inverse elastic modulus. We will sometimes denote it by J_e. For certain materials, creep continues indefinitely. Such behaviour is akin to that of a liquid. Certain classes of viscoelastic materials are in fact referred to as viscoelastic liquids. There are materials with the property that the limit

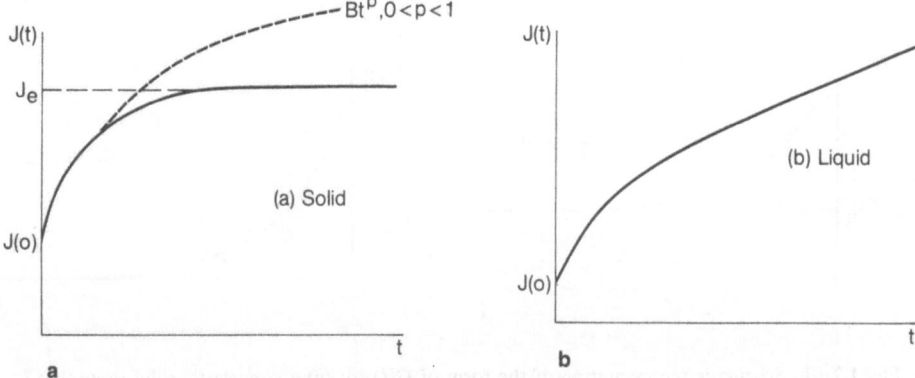

Fig. 1.1a,b. Schematic representation of the form of $J(t)$ for (**a**) a viscoelastic solid and (**b**) a viscoelastic liquid

$$\dot{J}(\infty) = \lim_{t \to \infty} \frac{dJ(t)}{dt} \tag{1.4.3}$$

is a non-zero constant. If $\dot{J}(\infty)$ is zero, the material is referred to as a viscoelastic solid. This distinction will be rendered physically plausible later. At large times, $J(t)$, for a viscoelastic liquid, behaves as

$$J(t) = t/\eta \tag{1.4.4}$$

$$\eta = [\dot{J}(\infty)]^{-1} \ .$$

Inserting (1.4.4) into (1.4.1) and differentiating with respect to time shows that η is the effective coefficient of viscosity at large times. Note that $J(t)$ may diverge for large t, in the case of a viscoelastic solid, as t^a, $a<1$, for example.

It must be remarked that viscoelastic liquids will behave quite differently to viscous liquids over short time periods.

These various observations allow us to form a rough qualitative picture of the possible shapes of $J(t)$, which are summarized on Fig. 1.1. Note that the option that $\lim_{t \to \infty} \dot{J}(t)$ may diverge is not included. It will be shown later that this is not possible on physical grounds.

It may not be so easy to decide experimentally whether a viscoelastic material is a solid or a liquid. In particular, it may not be apparent when a truly asymptotic state has been reached. For polymeric materials, the distinction between a liquid and a solid is essentially that between cross-linked and uncross-linked polymers. Further discussion of molecular structure and its consequences may be found in Ferry (1970).

The instantaneous strain is given by $J(0)\sigma$. If a material exhibits negligible instantaneous deformation then $J(0)$ is small or zero. Special models of such materials are noted in Sect. 1.6.

Consider a material subject to a constant shear stress from $t = t_1$, which is released at $t = t_2$. From (1.2.30), we have, instead of (1.4.1):

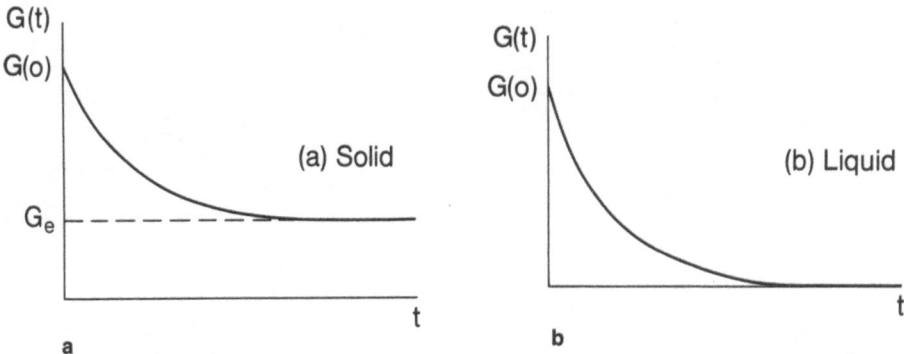

Fig. 1.2a, b. Schematic representation of the form of $G(t)$ for (a) a viscoelastic solid (note that G_e may be zero, corresponding to $J(t)$ becoming infinite as t^p, $0 < p < 1$) and (b) a viscoelastic liquid

$$\varepsilon(t) = \sigma\{J(t-t_1) - J(t-t_2)\} \approx \sigma \dot{J}(t)(t_2-t_1) \qquad (1.4.5)$$

at large t. If $\dot{J}(t)$ tends to zero, $\varepsilon(t)$ finally vanishes. This behaviour justifies the term viscoelastic solid. The material recovers completely from applied stress, like an elastic body, but with time delays caused by internal frictional losses.

If $\dot{J}(\infty)$ is finite, then permanent deformation occurs as a result of the application and removal of stress. This justifies the term viscoelastic liquid. Such materials also exhibit recovery to the extent that there exists a portion of $J(t)$, namely $\{J(t) - t/\eta\}$, the time derivative of which vanishes at large times.

Consider now a sudden, constant strain ε applied at time t_1 to an undeformed body. From (1.2.28), the resulting stress will be given by

$$\sigma(t) = G(t-t_1)\varepsilon \ . \qquad (1.4.6)$$

Experimental observation tells us that the initial stress is high and then relaxes to some final value, from which we deduce the qualitative behaviour of $G(t)$ (Fig. 1.2). Physically, the high initial resistance to deformation is due to frictional forces opposing the reorganization of molecules or particles required for optimal compliance with the change of shape. These internal frictional forces are gradually overcome, leading to a relaxation of stress. Hence the name relaxation for $G(t)$.

At first sight, the phenomena of creep and relaxation are unrelated. However, we know that $G(t)$ is determined, once $J(t)$ is given, and vice versa. Therefore, the two phenomena must in fact be intimately connected. In particular, the initial high level of stress, after a suddenly imposed strain, is given by

$$G(0)\varepsilon = \varepsilon/J(0) \qquad (1.4.7)$$

from (1.2.37a). In other words, it is determined by the level of instantaneous strain resulting from a suddenly imposed stress, since this is what $J(0)$ characterizes. In materials for which this quantity is small, $G(0)$ is very large and a sudden finite strain is difficult or impossible to impose.

The final value of $\sigma(t)$ is determined by $G(\infty)$, which is a natural generalization of the elastic modulus and will sometimes be referred to as G_e. It has been shown that $G(\infty), J(\infty)$ are inverses of each other, if they are finite (see (1.2.41)). The physical content of this is that if a final equilibrium state exists, it is irrelevant how it was arrived at, whether by a creep or relaxation process. This may be seen also by the following argument. Consider the constitutive equations (1.2.28 b, 30 b) in the limit of large t. If this limit can be interchanged with the integration, it is clear that the final condition of the material is independent of stress or strain history.

If $J(t)$ continues to increase indefinitely, $G(t)$ must tend to zero, according to (1.2.40). In particular, this is true for a viscoelastic liquid. It corresponds to the physical statement that the stress resulting from a suddenly applied strain relaxes to zero, even though the strain is maintained.

The concept of a viscoelastic liquid can also be characterized in terms of the relaxation function. Consider (1.2.29 b), letting the rate of strain (taken to be in shear) $\dot{\varepsilon}(t)$ be constant, for all time, and equal to s. Also, let $\varepsilon(-\infty)$ be zero. Then

$$\sigma = \eta s \; , \quad \eta = \int_0^\infty dt\, G(t) \; . \tag{1.4.8}$$

If $G(\infty)$ is not zero, the integral over $G(t)$ diverges, so that η is infinite for a viscoelastic solid. This definition of η may be shown to be equivalent to that given by (1.4.4). Consider the left-hand side of (1.2.38) for a viscoelastic liquid. Carrying out a partial integration, and taking the limit of large t, we obtain

$$\dot{J}(\infty) \int_0^\infty dt\, G(t) = 1 \tag{1.4.9}$$

by the same argument used to derive (1.2.40), since $G(t)$ goes to zero at large t. This gives the required consistency. We stated earlier that $\lim\limits_{t\to\infty} \dot{J}(t)$ could not diverge. Note that (1.4.9) would indicate that if this occurs, $\int\limits_0^\infty G(t)dt$ must be zero which, given the positive nature of $G(t)$, will occur only if $G(t)$ is zero everywhere.

Finally, we observe that since $G(t)$ is decreasing and bounded below, its slope $\dot{G}(t)$ must tend to zero at large t.

1.4.1 The Boltzmann Superposition Principle

Consider two strain increments $\Delta\varepsilon_1, \Delta\varepsilon_2$ applied to an initially undeformed body, at times t_1, t_2. From (1.2.29) the resulting stress at a subsequent time t is given by

$$\sigma(t) = G(t-t_1)\Delta\varepsilon_1 + G(t-t_2)\Delta\varepsilon_2 = \sigma_1(t) + \sigma_2(t) \; ,$$

where $\sigma_1(t)$ is the stress at time t due to $\Delta\varepsilon_1$ alone and $\sigma_2(t)$ is the stress at time t due to $\Delta\varepsilon_2$ alone. This result, which can be extended to any strain history, is the content of Boltzmann's Superposition Principle. It is essentially an expression of the linearity of the theory. In fact, by taking this principle as the starting point, it is possible to derive the constitutive equation (1.2.29).

We finally observe that delayed response phenomena akin to creep and relaxation occur in other areas of Mechanics and Physics, and are attributable to the same fundamental cause, namely (usually internal) frictional losses. The mathematical techniques used for analyzing such phenomena are similar to those used in analyzing the properties of the viscoelastic functions. Such a close analogy exists between certain phenomena in the theory of Dielectrics and Linear Viscoelasticity, as emphasized by Gross (1953).

1.5 The Frequency Representation

As noted in Sect. 1.2, the right-hand side of (1.2.32a) has the form of a convolution integral, given by (A3.1.17), since $\mu(t)$ is zero for negative values of t. Its Fourier transform (FT) is given by

$$\hat{\sigma}(\omega) = \hat{\mu}(\omega)\hat{\varepsilon}(\omega) , \tag{1.5.1}$$

where $\hat{\sigma}(\omega)$, $\hat{\varepsilon}(\omega)$ are the FTs of $\sigma(t)$, $\varepsilon(t)$. Using (1.2.33), we obtain

$$\hat{\mu}(\omega) = \int_0^\infty dt\, \mu(t)\mathrm{e}^{-\mathrm{i}\omega t} \tag{1.5.2a}$$

$$= G(0) + \int_0^\infty dt\, \dot{G}(t)\mathrm{e}^{-\mathrm{i}\omega t} \tag{1.5.2b}$$

$$= G(\infty) + \mathrm{i}\omega \int_0^\infty dt\, [G(t) - G(\infty)]\mathrm{e}^{-\mathrm{i}\omega t} . \tag{1.5.2c}$$

The form (1.5.2c) may be demonstrated by deriving from it, with the aid of a partial integration, the form (1.5.2b). The subtraction constant $G(\infty)$ has been introduced into the integrand for convergence. Relation (1.5.1) has the same form as the linear elastic constitutive relations, apart from the frequency dependence of the modulus. This fundamental observation will be extremely important in later chapters. We write

$$\hat{\mu}(\omega) = \hat{\mu}_1(\omega) + \mathrm{i}\hat{\mu}_2(\omega) \quad \text{where}$$

$$\hat{\mu}_1(\omega) = G(0) + \int_0^\infty dt\, \dot{G}(t) \cos \omega t \tag{1.5.3}$$

$$= G(\infty) + \omega \int_0^\infty dt\, [G(t) - G(\infty)] \sin \omega t \tag{1.5.4a}$$

$$\hat{\mu}_2(\omega) = -\int_0^\infty dt\, \dot{G}(t) \sin \omega t = \omega \int_0^\infty dt\, [G(t) - G(\infty)] \cos \omega t . \tag{1.5.4b}$$

There is no reason in general to believe that $\hat{\mu}_2(\omega)$ is zero, so that $\hat{\mu}(\omega)$ must be taken to be a complex quantity. This statement is validated more conclusively, later. It is generally known as the complex modulus. The quantity $\hat{\mu}_1(\omega)$ is referred to as the storage modulus and $\hat{\mu}_2(\omega)$ as the loss modulus, for reasons which will emerge later from the discussion of energy loss. Our object, in this section, is to examine to properties of these quantities.

In the first place, let us discuss, in a more concrete manner, the physical significance of $\hat{\mu}(\omega)$. Consider the simple but very important sinusoidal strain history given by

$$\varepsilon(t) = \varepsilon_0 \sin \omega t \ . \tag{1.5.5}$$

Substituting this form into (1.2.32a) gives

$$\sigma(t) = \mathrm{Im}\{\hat{\mu}(\omega)\varepsilon_0 e^{i\omega t}\} = \varepsilon_0[\hat{\mu}_1(\omega) \sin \omega t + \hat{\mu}_2(\omega) \cos \omega t] \ . \tag{1.5.6}$$

If we write $\hat{\mu}(\omega)$ in polar form, as

$$\hat{\mu}(\omega) = |\hat{\mu}(\omega)| \, e^{i\phi} \tag{1.5.7a}$$

$$\tan\phi = \hat{\mu}_2(\omega)/\hat{\mu}_1(\omega) \ , \quad \text{then} \tag{1.5.7b}$$

$$\sigma(t) = \varepsilon_0|\hat{\mu}(\omega)|\sin(\omega t + \phi) \ . \tag{1.5.8}$$

We can summarize the physical significance of $\hat{\mu}(\omega)$ in the following way. If the material is subjected to a long term sinusoidal strain of amplitude ε_0, the resulting stress is also sinusoidal, of amplitude $\varepsilon_0|\hat{\mu}(\omega)|$. There is a phase difference between the stress and strain oscillations, given by $\phi(\omega)$. It will be shown that this angle must be positive so that strain will always lag behind stress in phase. This phase lag is a characteristically viscoelastic effect. The angle $\phi(\omega)$ is often referred to as the loss angle and $\tan\phi$ as the loss tangent. We choose ϕ as the solution to (1.5.7b) in the first quadrant, that is, in $[0,\frac{\pi}{2}]$.

These results apply to the steady-state response which prevails after a long time. If we replace the lower limit in (1.2.32a) by some finite time, there will be other, transient, terms.

Problem 1.5.1: If

$$\sigma(t) = \int_0^t dt' \mu(t-t')\varepsilon(t') \tag{1.5.1p}$$

where $\varepsilon(t)$ is given by (1.5.5), show that if

$$\mu(t) \sim -\frac{G}{\tau} e^{-t/\tau} \tag{1.5.2p}$$

at large t, then the dominant transient contribution to $\sigma(t)$ has the form

$$\sigma_t(t) \sim -\frac{G\omega\tau\varepsilon_0}{1+\omega^2\tau^2} e^{-t/\tau} = -\hat{\mu}_{2a}(\omega)\varepsilon_0 e^{-t/\tau} \ , \tag{1.5.3p}$$

where $\hat{\mu}_{2a}(\omega)$ is the loss modulus corresponding to the asymptotic form (1.5.2p).

From (1.5.2), [see (A3.1.13)] it is clear that the complex conjugate of $\hat{\mu}(\omega)$ is given by

$$\bar{\hat{\mu}}(\omega) = \hat{\mu}(-\omega) \ , \tag{1.5.9}$$

which implies that the real part of $\hat{\mu}(\omega)$ is even in ω and the imaginary part odd. This relation can be used to attach meaning to $\hat{\mu}(\omega)$ at negative frequencies, which are physically meaningless.

Note that, from (1.5.2c), we have

$$\hat{\mu}(0) = G(\infty) = G_e \ . \tag{1.5.10}$$

At zero frequency, $\hat{\mu}(\omega)$ is therefore real. Also, from (1.5.2c) and (A3.1.19), we have

$$\lim_{\omega \to \infty} \hat{\mu}(\omega) = G(0) \ . \tag{1.5.11}$$

Since $\hat{\mu}(\omega)$ is in general non-zero at large ω, it does not have a finite inverse transform. This is of course no surprise since we expect that, formally, its inverse transform is equal to $\mu(t)$, which contains a delta function, i.e.

$$\mu(t) = \frac{1}{2\pi} \int_{-\infty}^{\infty} d\omega \, \hat{\mu}(\omega) e^{i\omega t} \ , \qquad t > 0 \ . \tag{1.5.12}$$

More rigorously we see from (1.5.2c, 10), that

$$G(t) = G(\infty) + \frac{1}{2\pi} \int_{-\infty}^{\infty} \frac{d\omega}{i\omega} \, [\hat{\mu}(\omega) - \hat{\mu}(0)] e^{i\omega t} \ , \qquad t > 0 \ . \tag{1.5.13}$$

For both very large and very small frequencies, $\hat{\mu}(\omega)$ becomes real, so that

$$\lim_{\substack{\omega \to \infty \\ \omega \to 0}} \hat{\mu}_2(\omega) = 0 \ . \tag{1.5.14}$$

We know, from the behaviour of $G(t)$, that, in general, $G(\infty) < G(0)$, so that $\hat{\mu}(0) < \hat{\mu}(\infty)$. The quantity $\hat{\mu}_1(\omega)$ generally increases monotonically from $\hat{\mu}(0)$ to $\hat{\mu}(\infty)$. Also, as suggested by (1.5.14), $\hat{\mu}_2(\omega)$ characteristically has a hump shape. These forms are shown schematically on Fig. 1.3. In practice, the loss modulus may have more than one maximum.

1.5.1 Dispersion Relations

The real function $G(t)$ determines the complex function $\hat{\mu}(\omega)$ so we expect that, in general, a relationship will exist between the real and imaginary parts of the latter. This is in fact the case, and may be demonstrated by using the analytic properties of $\hat{\mu}(\omega)$. Consider the expressions (1.5.2) for $\hat{\mu}(\omega)$. The integrals are over positive time only. This is an expression of the Principle of Causality (see Sect. 1.10). It is equivalent to the requirement that $\hat{\mu}(\lambda)$, $\lambda = \omega + i\alpha$, is analytic in the lower half-plane, that is for $\alpha < 0$ (Sect. A3.2). If follows from (A2.1.6) that

$$\hat{\mu}(\omega) - \hat{\mu}(\infty) = -\frac{1}{i\pi} \int_{-\infty}^{\infty} d\omega' \, \frac{\hat{\mu}(\omega') - \hat{\mu}(\infty)}{\omega' - \omega} \ , \tag{1.5.15}$$

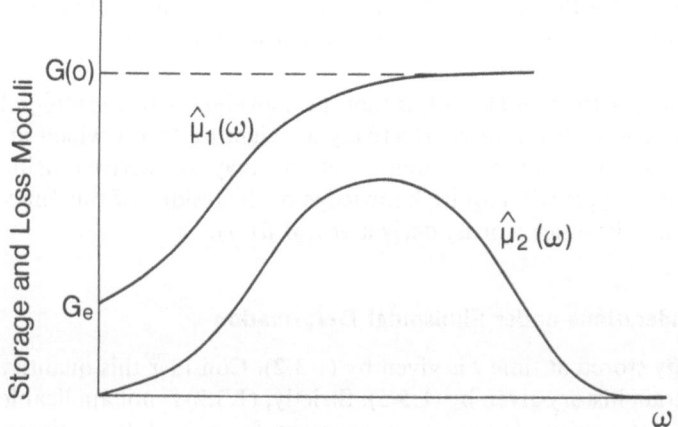

Fig. 1.3. Schematic representation of the form of the storage and loss moduli

where the integral is to be interpreted as a Cauchy principal value. We may drop the constant $\hat{\mu}(\infty)$ in the integrand because the $(\omega' - \omega)^{-1}$ kernel is odd. Since only positive frequencies are meaningful, we use (1.5.9) to re-express the integral in the form

$$\hat{\mu}(\omega) - \hat{\mu}(\infty) = -\frac{1}{i\pi} \int\limits_0^\infty d\omega' \left\{ \frac{\hat{\mu}(\omega')}{\omega' - \omega} - \frac{\bar{\mu}(\omega')}{\omega' + \omega} \right\} . \qquad (1.5.16)$$

The physical assumption underlying (1.5.9), being used here, is the fact that $G(t)$ is real. Splitting into real and imaginary parts, we find that

$$\hat{\mu}_1(\omega) = \hat{\mu}(\infty) - \frac{2}{\pi} \int\limits_0^\infty d\omega' \frac{\hat{\mu}_2(\omega')\omega'}{\omega'^2 - \omega^2} \quad \text{and} \qquad (1.5.17)$$

$$\hat{\mu}_2(\omega) = \frac{2\omega}{\pi} \int\limits_0^\infty d\omega' \frac{\hat{\mu}_1(\omega')}{\omega'^2 - \omega^2} . \qquad (1.5.18)$$

From (1.5.17) we see that if $\hat{\mu}_2(\omega)$ is zero everywhere, so that $\hat{\mu}(\omega)$ is real, then it must also be a constant and the theory degenerates to elastic theory. It is clear therefore that if the theory is to be a non-trivial extension of elastic theory, $\hat{\mu}(\omega)$ is essentially complex.

Relations of the form (1.5.17, 18) are often referred to as dispersion relations since they were first derived in the theory of optical dispersion by Kronig (1926) and Kramers (1927). Similar relations have been derived in other contexts, notably the Theory of Dielectrics as remarked by Gross (1953) [see for example Scaife (1971)] and the Quantum Theory of Scattering [Goldberger and Watson (1964)]. The main physical input to these relations is Causality. Gross (1953) first gave a relation of this kind for viscoelasticity, not in fact the same as either (1.5.17, 18), but of the form

$$\hat{\mu}_1(\omega) = \hat{\mu}(0) - \frac{2\omega^2}{\pi} \int\limits_0^\infty \frac{d\omega'}{\omega'} \frac{\hat{\mu}_2(\omega')}{\omega'^2 - \omega^2} . \qquad (1.5.19)$$

Problem 1.5.2: Derive relation (1.5.19), using a similar argument to that used above, but applied to $[\hat{\mu}(\omega) - \hat{\mu}(0)]/\omega$. Recall that $\hat{\mu}_2(0)$ is zero.

The advantage of (1.5.19) over (1.5.17) is that the convergence of the integral is better at large ω. The factor $1/\omega$ is effectively a weighting factor which attenuates the large ω contribution. Similar relations may be derived using $\hat{\mu}(\omega)/\omega^n$, $n > 1$, though these will require knowledge of the residue of this function at zero frequency, in other words, derivatives of $\hat{\mu}(\omega)$.

1.5.2 Energy Considerations under Sinusoidal Deformation

The elastic energy stored at time t is given by (1.3.2). Consider this quantity for the sinusoidal strain history given by (1.5.5). Strictly, (1.3.2) is not applicable since $\varepsilon(-\infty)$ does not vanish. However, it averages to zero at large times. Transformation of the time variable gives

$$V(t) = \frac{\varepsilon_0^2 \omega^2}{2} \int_0^\infty dt_1 \int_0^\infty dt_2 K(t_1, t_2) \cos \omega(t - t_1) \cos \omega(t - t_2) . \tag{1.5.20}$$

On decomposing the trigonometric product into a sum and averaging over a period $2\pi/\omega$, we find that the average level of stored elastic energy is

$$\langle V(t) \rangle = \frac{\varepsilon_0^2 \omega^2}{4} \int_0^\infty dt_1 \int_0^\infty dt_2 K(t_1, t_2) \cos \omega(t_1 - t_2) . \tag{1.5.21}$$

If (1.3.9) holds, then putting $K(t_1, t_2) = G(t_1 + t_2)$, we have

$$\langle V(t) \rangle = \frac{\varepsilon_0^2 \omega^2}{4} \int_0^\infty dt_1 \int_0^\infty dt_2 G(t_1 + t_2) \cos \omega(t_1 - t_2) . \tag{1.5.22}$$

Changing variables to

$$u = t_1 + t_2 \qquad v = t_1 - t_2 , \tag{1.5.23}$$

one obtains

$$\langle V(t) \rangle = \frac{\varepsilon_0^2 \omega^2}{8} \int_0^\infty du \int_{-u}^u dv \, G(u) \cos (\omega v)$$

$$= \frac{\varepsilon_0^2 \omega}{4} \int_0^\infty du \, G(u) \sin (\omega u) . \tag{1.5.24}$$

With the aid of (A3.1.5p) and (1.5.4a), we see that

$$\langle V(t) \rangle = \frac{\varepsilon_0^2}{4} \hat{\mu}_1(\omega) , \tag{1.5.25}$$

so that the average level of stored elastic energy per unit volume is proportional to $\hat{\mu}_1(\omega)$, motivating the term storage modulus. Similarly the *rate* of dissipation of mechanical energy into heat, given by (1.3.6), in this case has the form

$$\dot{D} = \frac{-\varepsilon_0^2\omega^2}{2} \int_0^\infty dt_1 \int_0^\infty dt_2 \dot{K}(t_1,t_2) \cos \omega(t-t_1) \cos \omega(t-t_2)$$

$$\dot{K}(t_1,t_2) = \left(\frac{\partial}{\partial t_1} + \frac{\partial}{\partial t_2}\right) K(t_1,t_2) \ . \tag{1.5.26}$$

Applying essentially the same argument and using (1.3.9) and (1.5.4b), it may be shown that

$$\langle \dot{D} \rangle = \frac{\varepsilon_0^2\omega}{2} \hat{\mu}_2(\omega) \ , \tag{1.5.27}$$

thus justifying the description loss modulus.

Since $\langle \dot{D} \rangle$ is non-negative, it follows that $\hat{\mu}_2(\omega)$ is also. Furthermore, the average level of stored energy in the body under sinusoidal deformation will be non-negative, so that the same is true for $\hat{\mu}_1(\omega)$, from (1.5.25). It follows that the loss tangent and loss angle are positive, as claimed earlier.

1.5.3 Creep

If, instead of starting from (1.2.32a), we start from (1.2.32b) we obtain the inverted form of (1.5.1), namely

$$\hat{\varepsilon}(\omega) = \hat{\gamma}(\omega)\hat{\sigma}(\omega) \quad \text{where} \tag{1.5.28}$$

$$\hat{\gamma}(\omega) = \int_0^\infty dt\,\gamma(t)e^{-i\omega t} \tag{1.5.29a}$$

$$= J(0) + \int_0^\infty dt\,\dot{J}(t)e^{-i\omega t} \ . \tag{1.5.29b}$$

In the case of a viscoelastic liquid, $\dot{J}(t) \to 1/\eta$ at large t and we obtain, with the help of (A3.1.4p):

$$\hat{\gamma}(\omega) = J(0) - \frac{i}{\eta\omega} + \int_0^\infty dt \left(\dot{J}(t) - \frac{1}{\eta}\right) e^{-i\omega t} \ . \tag{1.5.30}$$

Therefore, if the material is a liquid, $\hat{\gamma}(\omega)$ has an imaginary pole – contributing to $\hat{\gamma}_2(\omega)$ – at $\omega = 0$.

If $J(\infty)$ is finite, then, corresponding to (1.5.2c),

$$\hat{\gamma}(\omega) = J(\infty) + i\omega \int_0^\infty dt\,[J(t) - J(\infty)]e^{-i\omega t} \ . \tag{1.5.31}$$

Corresponding to (1.5.3, 4), we put

$$\hat{\gamma}(\omega) = \hat{\gamma}_1(\omega) + i\hat{\gamma}_2(\omega)$$

$$\hat{\gamma}_1(\omega) = J(0) + \int_0^\infty dt\,\dot{J}(t) \cos \omega t \ , \quad \hat{\gamma}_2(\omega) = -\int_0^\infty dt\,\dot{J}(t) \sin \omega t \ , \tag{1.5.32}$$

which can be made explicitly convergent, in the case of a liquid, by adding and subtracting $1/\eta$ in the integrand and using (A3.1.5p).

If $J(\infty)$ is finite,

$$\hat{\gamma}_1(\omega) = J(\infty) + \omega \int_0^\infty dt\,[J(t) - J(\infty)] \sin \omega t \; ,$$

$$\hat{\gamma}_2(\omega) = \omega \int_0^\infty dt\,[J(t) - J(\infty)] \cos \omega t \; . \tag{1.5.33}$$

The really interesting relation is

$$\hat{\gamma}(\omega) = 1/\hat{\mu}(\omega) \; . \tag{1.5.34}$$

This provides a simple but powerful method for determining the creep from the relaxation function, and vice versa, as we shall see in the next section. From (1.5.11, 34) and (1.2.37a), we have

$$\hat{\gamma}(\infty) = J(0) \; . \tag{1.5.35}$$

Also, from (1.5.33)

$$\lim_{\omega \to 0} \hat{\gamma}(\omega) = \lim_{t \to \infty} J(t) \; . \tag{1.5.36}$$

Dispersion relations analogous to (1.5.17 – 19) may be written down for the part of $\hat{\gamma}(\omega)$ that is regular at small ω.

From (1.5.30, 34), we deduce that at low frequencies the relaxation complex modulus for a liquid behaves as

$$\hat{\mu}(\omega) = i\omega\eta(1 - i\omega\tau + \dots) \; , \tag{1.5.37}$$

where τ will be defined below. This expression assumes that, apart from the pole in (1.5.30), a Taylor expansion can be constructed about $\omega = 0$. Now, for a liquid, $G(\infty)$ is zero and we can write (1.5.2c) as

$$\hat{\mu}(\omega) = i\omega \int_0^\infty dt\, e^{-i\omega t} G(t) \; . \tag{1.5.38}$$

Expanding the exponential and comparing with (1.5.37), we deduce that [cf. (1.4.8)]

$$\eta = \int_0^\infty dt\, G(t) \; , \qquad \tau = \frac{1}{\eta} \int_0^\infty dt\, t\, G(t) \; , \tag{1.5.39}$$

where τ is referred to as the mean relaxation time. The quantities η, τ are the zeroth and first moments of $G(t)$. For a further discussion of low frequency expansions, see Pipkin (1972).

Note that Causality implies that $\hat{\gamma}(\omega)$ cannot have singularities in the lower half-plane, just as was observed for $\hat{\mu}(\omega)$. Relation (1.5.34) therefore implies that neither of these two quantities can have zeros in the lower half-plane either.

1.6 Special Forms of the Viscoelastic Functions

In this section, we discuss some explicit forms of the viscoelastic functions that have been found useful in practice. There are two categories: (a) exponential decay models and degenerate limits of these, and (b) power law models. The bulk of the section is devoted to the first category, partly because exponential decay models are used often in later chapters. Power law models, however, are of considerable importance in that they are both simple and physically valid to a surprising degree.

The traditional discussion of mechanical (spring and dashpot) models and the related topic of differential forms of the constitutive equations will not be included here, but are treated extensively in several older references, Gross (1953), Ferry (1970), Bland (1960) for example. See also Nowacki (1965), Flügge (1967) and Lockett (1972). A consistent development of the theory is possible without these concepts. However, they do provide insights into the nature of viscoelastic behaviour and physically motivate exponential decay models.

1.6.1 Standard Linear Solid

This model is a convenient starting point in that it represents the simplest material which contains no special, degenerate features in a sense that will become clear when we discuss limiting cases of it — namely the Maxwell and Voigt materials.

A standard linear solid has a relaxation function of the form

$$G(t) = G_0 + G_1 e^{-t/\tau} , \tag{1.6.1}$$

where G_0, G_1 are positive constants and τ is the positive decay constant. Clearly,

$$G(0) = G_0 + G_1 , \qquad G(\infty) = G_0 < G(0) . \tag{1.6.2}$$

The elastic limit is obtained by letting G_1 tend to zero. The function $\mu(t)$, defined by (1.2.33a), has the form

$$\mu(t) = g_0 \delta(t) + g_1 e^{-t/\tau} H(t)$$

$$g_0 = G_0 + G_1 = G(0) , \qquad g_1 = -G_1/\tau < 0 \tag{1.6.3}$$

so that the complex modulus is given by

$$\hat{\mu}(\omega) = g_0 + \frac{g_1 \tau}{1 + i\omega\tau} = G_0 + \frac{G_1 i\omega\tau}{1 + i\omega\tau} = G_0 \frac{1 + i\omega\tau'}{1 + i\omega\tau} , \tag{1.6.4}$$

where τ' is given by

$$\tau' = \tau \frac{G_0 + G_1}{G_0} > \tau . \tag{1.6.5}$$

Its significance will emerge shortly. The storage and loss moduli have the form

$$\hat{\mu}_1(\omega) = G_0 \frac{1 + \omega^2 \tau\tau'}{1 + \omega^2 \tau^2} , \qquad \hat{\mu}_2(\omega) = \frac{G_1 \omega\tau}{1 + \omega^2 \tau^2} . \tag{1.6.6}$$

Also, the loss tangent has the form

$$\tan \phi = \frac{G_1 \omega \tau}{G_0(1 + \omega^2 \tau \tau')} . \tag{1.6.7}$$

It is interesting to remark on the singularity structure of $\hat{\mu}(\omega)$. It has one pole in the upper half-plane at i/τ. There is no singularity in the lower half-plane as required by Causality (Sect. 1.5). Also, it has a zero at $\omega = i/\tau'$.

It is readily perceived that if the curves of $\hat{\mu}_1(\omega)$, $\hat{\mu}_2(\omega)$, given by (1.6.6), were plotted, the results would be very much as indicated on Fig. 1.3.

The simplest way to determine the form of the creep function $J(t)$ is to use (1.5.34) giving that

$$\hat{\gamma}(\omega) = \frac{1}{G_0} \frac{1 + i\omega\tau}{1 + i\omega\tau'} . \tag{1.6.8}$$

Remembering that $J(t)$ is related to $\gamma(t)$ by (1.2.33b), which is identical in form to (1.2.33a), and comparing (1.6.8) and (1.6.4) gives that

$$J(t) = A + Be^{-t/\tau'} , \quad A = 1/G_0 , \quad \tau = \tau' \frac{A+B}{A} . \tag{1.6.9}$$

The last relation emerges by considering (1.6.5). Also from (1.6.5), it follows that $B < 0$, so that it is more convenient to write

$$J(t) = J_0 + J_1(1 - e^{-t/\tau'}) , \quad J_0 + J_1 = \frac{1}{G_0} , \quad \tau = \tau' \frac{J_0}{J_0 + J_1} \tag{1.6.10}$$

in terms of the positive quantities J_0, J_1. Comparing this last equation with (1.6.5) gives that

$$\frac{G_0}{G_0 + G_1} = \frac{J_0}{J_0 + J_1} \tag{1.6.11}$$

so that, from (1.6.10):

$$J_0 = \frac{1}{G_0 + G_1} \tag{1.6.12a}$$

$$J_1 = \frac{1}{G_0} - J_0 = \frac{G_1}{G_0(G_0 + G_1)} . \tag{1.6.12b}$$

It emerges therefore that the creep function also has the exponential decay form, with decay constant τ', given by (1.6.5). This quantity is always greater than the relaxation decay constant, except in the elastic limit where both parameters are redundant anyway.

1.6.2 Maxwell and Voigt Models

These models can be extracted as limiting cases of the standard linear solid. Let

$$G_0 \to 0 , \quad G_1 = G \tag{1.6.13}$$

giving

$$G(t) = Ge^{-t/\tau}$$

for the relaxation function. From (1.6.5, 12) it follows that J_1, τ' become infinite, though

$$\frac{J_1}{\tau'} \to \frac{1}{G\tau} \ . \tag{1.6.14}$$

Expansion of the exponential in (1.6.10) gives

$$J(t) = J + t/\eta \ , \quad J = 1/G \ , \quad \eta = G\tau \ . \tag{1.6.15}$$

This is the Maxwell (1867) model. Referring to the discussion in Sect. 1.4, we infer that it is a (very simple) model of a viscoelastic fluid. It is interesting to note that the large time Newtonian viscous fluid term in (1.6.15) emerged by taking the limit of a large creep decay constant and large coefficient J_1, starting from a model of a solid. The degenerate feature of this material is that $J(t) - t/\eta$ is a constant. The recoverable portion is the instantaneous term J.

The real and imaginary parts of the complex moduli for the Maxwell model are given by

$$\hat{\mu}_1(\omega) = \frac{G\omega^2\tau^2}{1+\omega^2\tau^2} \ , \quad \hat{\mu}_2(\omega) = \frac{G\omega\tau}{1+\omega^2\tau^2} \tag{1.6.16}$$

and the loss tangent is given by

$$\tan\phi = 1/\omega\tau \ . \tag{1.6.17}$$

These may be derived directly or deduced from (1.6.5) and (1.6.6).

Consider next the Voigt model, also referred to as the Kelvin model since it was in fact considered earlier by Lord Kelvin (and even earlier by Meyer – see Preface). Let

$$J_0 \to 0 \ , \quad J_1 = J \tag{1.6.18}$$

in the creep function, giving

$$J(t) = J(1 - e^{-t/\tau'}) \ . \tag{1.6.19}$$

We have from (1.6.12b)

$$G_0 = 1/J \ , \tag{1.6.20}$$

which is finite. Let

$$G_0 = G \ . \tag{1.6.21}$$

From (1.6.12a) it follows that $G_0 + G_1$ tends to infinity so that G_1 must become large. Inspection of (1.6.5) gives that τ must therefore become small to keep τ' finite. Formula (A3.1.6p) allows us to write

$$G(t) = G + \eta\delta(t) \ , \quad \eta = \lim_{G_1 \to \infty} G_1\tau = G\tau' \tag{1.6.22}$$

by virtue of (1.6.5). The fact that $J(0)$ is zero, or $G(0)$ is infinite, in this model means that there is no instantaneous response to a suddenly applied stress. Furthermore, stress relaxation under an applied strain is instantaneous since τ is zero.

The storage and loss moduli and the loss tangent are given by

$$\hat{\mu}_1(\omega) = G \ , \quad \hat{\mu}_2(\omega) = \eta \omega \ , \quad \tan \phi = \frac{\eta \omega}{G} = \omega \tau' \ , \tag{1.6.23}$$

the first term of which may be derived using (1.5.4c), or deduced by taking the limit of (1.6.6).

We remark that the Maxwell and Voigt models are equivalent to simple differential constitutive relations [for example, Christensen (1982)].

The standard linear solid is a convenient non-trivial but simple model, frequently used in theoretical analysis for purposes of illustrating techniques. However, very few materials exhibit such simple behaviour. In general, to fit experimental results, one must go to spectrum models discussed below.

The Maxwell and Voigt models are of historical interest. They are also occasionally useful in theoretical analysis as examples for which particularly complex general results simplify drastically. They have some theoretical importance deriving from the fact that by combining Maxwell or Voigt elements in the sense of spring and dashpot models, one generates spectrum models [Ferry (1970)].

1.6.3 Spectrum Models

These are models for which the relaxation function is a sum of decaying exponentials with different decay times. As noted above, such expressions emerge by combining Maxwell or Voigt spring and dashpot elements. More fundamentally, they also emerge from molecular models of polymeric materials. The theory of Stochastic Processes applied to long polymers leads naturally to the "bead-spring" model of Rouse (1953) and others [see Ferry (1970) for reference to the original papers and for example Doi and Edwards (1986); also Golden (1975)], which in simplest terms amounts to beads connected by springs with temperature-dependent moduli and subject to friction forces from the surrounding medium. The decay constants in models of this kind are determined from the eigenvalue spectrum of a particular matrix.

Specific predictions of molecular models of this type will not be discussed here. Instead, we simply consider the general form

$$G(t) = G_0 + \sum_{i=1}^{N} G_i e^{-t/\tau_i} \ , \tag{1.6.24}$$

where the N parameters τ_i are the (positive) decay times and the G_i are constants, also taken to be positive, although it is possible that certain of them could be negative without violating the fundamental requirements on $G(t)$ and its derivatives discussed at the end of Sect. 1.3. However, in expressions derived from simple mechanical and molecular models they are positive.

Observe that the constant term G_0, which is in fact a generalization of the elastic modulus (it is G_e mentioned in Sect. 1.4), may be regarded as a limiting case of the exponential decay form where one of the decay constants is infinite. The presence of this term implies that the material is a solid.

For this model we have

$$\mu(t) = g_0\delta(t) + \sum_{i=1}^{N} g_i e^{-t/\tau_i} ,$$

$$g_0 = \sum_{i=0}^{N} G_i = G(0) > 0 , \qquad g_i = -G_i/\tau_i < 0 , \qquad i = 1, 2, \ldots, N , \tag{1.6.25}$$

so that

$$\hat{\mu}(\omega) = g_0 + \sum_{i=1}^{N} \frac{g_i\tau_i}{1 + i\omega\tau_i} = G_0 + \sum_{i=1}^{N} \frac{G_i i\omega\tau_i}{1 + i\omega\tau_i} . \tag{1.6.26}$$

Taking real and imaginary parts gives

$$\hat{\mu}_1(\omega) = G_0 + \sum_{i=1}^{N} \frac{G_i\omega^2\tau_i^2}{1 + \omega^2\tau_i^2} , \qquad \hat{\mu}_2(\omega) = \sum_{i=1}^{N} \frac{G_i\omega\tau_i}{1 + \omega^2\tau_i^2} . \tag{1.6.27}$$

Relation (1.6.26) shows that $\hat{\mu}(\omega)$ is the ratio of two polynomials of degree N in ω, so that $\hat{\gamma}(\omega) = 1/\hat{\mu}(\omega)$ is also, which leads us to expect that $J(t)$ is a series of decaying exponentials. This step is not rigorous. All that we are entitled to deduce is that $\gamma(t)$ and therefore $J(t)$ will be a series of exponentials. The decay constants for creep are determined by the zeros of the numerator polynomial in $\hat{\mu}(\omega)$. On physical grounds, however, it is to be expected that the exponentials are real and decaying rather than increasing. In anticipation of this, we write

$$\gamma(t) = h_0\delta(t) + \sum_{i=1}^{N} h_i e^{-t/\tau_i'} , \qquad J(t) = J_0 + \sum_{i=1}^{N} J_i(1 - e^{-t/\tau_i'})$$

$$h_0 = J(0) = J_0 > 0 , \qquad h_i = J_i/\tau_i' > 0 \tag{1.6.28}$$

where $J(t)$ is put in this form because it is an increasing function of t.

It is convenient to use a slightly different form of $\mu(t)$, $\gamma(t)$ in order to discuss the relationships between their parameters. We put

$$\mu(t) = g_0\delta(t) + \sum_{i=1}^{N} g_i e^{-a_i t} , \qquad \gamma(t) = h_0\delta(t) + \sum_{i=1}^{N} h_i e^{-\beta_i t} , \tag{1.6.29}$$

where

$$a_i = 1/\tau_i , \qquad \beta_i = 1/\tau_i' , \tag{1.6.30}$$

giving

$$\hat{\mu}(\omega) = g_0 + \sum_{i=1}^{N} \frac{g_i}{a_i + i\omega} , \qquad \hat{\gamma}(\omega) = h_0 + \sum_{i=1}^{N} \frac{h_i}{\beta_i + i\omega} . \tag{1.6.31}$$

Problem 1.6.1: Show, using (1.6.29) and (1.2.36), that

$$h_0 g_0 = 1 \ ,$$

$$g_0 + \sum_{i=1}^{N} \frac{g_i}{a_i - \beta_j} = 0, \quad j = 1, 2, \ldots, N \tag{1.6.1p}$$

$$h_0 + \sum_{j=1}^{N} \frac{h_j}{\beta_j - a_i} = 0, \quad i = 1, 2, \ldots, N \ .$$

Consider the functions

$$\hat{\mu}(\mathrm{i}z) = g_0 + \sum_{i=1}^{N} \frac{g_i}{a_i - z} \ , \quad \hat{\gamma}(\mathrm{i}z) = h_0 + \sum_{i=1}^{N} \frac{h_i}{\beta_i - z} \tag{1.6.32}$$

for general complex values of z. These are analytic except at the poles at a_i *for* $\hat{\mu}(\mathrm{i}z)$ and β_i for $\hat{\gamma}(\mathrm{i}z)$. They are clearly analytic continuations of $\hat{\mu}(\omega)$, $\hat{\gamma}(\omega)$ into the complex plane. Referring to the brief discussion of this topic in Sect. A2.1, we see, from (1.5.34), that

$$\hat{\gamma}(\mathrm{i}z) = \frac{1}{\hat{\mu}(\mathrm{i}z)} \tag{1.6.33}$$

over the whole complex plane — even at poles, since by inverting both sides, these become zeros. As z approaches β_i, the left-hand side has a pole. It follows that the right-hand side will have a zero in the denominator. Therefore,

$$h_i = \lim_{z \to \beta_i} (\beta_i - z)\hat{\gamma}(\mathrm{i}z) = -\left\{ \frac{d}{dz} \hat{\mu}(\mathrm{i}z) \right\}_{z=\beta_i}^{-1} = \left\{ -\sum_{j=1}^{N} \frac{g_j}{(a_j - \beta_i)^2} \right\}^{-1} , \tag{1.6.34}$$

$$i = 1, 2, \ldots, N \ .$$

If any of the a_j are equal to β_i, then h_i is zero. It follows that this cannot be the case. The inverse decay constants β_i are solutions of the equations

$$\hat{\mu}(\mathrm{i}z) = 0 \tag{1.6.35}$$

or, more explicitly, the second relation of (1.6.1p). On physical grounds, these must be real and positive. Equations (1.6.34) and the first two relations of (1.6.1p) determine the creep parameters in terms of those of the relaxation function. These results were given by Gross (1953). Note that h_i have the opposite sign to the g_i, from (1.6.34). The g_i are negative, from (1.6.25), so that the h_i, and therefore the J_i of (1.6.28), must be positive, as has been indicated.

Problem 1.6.2: Show that for a model with one decay time, or a standard linear solid, relations (1.6.1p) may be put in the form

$$h_0 = \frac{1}{g_0} \ , \quad \beta = a + \frac{g_1}{g_0} \ , \quad h_1 = -\frac{(a - \beta)^2}{g_1} = -\frac{g_1}{g_0^2} \ , \tag{1.6.2p}$$

where

$$\mu(t) = g_0\delta(t) + g_1 e^{-at} , \qquad \gamma(t) = h_0\delta(t) + h_1 e^{-\beta t} . \qquad (1.6.3\,\mathrm{p})$$

Show that these relations are equivalent to (1.6.10 − 12).

1.6.4 Continuous Spectra

A generalization of (1.6.24) is given by

$$G(t) = G_0 + \int_0^\infty da\, G_1(a) e^{-at} , \qquad G_0 \geq 0 , \qquad G_1(a) \geq 0 . \qquad (1.6.36)$$

If $G_1(a)$ is a series of delta functions, one recovers the discrete spectrum. We write

$$\mu(t) = g_0\delta(t) + H(t) \int_0^\infty da\, g(a) e^{-at}$$

$$(1.6.37)$$

$$g_0 = G_0 + \int_0^\infty da\, G_1(a) , \qquad g(a) = -a\, G_1(a) ,$$

so that

$$\hat{\mu}(\omega) = g_0 + \int_0^\infty da\, \frac{g(a)}{a + i\omega} . \qquad (1.6.38)$$

The analytic continuation of $\hat{\mu}(\omega)$ is

$$\hat{\mu}(iz) = g_0 + \int_0^\infty da\, \frac{g(a)}{a - z} \qquad (1.6.39)$$

for general complex values of z, off the positive real axis. Note that $\hat{\mu}(iz)$ has a discontinuity or cut from $z = a_0$ to infinity where a_0 is the lowest value of a such that $g(a)$ is non-zero, so that a_0 is in general a branch point. Consider the limits $\hat{\mu}^\pm(ia)$ as $z \to a_1$ on the real axis from above and below. The Plemelj Formula (A2.2.10) gives that

$$g(a_1) = \frac{1}{2\pi i} [\hat{\mu}^+(ia_1) - \hat{\mu}^-(ia_1)] = \pm\frac{1}{\pi} \operatorname{Im}\{\hat{\mu}^\pm(ia_1)\} , \qquad a_1 > a_0 , \qquad (1.6.40)$$

since the function $g(a_1)$ is real so that, from (1.6.39), $\hat{\mu}^+(ia_1) = \bar{\hat{\mu}}^-(ia_1)$. Equation (1.6.40) gives the function $g(a)$ in terms of the complex modulus − though evaluated at imaginary frequencies, which limits its usefulness. In principle, it might be possible to obtain estimates of $\hat{\mu}(i\omega)$ from experiment by means of analytic continuation techniques. However, in practice, this is rarely attempted [Ferry (1970)]. Gross (1953) discusses this type of relation in more detail, and gives references to the original papers.

Guided by the discrete case, we look for a representation of $J(t)$ also in the form

$$J(t) = J_0 + \int_0^\infty da\, J_1(a)(1 - e^{-at}) \qquad (1.6.41)$$

and seek an expression for $J_1(a)$, in terms of $G_1(a)$. As before,

$$\gamma(t) = h_0\delta(t) + H(t) \int_0^\infty da\, h(a)e^{-at}$$

(1.6.42)

$$h_0 = J(0) = J_0 = 1/g_0 , \qquad h(a) = aJ_1(a)$$

and

$$\hat{\gamma}(\omega) = h_0 + \int_0^\infty da\, \frac{h(a)}{a+i\omega} .$$

(1.6.43)

Let us consider the analytic continuations $\hat{\gamma}(iz) = 1/\hat{\mu}(iz)$ and take the limit of this relationship as $z \to \beta$ from above. The Plemelj Formulae (A2.2.9) give that

$$h_0 + \pi i h(\beta) + \int_0^\infty da\, \frac{h(a)}{a-\beta} = \left[g_0 + \pi i g(\beta) + \int_0^\infty da\, \frac{g(a)}{a-\beta} \right]^{-1} ,$$

(1.6.44)

where the integrals are interpreted as principle values.

Taking the imaginary part of both sides gives

$$h(\beta) = -\frac{g(\beta)}{\Lambda(\beta)} , \qquad \Lambda(\beta) = \left(g_0 + \int_0^\infty da\, \frac{g(a)}{a-\beta} \right)^2 + \pi^2 g^2(\beta) .$$

(1.6.45)

This relationship determines $h(\beta)$ and therefore the creep function in terms of the relaxation function. A similar equation obtained by interchanging the creep and relaxation parameters expresses $g(a)$ in terms of $h(a)$. This relationship, in somewhat different, and slightly more general, form, was given by Gross (1953). Ferry (1970) discusses theoretically based forms of the function $J_1(a)$. In many cases, (1.6.45) does not give analytic expressions and so it may be that approximate methods are more useful, for example than that of Rapp and Romas'ko (1972) who assume that (1.2.1p) is an equality.

1.6.5 Power Law Viscoelastic Functions

On a log-log plot, many viscoelastic functions are approximately linear over significant time ranges. This suggests the use of a relaxation function of the form

$$G(t) = At^{-p} ,$$

(1.6.46)

where p is a positive real number, less than unity. This may be a good representation at intermediate times but at short and long times there are problems in that it is infinite at $t = 0$ and zero at infinite time. Therefore, the material has zero instantaneous response to a suddenly applied stress, and the stress relaxes to zero at large times. These features will be approximately valid, at best, for real materials. However, in a given problem for which typical times are neither long nor short, if p is chosen to fit the experimental behaviour reasonably well over the range, then (1.6.46) may well be a valid choice for theoretical analysis of the problem. Note that by adding a constant we render $G(\infty)$ finite. However, this also has the effect of making the theoretical properties of the function less manageable.

Observe that $\mu(t)$, given by (1.2.33a) is not defined in this case. However, the complex modulus can be defined by (1.5.2c), giving

$$\hat{\mu}(\omega) = i\omega A \int_0^\infty dt\, t^{-p} e^{-i\omega t} . \tag{1.6.47}$$

Putting $z = i\omega$ and rotating it to a point x on the positive real axis, the integral can be evaluated in terms of the Gamma function (A1.4.1). Rotating back, we find the

$$\hat{\mu}(\omega) = A\Gamma(1-p)\,\omega^p \exp\left(i\,\frac{p\pi}{2}\right) , \tag{1.6.48}$$

so that

$$\hat{\mu}_1(\omega) = A\Gamma(1-p)\,\omega^p \cos p\pi/2 , \qquad \hat{\mu}_2(\omega) = A\Gamma(1-p)\,\omega^p \sin p\pi/2 \tag{1.6.49}$$

and the loss angle is given by

$$\phi = p\pi/2 . \tag{1.6.50}$$

The fact that ϕ is frequency-independent is an interesting feature of the model. Experimentally, this is found to be a good approximation for nearly all high polymers over a broad range of frequencies [Kolsky (1956), Pipkin (1972)].

From (1.5.34) we obtain

$$\hat{\gamma}(\omega) = \frac{\omega^{-p}}{A\Gamma(1-p)} \exp\left(-i\,\frac{p\pi}{2}\right) . \tag{1.6.51}$$

It follows that

$$J(t) = B\,t^p , \tag{1.6.52a}$$

$$B = \frac{1}{A\Gamma(1+p)\Gamma(1-p)} = \frac{\sin p\pi}{A p\pi} , \tag{1.6.52b}$$

which is most easily shown by going backwards, by means of [see (1.5.29b)]

$$\hat{\gamma}(\omega) = \int_0^\infty dt\, \dot{J}(t) e^{-i\omega t} . \tag{1.6.53}$$

Problem 1.6.3: Show that (1.2.39) holds for the forms of $G(t)$ and $J(t)$, given above. See (A1.4.3).

The power law form was apparently first proposed by Nutting (1921). In recent years [for example Torvik and Bagley (1984) and earlier papers by those authors; and Koeller (1984)] this form [specifically (1.6.46) or (1.6.52a) plus constant] has inspired the application of the fractional calculus (calculus of fractional derivatives) to the description of viscoelastic phenomena, a possibility which had been suggested by several authors, notably Rabotnov (1948, 1969,

1980), whose work is generalized and phrased in fractional calculus language by Koeller (1984). By replacing the derivatives in differential constitutive laws [see Christensen (1982) for example] by fractional derivatives, it is possible to generate a rich variety of relaxation and creep functions of far greater complexity than the simple power law form which provided the original stimulus for the approach.

1.7 Temperature Dependence of the Viscoelastic Functions

The viscoelastic functions of polymeric materials are strongly temperature-dependent. Molecular models suggest that the source of this dependence is a factor in the decay constants, the same factor in all of them, which is very sensitive to temperature. This would imply that if the factor is denoted by $a(T)$, where T is the absolute temperature, and

$$\xi = t/a(T) , \tag{1.7.1}$$

then $G(t)$, $J(t)$, expressed as functions of ξ, have no other temperature dependence apart from through ξ, so that

$$G_0(\xi) \equiv G(t,T) , \quad J_0(\xi) \equiv J(t,T) \tag{1.7.2}$$

are functions of the single variable ξ rather than the two variables t, T. So, for example, a change in temperature, leaving t fixed, is equivalent to a time change, leaving T fixed, in such a way as to bring about the same value of ξ. A temperature change effectively scales the time dependence of the material functions.

This observation is totally confirmed by experimental observation, for a wide range of polymers, as discussed by Ferry (1970) who also refers to the original literature. Apart from the temperature dependence through $a(T)$, there is an overall factor in the viscoelastic functions which depends on temperature [Ferry (1970)]. However, it is frequently the case that this overall variation can be neglected compared with the effect of $a(T)$. An exception to this may occur at times for which the viscoelastic functions are very slowly varying. In the present discussion, we shall ignore it.

In practice, $a(T)$ is determined empirically rather than from any fundamental theoretical considerations. A reference temperature T_0 is selected and $a(T_0)$ is set equal to unity. If follows that $G(t,T_0)$, $J(t,T_0)$ are the universal curves $G_0(\xi)$, $J_0(\xi)$ with $t = \xi$. One may then fill in the detailed structure of $a(T)$ by taking measurements at different temperatures.

Williams, Landel and Ferry (1955) have given an empirical expression for $a(T)$, applicable to a wide variety of polymeric materials, for temperatures not too different from the glass transition temperature T_g of the polymer. Their expression contains material-dependent parameters. In its simplest form, there is only one such parameter, the glass transition temperature T_g. This approximate universal expression is given by

$$\log_{10} a(T) = \frac{-17.44(T - T_g)}{51.6 + T - T_g} \tag{1.7.3}$$

for T_g taken to be the reference temperature. It may be checked that the variation of $a(T)$ with temperature is very rapid, particularly for values of T in the vicinity of T_g. This has the consequence that it may not be possible to adopt an isothermal approximation under conditions where relatively small changes of temperature occur. It also restricts the validity of the fully linearized non-isothermal theory, which we discuss briefly below.

The combining of time and temperature dependence in the manner described above is referred to as the time-temperature superposition principle. Materials which have this property are sometimes referred to as thermorheologically simple materials.

Problem 1.7.1: Show that the complex moduli of thermorheologically simple materials have frequency and temperature dependence combined into the variable $u = \omega a(T)$.

1.7.1 Variable Temperature History

Formulae (1.7.2) apply only to the case where temperature does not vary with time. If the temperature is time-dependent, then the first point to be observed is that the non-aging hypothesis may break down. Consider a sudden constant strain ε applied at $t = t_0$. Then at a later time t,

$$\sigma(t) = G(t, t_0)\varepsilon . \tag{1.7.4}$$

Let us divide the time interval $[t_0, t]$ into small sub-intervals Δt_i and make the approximation that the temperature has a constant value T_i during each of these. We now assume that each sub-interval of time is scaled according to the temperature prevailing at that time, so that

$$G(t, t_0) = G_0(\xi(t, t_0))$$
$$\xi(t, t_0) = \sum \frac{\Delta t_i}{a(T_i)} \rightarrow \int_{t_0}^{t} \frac{dt'}{a(T(t'))} = \xi(t) - \xi(t_0) \tag{1.7.5}$$

where

$$\xi(t) = \int_0^t \frac{dt'}{a(T(t'))} . \tag{1.7.6}$$

This result was first given by Morland and Lee (1960), who refer to ξ as the reduced time or the pseudo-time. Note that

$$G(t, t_0) = G_0(\xi - \xi_0) , \tag{1.7.7}$$

where $\xi = \xi(t)$, $\xi_0 = \xi(t_0)$. The isothermal limit is extracted by putting $a(T_0) = 1$ where T_0 is the prevailing temperature, so that $\xi(t)$ reduces to t. Thus $G_0(t)$ is the isothermal viscoelastic function. It is clear that if the temperature is variable, the non-aging structure breaks down. However, by using the pseudo-time variables, it can be re-established, at least if the temperature field is independent of the space variables. For example, the constitutive equations (1.2.28) can be rewritten as

$$\sigma(\xi) = G_0(0)\varepsilon(\xi) + \int_{\xi_1}^{\xi} d\xi' \frac{d}{d\xi} G_0(\xi - \xi')\varepsilon(\xi')$$

$$= G_0(\xi - \xi_1)\varepsilon(\xi_1) + \int_{\xi_1}^{\xi} d\xi' G_0(\xi - \xi') \frac{d}{d\xi'} \varepsilon(\xi') \ . \tag{1.7.8}$$

The strain in these equations should be considered to include a contribution due to thermal expansion of the material. If the temperature depends on position, the viscoelastic function will also, and space homogeneity is lost. Equation (1.7.8) remains valid in fact but the dynamical equations, which will be introduced in Sect. 1.8, are altered in a significant manner.

Christensen (1982) derives the result of Morland and Lee (1960) by first considering non-linear constitutive equations and then linearizing with respect to strain but not temperature. His derivation is based on work by Crochet and Naghdi (1969, 1970). Stouffer and Wineman (1971) give a generalized form which includes that of Morland and Lee as a special case. Stouffer and Wineman (1972) give a general form for the constitutive equation of an aging material, the properties of which are dependent upon environmental variables. This is intended to cover cases where the material is influenced by temperature or other environmental variables such as humidity, water content and so on. Stouffer (1972) also presents an alternative method of taking account of temperature history. F. Williams (1975) discusses and uses these approaches.

The dependence on temperature of the viscoelastic functions is clearly non-linear, though the mechanical constitutive relation, discussed in Sect. 1.2, is linear. If the temperature variation is small, the theory may be completely linearized. Let

$$T = T_0 + \theta(t) \tag{1.7.9}$$

where $\theta(t)$ is a small quantity and T_0 is the average background temperature. Then, choosing the reference temperature to be T_0,

$$\xi(t) = t + c\int_0^t \theta(t')dt' = t + \Delta(t)$$

$$c = \frac{d}{dT}\left(\frac{1}{a(T)}\right)_{T=T_0} . \tag{1.7.10}$$

Therefore

$$G(t, t_0) = G_0(t - t_0) + G_0'(t - t_0)[\Delta(t) - \Delta(t_0)] \ , \tag{1.7.11}$$

where $G_0(t - t_0)$ is the limit of $G(t, t_0)$ as $\Delta(t)$ vanishes, namely, the isothermal relaxation function. If $\Delta(t)$ is regarded as being of the same order as the strain $\varepsilon(t)$, then in fact, the second term on the right of (1.7.11) is neglected in a linear theory and one simply has $G_0(t - t_0)$ replacing $G(t, t_0)$. However, as remarked previously, $a(T)$ varies very rapidly with T in certain temperature ranges, so that the temperature variation must be very slight in such cases for $\Delta(t)$ to be small. A systematic derivation of the equations of such a fully linearized theory is given by Christensen (1982). These are not discussed here since they are not needed in later chapters.

1.8 Three-dimensional Constitutive and Dynamical Equations

We now move on to the full three-dimensional theory of linear viscoelasticity. The isothermal case will be discussed first, using essentially the approach developed in Sect. 1.3. Adjustments will then be made to incorporate non-isothermal effects. The material is assumed to be space-homogeneous and non-aging, though not necessarily isotropic. In later chapters, only isotropic materials are considered. However, the anisotropy assumption allows greater notational compactness and is therefore convenient for this initial, general discussion. The corresponding isotropic equations are given in Sect. 1.9.

1.8.1 Isothermal Theory

Let the stored energy density at a point be given by the quadratic form (see Sect. 1.3):

$$V(t) = \frac{1}{2} \int_{-\infty}^{t} dt_1 \int_{-\infty}^{t} dt_2 K_{ijkl}(t - t_1, t - t_2) \dot{\varepsilon}_{ij}(t_1) \dot{\varepsilon}_{kl}(t_2) \ , \tag{1.8.1}$$

where the space dependence is omitted for convenience. The quantity K_{ijkl} is symmetric in the interchange of i, j and also k, l. Furthermore,

$$K_{ijkl}(t, t') = K_{klij}(t, t') \ . \tag{1.8.2}$$

We differentiate $V(t)$ with respect to t and, following the arguments of Sect. 1.3, deduce the constitutive relationship

$$\sigma_{ij}(t) = \int_{-\infty}^{t} dt_2 K_{ijkl}(0, t - t_2) \dot{\varepsilon}_{kl}(t_2) \tag{1.8.3}$$

and that the rate of dissipation of mechanical energy into heat per unit volume is given by

$$\dot{D}(t) = -\frac{1}{2} \int_{-\infty}^{t} dt_1 \int_{-\infty}^{t} dt_2 \frac{\partial}{\partial t} K_{ijkl}(t - t_1, t - t_2) \dot{\varepsilon}_{ij}(t_1) \dot{\varepsilon}_{kl}(t_2) \ . \tag{1.8.4}$$

For the reasons discussed in Sect. 1.3, we adopt (1.3.9) without further ado, and put

$$K_{ijkl}(t, t') = G_{ijkl}(t + t') \tag{1.8.5}$$

so that (1.8.3, 4) become

$$\sigma_{ij}(t) = \int\limits_{-\infty}^{t} dt_1 \, G_{ijkl}(t-t_1)\dot{\varepsilon}_{kl}(t_1) \quad \text{and} \tag{1.8.6}$$

$$\dot{D}(t) = -\int\limits_{-\infty}^{t} dt_1 \int\limits_{-\infty}^{t} dt_2 \dot{G}_{ijkl}(2t-t_1-t_2)\dot{\varepsilon}_{ij}(t_1)\dot{\varepsilon}_{kl}(t_2) \; . \tag{1.8.7}$$

Note that if (1.8.5) holds, G_{ijkl} is symmetric in the interchange of (ij) and (kl). In the shorthand notation of (1.2.28) we can write (1.8.6) as

$$\sigma_{ij}(r, t) = [G_{ijkl} * d\varepsilon_{kl}](r, t) \; . \tag{1.8.8}$$

In the alternative notation, corresponding to (1.2.32), (1.8.6) becomes

$$\sigma_{ij}(r, t) = \int\limits_{-\infty}^{t} dt' \mu_{ijkl}(t-t')\varepsilon_{kl}(r, t') = [\mu_{ijkl} * \varepsilon_{kl}](t) \tag{1.8.9}$$

where, as in (1.2.33a)

$$\mu_{ijkl}(t) = G_{ijkl}(0)\delta(t) + \dot{G}_{ijkl}(t)H(t) \; . \tag{1.8.10}$$

The dynamical equations may be deduced from (1.1.3, 4) and (1.8.9) to have the form

$$\varrho \frac{\partial^2}{\partial t^2} u_i(r, t) = \frac{1}{2} \int\limits_{-\infty}^{t} dt' \mu_{ijkl}(t-t')(u_{l,jk} + u_{k,jl})(r, t') \tag{1.8.11}$$

where the subscripts after the comma indicate differentiation with respect to the corresponding space variable.

Finally, it is necessary to supply boundary and initial conditions on the field variables. The initial conditions are often of the form

$$u_i(r, t) = 0, \quad t < t_1 \tag{1.8.12}$$

for all r. In other words, the medium was undisturbed before a certain time t_1. For an important class of problems, those involving a steady state assumption, condition (1.8.12) does not apply. Instead of an initial condition in such problems, one has the steady-state assumption itself.

Before considering the various types of boundary conditions that will be of interest, we note that, apart from some passing comments, the discussion here and in later chapters will be confined to bodies that occupy fixed regions of space (in their undeformed state). This excludes, for example, bodies that ablate with time.

There are two simple but very important types of boundary conditions which will be noted before moving on to the general case. The so-called first boundary value problem is where the stresses $\sigma_{ij}(r, t)n_j(r)$ [see (1.1.1)] are specified all over the boundary B of the body. Specifically:

$$\sigma_{ij}(r, t)n_j(r) = c_i(r, t), \quad r \in B, \quad i = 1, 2, 3 \; , \tag{1.8.13}$$

where $c_i(r, t)$ is a known function. The second boundary value problem is where the displacements $u_i(r, t)$ are specified over the whole boundary:

$$u_i(r, t) = d_i(r, t), \quad r \in B, \quad i = 1, 2, 3 \tag{1.8.14}$$

where $d_i(r, t)$ is a known function. For viscoelastic, as for elastic materials, these two boundary value problems are generally easy to solve, at least for simple shapes.

In the general case, termed the mixed boundary value problem, the boundary is split into two groups of complementary regions $B_u^{(i)}(t)$, $B_\sigma^{(i)}(t)$ such that the displacements $u_i(r, t)$ are specified on $B_u^{(i)}(t)$ and the boundary stresses $\sigma_{ij}(r, t) n_j(r)$ are specified on $B_\sigma^{(i)}(t)$. Explicitly, we have

$$u_i(r, t) = d_i(r, t), \quad r \in B_u^{(i)}(t) \tag{1.8.15}$$

$$\sigma_{ij}(r, t) n_j(r) = c_i(r, t), \quad r \in B_\sigma^{(i)}(t)$$

where $d_i(r, t)$, $c_i(r, t)$ are known functions. The complementary property of $B_u^{(i)}(t)$, $B_\sigma^{(i)}(t)$ means that

$$B_u^{(i)}(t) \cup B_\sigma^{(i)}(t) = B, \quad i = 1, 2, 3 . \tag{1.8.16}$$

Note that we allow the boundary regions $B_u^{(i)}(t)$, $B_\sigma^{(i)}(t)$ to be time-dependent. In fact, it emerges that if they are not, the boundary value problem is relatively trivial in the sense of being closely related to the corresponding elastic problem (Sect. 1.2.1). The first and second boundary value problems are particular examples of this.

We have also allowed the boundary regions to depend on the component being specified. This is required to cover examples such as frictionless contact. These are not the most general boundary conditions that can be conceived. For example the case of frictional contact is not even covered by this scheme. However, we shall see in Chap. 3 that, at least in the plane case, such problems can be handled by methods similar to those used in the frictionless case. The fact is that the methods outlined in later chapters for attacking viscoelastic boundary value problems are all indirect, in the sense that they focus on the boundary quantities, with the aim of determining these quantities everywhere on B. Once this is done, the task of determining any quantity in the interior of the medium is in principle easy. Indeed, it is either a first or second boundary value problem as defined above.

It is possible also to conceive of bodies that (in their undeformed state) occupy time-dependent regions of space: material points may be removed (ablation) or added (accretion) to the body as time progresses. Examples of ablation are provided by the burning of a solid rocket grain or the erosion of the Earth's surface while the growth of an icicle is an everyday example of accretion. Boundary conditions for these problems may take, among others, any of the forms $(1.8.13 - 15)$ where the boundary B of the body is itself time-dependent.

1.8.2 The Non-inertial Approximation

In materials with high internal friction losses, it often happens that inertial effects, namely those depending on ϱ, the density of the material, may be neglected compared with viscous effects. This is analogous to the approximation in the theory of simple harmonic motion through a frictional medium, where the acceleration term is neglected compared with the frictional resistance term. In the particle mechanics context, such an approximation results in the reduction of the differential equation of motion from a second to a first order equation. This is a considerable simplification. In the continuum mechanics context, the non-inertial approximation results in major simplification also, though of a different kind. One drops the acceleration term in (1.1.3) to obtain

$$\sigma_{ij,j}(r, t) = 0 \tag{1.8.17}$$

in the absence of body forces, or from (1.8.11):

$$\int_{-\infty}^{t} dt' \mu_{ijkl}(t - t')(u_{l,jk} + u_{k,jl})(r, t') = 0 . \tag{1.8.18}$$

Except in Chap. 7, the non-inertial approximation is adopted throughout this work. Even for the very simple inertial problems considered in Chap. 7, it is apparent that there are significant difficulties associated with the retention of inertial terms.

1.8.3 Frequency Representation

We seek here to generalize the observation contained in (1.5.1) to the three-dimensional context. Consider (1.8.9 – 11). On taking Fourier transforms (FTs), these become, with the aid of the Faltung theorem [(A3.1.14 – 17)]:

$$\hat{\sigma}_{ij}(r, \omega) = \hat{\mu}_{ijkl}(\omega)\hat{\varepsilon}_{kl}(r, \omega)$$
$$- \varrho\omega^2 \hat{u}_i(r, \omega) = \tfrac{1}{2}\hat{\mu}_{ijkl}(\omega)[\hat{u}_{l,jk}(r, \omega) + \hat{u}_{k,jl}(r, \omega)] . \tag{1.8.19}$$

These are formally equivalent to the FT equations governing elastic problems, where the complex moduli $\hat{\mu}_{ijkl}(\omega)$, the properties of which are discussed in Sect. 1.5, replace the elastic moduli. This immediately suggests that elastic solutions can be modified so that they apply to corresponding viscoelastic problems. We have not mentioned boundary conditions however, and this is the catch. It will emerge that there is a simple correspondence between elastic and viscoelastic solutions only for certain types of boundary conditions. However, under quite general conditions, there is a connection between the two, and the exploration and utilization of this connection forms the subject manner of virtually all later chapters.

1.8.4 Proportionality Assumption

Under the assumption that (1.8.5) holds, there may be as many as 21 independent components of $G_{ijkl}(t)$. The presence of this many independent functions significantly complicates the mathematical analysis of many problems. There are also, naturally, major experimental problems involved in the determination of so many functions. Therefore, if approximate or exact interrelationships between the various functions can be established, considerable simplifications may result. A rather strong assumption, from the physical point of view, which greatly simplifies the mathematics is to put all the $G_{ijkl}(t)$ proportional to one function:

$$G_{ijkl}(t) = c_{ijkl}D(t) \ , \tag{1.8.20}$$

where the c_{ijkl} are constants. The relaxation characteristics of the medium are all contained in the function $D(t)$. This assumption will be discussed further in the next section, in the context of isotropic materials. Note that, in terms of $\mu_{ijkl}(t)$, it becomes

$$\mu_{ijkl}(t) = c_{ijkl}\mu(t) \ , \quad \mu(t) = D(0)\delta(t) + \dot{D}(t)H(t) \ . \tag{1.8.21}$$

Taking FTs gives

$$\hat{\mu}_{ijkl}(\omega) = c_{ijkl}\hat{\mu}(\omega) \ , \tag{1.8.22}$$

so that the same property holds for the complex moduli.

A consequence of the proportionality assumption, which will be of some importance, is that if the quantities

$$v_i(r, t) = \int\limits_{-\infty}^{t} dt' \mu(t - t')u_i(r, t') \tag{1.8.23}$$

are substituted for the displacements, the non-inertial governing field equations reduce to linear elastic form. In particular, if the quantities $v_i(r,t)$ can be determined on the boundary regions where displacements are specified for a given problem, then the solutions to that problem in the elastic case, if known, immediately given solutions to the corresponding viscoelastic problem with the same specified stresses. If the hereditary integral is cast over onto the stresses, then a similar observation applies, with displacements and stresses interchanged. This point is illustrated in a more concrete fashion in Graham (1969), Graham and Golden (1988) and in Sect. (2.4).

1.8.5. Non-isothermal Equations

Consider the constitutive and dynamical equations governing a thermorheologically simple material (Sect. 1.7). Firstly, the space and time homogeneity properties break down so that we write the material functions as $G_{ijkl}(r, t, t')$. We denote the temperature variation about an average background value T_0 as $\theta(r, t)$ [see (1.7.9)]. It is not however assumed that $\theta(r, t)$ is sufficiently small so that a fully linearized theory is justified.

In a thermally anisotropic medium, temperature variation will cause expansion or contraction and other distortions, thereby influencing the strain and stress tensors. We replace (1.8.9, 10) by

$$\sigma_{ij}(r, t) = \int\limits_{-\infty}^{t} dt' \mu_{ijkl}(r, t, t') \ [\varepsilon_{kl}(r, t') - \alpha_{kl}\theta(r, t')] \tag{1.8.24a}$$

where

$$\mu_{ijkl}(r, t, t') = G_{ijkl}(r, t, t)\delta(t - t') - \left[\frac{d}{dt'} G_{ijkl}(r, t, t')\right] H(t - t') \ , \tag{1.8.24b}$$

and α_{kl} is a constant matrix of coefficients of thermal expansion. Equation (1.8.24a) is chosen because it is a natural generalization of the thermoelastic constitutive relation. By putting σ_{ij} and ε_{ij} equal to zero in turn, we see that it gives the limiting behaviour that we would expect.

The same mechanical equations of motion hold, namely (1.8.11) or (1.8.17). In general one would require further constitutive and dynamical equations to determine $\theta(r, t)$ as well as σ_{ij}, ε_{ij}. However, we will have occasion to consider only a problem where $\theta(r, t)$ is specified as a function of space and time (see Chap. 6). In these circumstances, no further relations are necessary.

Homogeneity can be re-established, in terms of the pseudo-time variable ξ, given by (1.7.6), for a thermorheologically simple material, in the constitutive equation at least. In the light of (1.7.8), we rewrite (1.8.24a) as

$$\sigma_{ij}(r, \xi) = \int\limits_{-\infty}^{\xi} d\xi' \mu_{ijkl}(\xi - \xi') \left[\varepsilon_{kl}(r, \xi') - a_{kl}\theta(r, \xi') \right]$$

$$\mu_{ijkl}(\xi - \xi') = G_{ijkl}(0)\delta(\xi - \xi') + \left[\frac{d}{d\xi} G_{ijkl}(\xi - \xi') \right] H(\xi - \xi') \ .$$

(1.8.25)

However, it is clear that the dynamical equation (1.8.11) cannot be re-expressed in terms of ξ, so in fact, complete homogeneity is not re-established. In the non-inertial approximation, (1.8.17) becomes

$$\sigma_{ij,j}(r, \xi) = -\frac{\partial \sigma_{ij}}{\partial \xi} (r, \xi) \, \xi_{,j}(r, t)$$

(1.8.26)

if the temperature field, and hence the variable ξ, depends on position. If, on the other hand, the temperature field depends only on time, the right-hand side is zero and the dynamical equations, as well as constitutive equations can be re-expressed consistently in terms of ξ rather than t, and space and time homogeneity are recovered. This was first observed by Sternberg (1964). A position-independent temperature field is however a very particular case.

1.8.6. Uniqueness and Other Theorems

Rigorous proofs of the existence of unique solutions to linear viscoelastic boundary value problems have been given by various authors but will not be discussed here. Instead we refer to some of the literature. Gurtin and Sternberg (1962) revive a result for the isothermal case, due originally to Volterra. This theorem has been proved in an alternative manner by Onat and Breuer (1963). Sternberg and Gurtin (1963) extend Volterra's result to include thermal stresses in a thermorheologically simple, ablating body. More general isothermal theorems have been given by Edelstein and Gurtin (1964), Barberan and Herrera (1966), Lubliner and Sackman (1967) and Day (1970). Detailed discussion of this topic may be found in Christensen (1982) and also Leitman and Fisher (1973).

These theorems all apply to the boundary value problem (1.8.15) with B_u, B_σ independent of the component subscript i and, mainly speaking, of time also, though Sternberg and Gurtin (1963) do consider a restricted sort of time dependence. The fact that B_u, B_σ are the same for all components excludes application to contact problems. However, Boucher (1975) proves an existence and uniqueness theorem for contact problems, without friction. The restriction on the time dependence of the boundary regions really excludes applicability to most problems considered in the present volume. Another limitation is that, with the exception of Boucher's work, the theorems as formulated are primarily confined to finite bodies and in particular do not apply to half-spaces.

A review of many topics in viscoelasticity giving generalizations of results in classical elasticity to viscoelasticity is given by Leitman and Fisher (1973). Earlier papers on which this work is based include Gurtin and Sternberg (1962), Sternberg (1964) and Sternberg and Al Khozaie (1964).

1.9 Isotropic Media

If a medium is mechanically isotropic, the number of independent components of $G_{ijkl}(t)$ reduces to two. The argument is the same as that used in Elasticity Theory. Let us decompose the stress tensor into hydrostatic and deviatoric components σ, s given by (1.1.6, 7) and the strain tensors into volume strain ε, given by (1.1.5) and

$$e = \varepsilon - \tfrac{1}{3} I\varepsilon. \tag{1.9.1}$$

Then, in the isothermal approximation, we have

$$s(r, t) = 2\,[G*de]\,(r, t)\ , \qquad \sigma(r, t) = 3\,[K*d\varepsilon]\,(r, t)\ , \tag{1.9.2}$$

where $G(t)$, $K(t)$ are the two independent components characterizing shear and volumetric deformation, respectively. The numerical factors are conventional. Note that $G(t)$ here differs from that in Sect. 1.2 by a factor of 2. The purpose of that and subsequent sections was merely to establish general properties of the function, which are not influenced by the overall factor. We remark that the properties of many viscoelastic materials are very sensitive to volume change. This is true for example, in the case of polymers. Therefore the linear theory is valid only for very small volume changes. Combining the equations gives

$$\sigma(r, t) = 2\,[G*d\varepsilon]\,(r, t) + I\,[M*d\varepsilon]\,(r, t) \tag{1.9.3}$$

or, more explicitly:

$$\sigma_{ij}(r, t) = 2 \int_{-\infty}^{t} dt'\,G(t-t')\dot{\varepsilon}_{ij}(r, t) + \delta_{ij} \int_{-\infty}^{t} dt'\,M(t-t')\dot{\varepsilon}(r, t')$$

$$M(t) = K(t) - \tfrac{2}{3}\,G(t)\ . \tag{1.9.4}$$

This is the isotropic form of (1.8.8). It follows that

$$G_{ijkl}(t) = G(t)\,(\delta_{ik}\delta_{jl} + \delta_{il}\delta_{jk}) + M(t)\delta_{ij}\delta_{kl}\ . \tag{1.9.5}$$

We can rewrite (1.9.4) in the form

$$\sigma_{ij}(r, t) = 2 \int_{-\infty}^{t} dt'\,\mu(t-t')\varepsilon_{ij}(r, t') + \delta_{ij} \int_{-\infty}^{t} dt'\,\lambda(t-t')\varepsilon(r, t')\ , \tag{1.9.6}$$

where $\mu(t)$ is given by (1.2.33 a) and

$$\lambda(t) = M(0)\delta(t) + \dot{M}(t)H(t) . \tag{1.9.7}$$

Equation (1.9.6) will be the basis of many of the considerations of later chapters. It follows from (1.9.5) that $\mu_{ijkl}(t)$, given by (1.8.10), has the form

$$\mu_{ijkl}(t) = \mu(t)\,(\delta_{ik}\delta_{jl} + \delta_{il}\delta_{jk}) + \lambda(t)\delta_{ij}\delta_{kl} \tag{1.9.8}$$

in the isotropic case.

The dynamical equation (1.8.11) becomes

$$\varrho\,\frac{\partial^2}{\partial t^2}\,u_i(r,t) = \int\limits_{-\infty}^{t} dt'\mu(t-t')\nabla^2 u_i(r,t')$$

$$+ \int\limits_{-\infty}^{t} dt'\,[\mu(t-t') + \lambda(t-t')]\,\frac{\partial}{\partial x_i}\,\boldsymbol{\nabla}\cdot\boldsymbol{u}(r,t') \tag{1.9.9}$$

under suitable initial and boundary conditions, as discussed in the previous section. In the non-inertial approximation, this becomes

$$\sigma_{ij,j}(r,t) = \int\limits_{-\infty}^{t} dt'\mu(t-t')\nabla^2 u_i(r,t')$$

$$+ \int\limits_{-\infty}^{t} dt'\,[\mu(t-t') + \lambda(t-t')]\,\frac{\partial}{\partial x_i}\,\boldsymbol{\nabla}\cdot\boldsymbol{u}(r,t') \tag{1.9.10}$$

$$= 0.$$

From (1.8.1, 5) we have for the stored energy density, omitting explicit space dependence:

$$V(t) = \int\limits_{-\infty}^{t} dt_1 \int\limits_{-\infty}^{t} dt_2 G(2t - t_1 - t_2)\dot{\varepsilon}_{ij}(t_1)\dot{\varepsilon}_{ij}(t_2)$$

$$+ \frac{1}{2} \int\limits_{-\infty}^{t} dt_1 \int\limits_{-\infty}^{t} dt_2 M(2t - t_1 - t_2)\dot{\varepsilon}(t_1)\dot{\varepsilon}(t_2). \tag{1.9.11}$$

The rate of dissipation of energy per unit volume, given by (1.8.7), has the form

$$\dot{D}(t) = -2 \int\limits_{-\infty}^{t} dt_1 \int\limits_{-\infty}^{t} dt_2 \dot{G}(2t - t_1 - t_2)\dot{\varepsilon}_{ij}(t_1)\dot{\varepsilon}_{ij}(t_2)$$

$$- \int\limits_{-\infty}^{t} dt_1 \int\limits_{-\infty}^{t} dt_2 \dot{M}(2t - t_1 - t_2)\dot{\varepsilon}(t_1)\dot{\varepsilon}(t_2). \tag{1.9.12}$$

From the considerations of Sect. 1.2, we know that we can invert (1.9.2) to obtain

$$e(r,t) = \tfrac{1}{2}\,[J*ds]\,(r,t) , \qquad \varepsilon(r,t) = \tfrac{1}{3}\,[Q*d\sigma]\,(r,t) , \tag{1.9.13}$$

where $J(t)$, $Q(t)$ describe creep in shear and bulk deformation, respectively. Defining $\gamma(t)$ by (1.2.33 b) and putting

$$\psi(t) = Q(0)\delta(t) + \dot{Q}(t)H(t) , \tag{1.9.14}$$

we obtain

$$e(r, t) = \frac{1}{2} \int_{-\infty}^{t} dt' \gamma(t-t') s(r, t') = \frac{1}{2} [\gamma * s](r, t) \tag{1.9.15a}$$

$$\varepsilon(r, t) = \frac{1}{3} \int_{-\infty}^{t} dt' \psi(t-t') \sigma(r, t') = \frac{1}{3} [\psi * \sigma](r, t) \tag{1.9.15b}$$

where $\gamma(t)$, $\mu(t)$ are related by (1.2.35), as are $\psi(t)$, $\pi(t)$, the latter quantity being related to $K(t)$ by

$$\pi(t) = K(0)\delta(t) + \dot{K}(t)H(t). \tag{1.9.16}$$

The inverted constitutive equations read:

$$\varepsilon_{ij}(r, t) = \frac{1}{2} \int_{-\infty}^{t} dt' J(t-t') \dot{\sigma}_{ij}(r, t')$$

$$+ \frac{\delta_{ij}}{9} \int_{-\infty}^{t} dt' \left[Q(t-t') - \frac{3}{2} J(t-t') \right] \dot{\sigma}(r, t') \tag{1.9.17a}$$

$$= \frac{1}{2} \int_{-\infty}^{t} dt' \gamma(t-t') \sigma_{ij}(r, t')$$

$$+ \frac{\delta_{ij}}{9} \int_{-\infty}^{t} dt' \left[\psi(t-t') - \frac{3}{2} \gamma(t-t') \right] \sigma(r, t') . \tag{1.9.17b}$$

1.9.1 Frequency Representation

Equations (1.8.19) become in the isotropic case

$$\hat{\sigma}_{ij}(r, \omega) = 2\hat{\mu}(\omega)\hat{\varepsilon}_{ij}(r, \omega) + \delta_{ij}\hat{\lambda}(\omega)\hat{\varepsilon}(r, \omega)$$

$$-\varrho\omega^2 \hat{u}_i(r, \omega) = \hat{\mu}(\omega)\nabla^2 \hat{u}_i(r, \omega) + [\hat{\mu}(\omega) + \hat{\lambda}(\omega)] \frac{\partial}{\partial x_i} \nabla \cdot \hat{u}(r, \omega) , \tag{1.9.18}$$

which are, as observed in a more general context, identical to the elastic equations with $\hat{\mu}(\omega)$, $\hat{\lambda}(\omega)$ replacing Lamé's constants.

The various moduli and other parameters that can be formed out of Lamé's constants in elastic theory may be generalized to the viscoelastic case by simply substituting $\hat{\mu}(\omega)$, $\hat{\lambda}(\omega)$ for μ, λ, and, if one wishes to return to the time representation, inverting the Fourier transform. An important example is $\hat{v}(\omega)$, given by

$$\hat{v}(\omega) = \frac{\hat{\lambda}(\omega)}{2[\hat{\lambda}(\omega) + \hat{\mu}(\omega)]} , \tag{1.9.19}$$

which can be regarded as a generalization of Poisson's ratio.

1.9.2 Proportionality Assumption

Even in the isotropic case, significant mathematical simplification of many problems results from taking $G(t)$ and $M(t)$ or $\mu(t)$ and $\lambda(t)$ proportional. From (1.8.22), we see that this is equivalent to putting $\hat{\lambda}(\omega)$ proportional to $\hat{\mu}(\omega)$. This, together with (1.9.19), gives immediately that $\hat{v}(\omega)$ is a constant v, so that its inverse transform is

$$v(t) = v\delta(t) \ . \tag{1.9.20}$$

This assumption will be made in many contexts in later chapters. Let us explore a little further its essential physical content. If $\pi(t)$ is given by (1.9.16) and $K(t)$ by (1.9.4), then

$$\hat{\pi}(\omega) = \tfrac{2}{3}\hat{\mu}(\omega) + \hat{\lambda}(\omega) \ , \tag{1.9.21}$$

so that $\hat{\pi}(\omega)$ is proportional to $\hat{\mu}(\omega)$ and $\pi(t)$ to $\mu(t)$. In other words, the time behaviour of the shear and bulk viscoelastic functions have similar shapes. However, this would not be expected to be true in general since the molecular mechanisms which determine the time dependence of the viscoelastic functions will have little in common in these two cases. In fact, viscoelastic effects tend to be less pronounced for volume than for shear deformation.

This would seem to indicate that the proportionality assumption has no physical basis. However, in many cases the bulk modulus is much larger than the shear modulus (except perhaps near $t = 0$) so that, for practical purposes, we can take $v = \tfrac{1}{2}$. In other words, $K(t)$ is so much larger than $G(t)$ that it does not matter what shape we assume for $K(t)$. This is the case for amorphous polymers at temperatures well above their glass transition temperatures. Also for many rigid plastics, for example, amorphous polymers in the glassy state, v is constant but less than $\tfrac{1}{2}$, typically having values in the range $0.35 - 0.41$ [Schapery (1974)]. In such cases, the proportionality assumption would seem to have approximate validity. In summary, therefore, this assumption, while motivated primarily by the need for mathematical simplicity, is a reasonable approximation for many materials.

Another simplifying assumption, which is sometimes made, is to neglect viscoelastic effects associated with bulk deformation, in other words to put $K(t)$ equal to a constant or $\pi(t)$ proportional to a delta function.

1.9.3 Non-isothermal Relations

We will assume thermal as well as mechanical isotropy, putting

$$a_{ij} = a\delta_{ij} \ , \tag{1.9.22}$$

so that (1.8.25) becomes

$$\sigma_{ij}(r, \xi) = 2 \int_{-\infty}^{\xi} d\xi' \mu(\xi - \xi')\varepsilon_{ij}(r, \xi') + \delta_{ij} \int_{-\infty}^{\xi} d\xi' \lambda(\xi - \xi')\varepsilon(r, \xi')$$

$$- \delta_{ij} a \int_{-\infty}^{\xi} d\xi' [2\mu(\xi - \xi') + 3\lambda(\xi - \xi')] \theta(r, \xi')$$

$$= 2[\mu * \varepsilon_{ij}](r, \xi) + \delta_{ij}[\lambda * \varepsilon](r, \xi) - \delta_{ij} a[(2\mu + 3\lambda) * \theta](r, \xi) \ , \tag{1.9.23}$$

where the pseudo-time variable ξ is given by (1.7.6); $\mu(\xi)$ and $\lambda(\xi)$ are given by (1.2.33a) and (1.9.7) with t replaced by ξ; and we have used the notation of (1.2.22), (with $t_1 = -\infty$).

The inverted equations (1.9.17b) are modified by replacing time by pseudo-time variables and adding a term $\delta_{ij}a\theta$ to the right-hand side, so that

$$\varepsilon_{ij}(r, \xi) = \frac{1}{2}\,[\gamma * \sigma_{ij}](r, \xi) + \frac{\delta_{ij}}{9}\left[\left(\psi - \frac{3}{2}\,\gamma\right)*\sigma\right](r, \xi) + \delta_{ij}a\theta(r, \xi),\quad (1.9.24)$$

where $\gamma(\xi)$ and $\psi(\xi)$ are given by (1.2.33b) and (1.9.14) with t replaced by ξ, and we are using the notation of (1.2.22).

1.9.4 Compatibility Equations

The equations of compatibility have the same form as in Elasticity, namely

$$\varepsilon_{ij, kl} + \varepsilon_{kl, ij} - \varepsilon_{ik, jl} - \varepsilon_{jl, ik} = 0\ ,\qquad i, j, k, l = 1, 2, 3\ ,\qquad (1.9.25)$$

which are satisfied identically for ε_{ij} given by (1.1.4). In a simply connected region, satisfaction of (1.9.25) is sufficient to ensure that (1.1.4) has solutions that are single-valued continuous displacements [Sokolnikoff (1956)]. It sometimes happens that one is primarily interested in stresses rather than displacements so that the dynamical equations (1.9.10), expressed in terms of displacements, are not particularly convenient. In the non-inertial approximation, an alternative set of equations may be derived for an elastic medium, known as the Beltrami-Michell compatibility equations. These equations are easily generalized to the case of a viscoelastic medium. Consider the inverted constitutive relations (1.9.17). If these are substituted into (1.9.25), one obtains the desired equations in terms of the stresses. On going into the frequency representation, carrying out manipulations identical to those in the elastic case, including the use of the equilibrium equations [Sokolnikoff (1956)] and reverting to the time variable, one obtains

$$\nabla^2 \sigma_{ij} + \int\limits_{-\infty}^{t} dt'\, v_1(t - t')\sigma_{kk, ij} = 0 \qquad\qquad (1.9.26)$$

in the absence of body forces, where $v_1(t)$ is defined by the relation

$$\hat{v}_1(\omega) = \frac{1}{1 + \hat{v}(\omega)}\,, \qquad\qquad (1.9.27)$$

where $\hat{v}(\omega)$ is given by (1.9.19). If the proportionality assumption is adopted, then (1.9.26) reduces identically to the elastic Beltrami-Michell equations. A consequence of this is that if the boundary value problem is of the first kind (purely stress), and the body occupies a simply connected region, then the stress tensor is identical to the elastic form [see Gurtin and Sternberg (1962)].

In the above argument, it is assumed that the material is homogeneous in space and time. Under non-isothermal conditions in particular, this assumption breaks down. The resulting equations are more complicated than (1.9.26). They will not be discussed in general. A special case is considered in Sect. 6.1, using polar coordinates.

1.10 Causality

We have seen, in Sect 1.8, 1.9, that the Fourier transform of the equations governing the behaviour of the medium are formally identical to the elastic equations, except that we replace moduli by complex moduli. This suggests that elastic solutions can be used to generate viscoelastic solutions. The procedure by which this is done is not straightforward, as we shall see in Chap. 2, if the boundary regions are time-dependent. However, the statement is valid for many types of problems. A consequence of this is that we shall meet expressions, for physical quantities at time t, of the form

$$h(r, t) = \frac{1}{2\pi} \int_{-\infty}^{\infty} d\omega\, e^{i\omega t} \hat{f}_i(\omega) \hat{g}_i(r, \omega) \ , \tag{1.10.1}$$

where $h(r, t)$ is the quantity of interest; the $g_i(r, t)$ usually depend upon the boundary quantities, while the $\hat{f}_i(\omega)$ are functions of the complex moduli. Typically, these factors $\hat{f}_i(\omega)$ are equal to, or clearly related to, combinations of moduli that occur in elastic Green's functions relevant to the problem under consideration.

The Faltung Theorem (A3.1.14 − 17) gives that

$$h(r, t) = \int_{-\infty}^{\infty} dt' f_i(t - t') g_i(r, t') \ . \tag{1.10.2}$$

Now, $h(r, t)$ cannot depend upon the values of boundary quantities in the future. This is an absolute physical requirement, which we refer to as the Principle of Causality. It implies that (1.10.2) must reduce to

$$h(r, t) = \int_{-\infty}^{t} dt' f_i(t - t') g_i(r, t') \qquad \text{or} \tag{1.10.3}$$

$$f_i(t) = 0, \quad t < 0 \ . \tag{1.10.4}$$

If $f_i(t)$ does not diverge exponentially at large (t), then (1.10.4) is equivalent to the requirement that $\hat{f}_i(\omega)$ be analytic in the lower half complex ω plane. The presence of a singularity in the lower half-plane means that, if (1.10.4) is to hold, then $f_i(t)$ must diverge exponentially at large t. This point is discussed in Sect. A3.2. Such a singularity would be expected to have a definite physical meaning if it is present. In this section we shall argue that no such singularity is present in non-inertial problems. In Chap. 7, we shall come to a similar conclusion for subsonic inertial problems, though not for problems where the velocity is in the range of possible sonic speeds.

Let us consider such a factor $\hat{f}(\omega)$ (we drop the subscript) for non-inertial, isotropic problems. Typically, it would be a rational function of the complex moduli $\hat{\mu}(\omega)$, $\hat{\lambda}(\omega)$, which we write

$$\hat{f}(\omega) = \frac{p(\hat{\lambda}(\omega), \hat{\mu}(\omega))}{q(\hat{\lambda}(\omega), \hat{\mu}(\omega))} \tag{1.10.5}$$

where $p(x, y), q(x, y)$ are polynomials in x, y. It will be assumed that these polynomials are factorizable into linear terms with real coefficients. This is true for all the standard solutions. Our object is to explore the singularity structure of $\hat{f}(\omega)$, based on the given fact that $\hat{\mu}(\omega), \hat{\lambda}(\omega)$ are causal in the sense of having no singularities in the lower half-plane. Specifically, we wish to show that $\hat{f}(\omega)$ also has this property.

The only way a singularity can arise in $\hat{f}(\omega)$, in the lower half-plane, given the properties of $\hat{\mu}(\omega), \hat{\lambda}(\omega)$, is that $q(\hat{\lambda}(\omega), \hat{\mu}(\omega))$ becomes zero for ω in the lower half-plane. Let us re-express $q(\hat{\lambda}(\omega), \hat{\mu}(\omega))$ as $q_1(\hat{\pi}(\omega), \hat{\mu}(\omega))$ where $\pi(t)$ is defined by (1.9.16). It is the inverse of $\psi(t)$ which must be causal. Similarly, $\mu(t)$ is the inverse of the causal function $\gamma(t)$. It follows that neither $\hat{\mu}(\omega), \hat{\pi}(\omega)$ can have zeros in the lower-half plane, since if they do, $\hat{\gamma}(\omega), \hat{\psi}(\omega)$ will have poles there. The only possibility for generating a singularity in the lower half-plane is that cancellations occur between the terms of $q_1(\hat{\pi}(\omega), \hat{\mu}(\omega))$. However, we place one restriction on this function $q_1(k, \mu)$, namely that for $k > 0$, $\mu > 0$, it cannot be zero. The point is that in the elastic limit, $\hat{\pi}(\omega), \hat{\mu}(\omega)$ reduce to the bulk and shear moduli which can, in principle, take any positive values. A zero in $q_1(k, \mu)$ for values in this range would correspond to a totally un-physical singularity for one particular choice of the moduli, which must be excluded as a possibility.

We will now show that the real and imaginary parts of $\hat{\mu}(\omega), \hat{\pi}(\omega)$ do not change sign in the lower half-plane, and in particular, the real part remains positive. This will be sufficient to prove the desired result because if $q_1(\hat{\pi}, \hat{\mu})$ became zero in the lower half-plane, it would imply the presence of an unphysical singularity, in the elastic case, of the type excluded above. This follows on recall-ing the factorized structure of q and therefore, q_1, which implies that any cancellation will occur separately in the real and imaginary parts of the moduli.

Let us discuss $\hat{\mu}(\omega)$. A similar argument applies to $\hat{\pi}(\omega)$. Note that the fact that these quantities cannot be zero for $\text{Im}\{\omega\} < 0$ does not mean that their real and imaginary parts cannot change sign at distinct values of ω. We need a specific model to make further progress. Let us adopt a continuous spectrum form for $\hat{\mu}(\omega)$, namely [see (1.6.38)]:

$$\hat{\mu}(\omega) = g_0 + \int_0^\infty da \, \frac{g(a)}{a + i\omega} , \qquad g(a) < 0 \qquad (1.10.6)$$

for ω real. Now let

$$u = \omega - iv , \qquad v > 0 , \qquad (1.10.7)$$

giving

$$\hat{\mu}(u) = g_0 + \int_0^\infty da \, \frac{g(a)}{a + v + i\omega} , \qquad v > 0$$

$$(1.10.8)$$

$$= G_0 + \int_0^\infty G_1(a) \left(\frac{(a+v)v + \omega^2}{(a+v)^2 + \omega^2} + \frac{ia\omega}{(a+v)^2 + \omega^2} \right)$$

in terms of the quantities in (1.6.36), where $G_1(a)$ is positive. It is clear therefore that, at least for the continuous spectrum model, which is fairly general, the real part of $\hat{\mu}(\omega)$ remains positive in the lower half-plane. A similar argument applies to $\hat{\pi}(\omega)$. It follows, from the above assumptions, that $f(\omega)$ will have no singularities in the lower half-plane.

Problem 1.10.1: Show that $\hat{v}(\omega)$, given by (1.9.19) has no singularities in the lower half-plane. Show also that $v(t)$ is real [see (A3.1.13)].

Problem 1.10.2: We will have occasion to consider functions of $\hat{\mu}(\omega)$ and $\hat{\lambda}(\omega)$, free of singularities for $\text{Im}\{\omega\}<0$, that are finite at infinite frequency. Denoting such a function by $\hat{l}(\omega)$, show that $l(t)$ may be represented in the form

$$l(t) = L(0)\delta(t) + \dot{L}(t)H(t) ,$$ (1.10.1p)

where [see (1.5.2), (1.5.10 – 13)]

$$L(t) = \hat{l}(0) + \frac{1}{2\pi} \int_{-\infty}^{\infty} d\omega\, e^{i\omega t}\, \frac{\hat{l}(\omega) - \hat{l}(0)}{i\omega} , \quad t>0$$

(1.10.2p)

$$L(0) = \hat{l}(\infty) , \quad L(\infty) = \hat{l}(0) .$$

1.11 Summary

The purpose of this first chapter has been to present the fundamental components of the linear theory of viscoelasticity, upon which the rest of the book relies. Before proceeding to the next chapter, we now highlight the main results.

I. Stress and Strain. Cartesian components of stress (σ_{ij}) and strain (ε_{ij}) are defined as in the classical theory of elasticity.

II. One-dimensional Constitutive Equations. (a) One-dimensional stress-strain relations for a non-aging linear viscoelastic material take the form

$$\sigma(t) = G(\infty)\varepsilon(-\infty) + \int_{-\infty}^{t} dt'\, G(t-t')\dot{\varepsilon}(t') = [G*d\varepsilon](t)$$ (1.11.1)

$$= \int_{-\infty}^{t} dt'\, \mu(t-t')\varepsilon(t') = [\mu*\varepsilon](t) ,$$ (1.11.2)

where, in terms of the *relaxation function* $G(t)$, $t\geq 0$,

$$\mu(t) = G(0)\delta(t) + \dot{G}(t)H(t) .$$ (1.11.3)

(b) In terms of the *creep function* $J(t), t\geq 0$, the inverses of (1.11.1, 2) are

$$\varepsilon(t) = [J*d\sigma](t) = [\gamma*\sigma](t) \quad \text{where}$$ (1.11.4)

$$\gamma(t) = J(0)\delta(t) + \dot{J}(t)H(t) .$$

(1.11.5)

(c) $G(t)$ and $J(t)$ are related through

$$\int_0^t dt' J(t-t')G(t') = \int_0^t dt' G(t-t')J(t') = t , \quad t \geq 0 ,$$

(1.11.6)

$$G(0)J(0) = G(\infty)J(\infty) = 1 ,$$

(1.11.7)

while $\mu(t)$ is related to $\gamma(t)$ through

$$\int_0^t dt' \mu(t-t')\gamma(t') = \int_0^t dt' \gamma(t-t')\mu(t') = \delta(t) .$$

(1.11.8)

(d) For a *standard linear solid*

$$G(t) = G_0 + G_1 e^{-t/\tau} ; \quad J(t) = J_0 + J_1(1 - e^{-t/\tau'}) , \quad \text{where}$$

(1.11.9)

$$J_0 = J_1 \frac{G_0}{G_1} = (G_0 + G_1)^{-1}; \quad \tau = \tau' G_0(G_0 + G_1)^{-1} .$$

(1.11.10)

(e) For *discrete spectrum models*

$$G(t) = G_0 + \sum_{i=1}^N G_i e^{-t/\tau_i} ; \quad J(t) = J_0 + \sum_{i=1}^N J_i(1 - e^{-t/\tau_i'})$$

(1.11.11)

where the constants G_0, J_0, G_i, J_i, τ_i, τ_i', $i = 1, 2, \ldots, N$, are related according to formulae given in Sect. 1.1.6.

(f) For *continuous spectrum models*

$$G(t) = G_0 + \int_0^\infty da\, G_1(a)e^{-at} ; \quad J(t) = J_0 + \int_0^\infty da\, J_1(a)(1 - e^{-at}) .$$

(1.11.12)

(g) For *power law models*

$$G(t) = A t^{-p}; \quad J(t) = B t^p , \quad 0 < p < 1 , \quad \text{where}$$

(1.11.13)

$$ABp\pi = \sin(p\pi) .$$

(1.11.14)

III. Definition of Solids and Liquids. It is a physical requirement of the theory that $\dot{J}(\infty)$ be finite. The material is a solid if and only if $\dot{J}(\infty) = 0$; otherwise it is a fluid. The (long-time) viscosity of the fluid is given by

$$\eta = [\dot{J}(\infty)]^{-1} = \int_0^\infty dt\, G(t) .$$

(1.11.15)

IV. Energy Formulae (one-dimensional case). (a) stored energy density:

$$V(t) = \frac{1}{2} \int_{-\infty}^t dt' \int_{-\infty}^t dt'' G(2t - t' - t'')\dot{\varepsilon}(t')\dot{\varepsilon}(t'') ;$$

(1.11.16)

(b) rate of dissipation of mechanical energy density into heat:

$$\dot{D}(t) = (-) \int\limits_{-\infty}^{t} dt' \int\limits_{-\infty}^{t} dt'' \dot{G}(2t - t' - t'')\dot{\varepsilon}(t')\dot{\varepsilon}(t'') \ . \tag{1.11.17}$$

V. Frequency Relationships. (a) The *complex modulus* (in relaxation) is

$$\hat{\mu}(\omega) = \int\limits_{0}^{\infty} dt\, \mu(t)\mathrm{e}^{-\mathrm{i}\omega t} = \hat{\mu}_1(\omega) + \mathrm{i}\hat{\mu}_2(\omega) \ , \tag{1.11.18}$$

where $\hat{\mu}_1(\omega)$ is the *storage modulus* and $\hat{\mu}_2(\omega)$ is the *loss modulus*. These satisfy the relations

$$\hat{\mu}_1(0) = G(\infty) \ , \hat{\mu}_1(\infty) = G(0) \ ; \quad \hat{\mu}_2(0) = \hat{\mu}_2(\infty) = 0 \ . \tag{1.11.19}$$

(b) A steady-state sinusoidal strain of amplitude ε_0 and period $2\pi/\omega$ gives rise to a sinusoidal stress of amplitude $\varepsilon_0 |\hat{\mu}(\omega)|$ at the same frequency. The strain lags behind the stress in phase by an amount (the loss angle) $\phi(\omega)$, which lies in $[0, \frac{\pi}{2}]$ and satisfies

$$\tan \phi = \frac{\hat{\mu}_2(\omega)}{\hat{\mu}_1(\omega)} \ . \tag{1.11.20}$$

In this deformation, the average value of the stored energy density is

$$\langle V(t) \rangle = \frac{\varepsilon_0^2 \hat{\mu}_1(\omega)}{4} \tag{1.11.21}$$

while the average value of the rate of dissipation of mechanical energy into heat is

$$\langle \dot{D}(t) \rangle = \frac{\varepsilon_0^2 \omega \hat{\mu}_2(\omega)}{2} \ . \tag{1.11.22}$$

VI. Dispersion Relations

$$\hat{\mu}_1(\omega) = \mu(\infty) - \frac{2}{\pi} \int\limits_{0}^{\infty} d\omega' \frac{\mu_2(\omega')\omega'}{\omega'^2 - \omega^2} \tag{1.11.23}$$

$$\hat{\mu}_2(\omega) = \frac{2\omega}{\pi} \int\limits_{0}^{\infty} d\omega' \frac{\hat{\mu}_1(\omega')}{\omega'^2 - \omega^2} \ , \tag{1.11.24}$$

and there are other forms.

VII. Thermorheologically Simple Viscoelastic Materials. For these materials, a temperature change has the effect of scaling the relaxation and creep functions with respect to time. Equations (1.11.1, 2) are replaced by

$$\sigma(\xi) = G_0(\infty)\varepsilon(-\infty) + \int\limits_{-\infty}^{\xi} d\xi' G_0(\xi - \xi')\dot{\varepsilon}(\xi') \ , \tag{1.11.25}$$

$$= \int\limits_{-\infty}^{\xi} d\xi' \mu_0(\xi - \xi')\varepsilon(\xi') \ , \quad \text{where} \tag{1.11.26}$$

$$\xi = \xi(t) = \int_0^t \frac{d\tau}{a(T(\tau))} \ , \qquad \xi' = \xi(t') \ , \qquad \text{and} \tag{1.11.27}$$

$$\mu_0(\xi) = G_0(0)\delta(\xi) + \dot{G}_0(\xi)H(\xi) \ ; \tag{1.11.28}$$

$G_0(t)$ is the relaxation function measured at temperature T_0; $a(T)$ is given in terms of material parameters in such a way that $a(T_0) = 1$. (It follows that if $T(\tau) \equiv T_0$, then $\xi \equiv t$ and $\xi' \equiv t'$ and (1.11.1) is recovered.)

VIII. Three-dimensional Linear Viscoelasticity (isotropic case). Define

$$s_{ij} = \sigma_{ij} - \tfrac{1}{3}\sigma_{kk}\delta_{ij} \ ; \qquad e_{ij} = \varepsilon_{ij} - \tfrac{1}{3}\varepsilon_{kk}\delta_{ij} \ . \tag{1.11.29}$$

Then the *non-isothermal* stress-strain relation takes the form

$$s_{ij}(\xi) = 2 \int_{-\infty}^{\xi} d\xi' \mu_0(\xi - \xi') e_{ij}(\xi')$$

$$\tag{1.11.30}$$

$$\sigma_{kk}(\xi) = \int_{-\infty}^{\xi} d\xi' (2\mu_0 + 3\lambda_0)(\xi - \xi')(\varepsilon_{kk} - 3a\theta)(\xi')$$

where μ_0, given by (1.11.28), and λ_0 are analogues in viscoelasticity of Lamé's constants in elasticity theory; ξ and ξ' are given by (1.11.27); a is the (constant) coefficient of thermal expansion and θ is the difference between the temperature at time t and T_0. If the *Proportionality Assumption* holds, then

$$\lambda_0(\xi) = \frac{2\nu}{1-2\nu} \mu_0(\xi) \ , \tag{1.11.31}$$

where ν is a constant analogue to Poisson's ratio for an elastic material. The *isothermal* case is recovered by setting ξ, ξ' equal to t, t'; $\mu_0(\xi)$ equal to $\mu(t)$, given by (1.11.3); and $\theta \equiv 0$.

Equations (1.11.30) are supplemented by stress dynamical equations and strain-displacement equations which take the same form as in classical elasticity. Likewise, boundary conditions may take any of the forms appropriate to the classical (dynamic) theory of elasticity: in particular the type of boundary condition specified at a point on the boundary of a body may change with time; and further, the boundary of the body may change with time, through the process of ablation, for example.

IX. Causality. The requirement of Causality, namely that the current situation can be influenced only by past and contemporaneous events, may be shown to impose a constraint on the analytic structure of the complex moduli in the complex ω plane, and also on combinations of the moduli multiplying Green's functions in the solution of non-inertial boundary value problems. These quantities can have no singularities in the lower half-plane. In a restricted sense, this can be shown directly for certain combinations of complex moduli using properties of the individual complex moduli.

2. General Theorems and Methods of Solution of Boundary Value Problems

In this chapter, we discuss methods of solution of viscoelastic boundary value problems in general terms, together with certain relevant theorems. The main emphasis is on non-inertial problems. Also, most of the discussion is confined to the isothermal case.

The general results of Sects. 2.1 – 6 apply to anisotropic as well as isotropic materials. However, as the results become more specific, and forms of solutions are discussed, attention will be confined to the isotropic case.

2.1 The Classical Correspondence Principle

Consider the isothermal equations governing an anisotropic, non-aging medium, given by (1.8.9, 11) and the boundary conditions (1.8.15). It will be assumed in this section that the boundary regions $B_\sigma^{(i)}$, $B_u^{(i)}$, $i = 1, 2, 3$ are time-independent. This includes the special cases of the first and second boundary value problems. The time Fourier transform (FT) of (1.8.9, 11) are given by (1.8.19). The boundary conditions (1.8.15) similarly transformed, read

$$\hat{\sigma}_{ij}(r, \omega)n_j(r) = \hat{c}_i(r, \omega), \qquad r \in B_\sigma^{(i)} ,$$
$$\hat{u}_i(r, \omega) = \hat{d}_i(r, \omega), \qquad r \in B_u^{(i)} ,$$

(2.1.1)

where, since $B_\sigma^{(i)}$, $B_u^{(i)}$ are time-independent, the quantities $\hat{c}_i(r, \omega)$, $\hat{d}_i(r, \omega)$ can be calculated without difficulty.

Equations (1.8.19), (2.1.1) are formally identical to the equations governing the elastic problem, in FT form, where the regions $B_\sigma^{(i)}$, $B_u^{(i)}$ and the specified functions $c_i(r, t)$, $d_i(r, t)$ are the same but where the elastic moduli are replaced by the complex moduli $\hat{\mu}_{ijkl}(\omega)$. If solutions to the elastic problem are known, one can obtain a solution of the corresponding viscoelastic problem by replacing the moduli by the complex moduli in the expressions for the displacements and stresses and calculating the inverse transforms. This observation is the content of the Classical Correspondence Principle, which allows us to use the vast catalogue of known elastic solutions to generate viscoelastic solutions.

We will now spell out this procedure in more detail. Let $\{u_i^{(e)}(r, t), \sigma_{ij}^{(e)}(r, t)\}$ be an elastic solution, with constant moduli, obeying the boundary conditions

$$\sigma_{ij}^{(e)}(r, t)n_j(r) = c_i(r, t), \qquad r \in B_\sigma^{(i)} ,$$
$$u_i^{(e)}(r, t) = d_i(r, t), \qquad r \in B_u^{(i)} .$$

(2.1.2)

Taking FTs, we obtain

$$\hat{\sigma}_{ij}^{(e)}(r,\omega)n_j(r) = \hat{c}_i(r,\omega), \qquad r \in B_\sigma^{(i)} \,,$$

$$\hat{u}_i^{(e)}(r,\omega) = \hat{d}_i(r,\omega), \qquad r \in B_u^{(i)} \,, \qquad (2.1.3)$$

which have the same form as (2.1.1). However, $\{\hat{u}_i^{(e)}(r,\omega), \hat{\sigma}_{ij}^{(e)}(r,\omega)\}$ do not obey the dynamical equations (1.8.19) unless the moduli are replaced by the complex moduli. Let us do this before taking FTs. Equation (2.1.2) becomes, assuming that the specified boundary functions do not depend on the moduli,

$$\sigma_{ij}^{(e)}(r,\omega,t)n_j(r) = c_i(r,t), \qquad r \in B_\sigma^{(i)} \,,$$

$$u_i^{(e)}(r,\omega,t) = d_i(r,t), \qquad r \in B_u^{(i)} \,, \qquad (2.1.4)$$

where the frequency dependence on the left indicates that the complex moduli have been substituted. The fact that these functions depend on both time and frequency is a little strange at first sight. The time variable is associated with the time dependence of the boundary values and with inertial effects, if these are not neglected, while the frequency dependence comes in through the complex moduli. The FT of the quantities $\sigma_{ij}^{(e)}(r,\omega,t)$, $u_i^{(e)}(r,\omega,t)$ obey both the dynamical equations and the same boundary conditions as $\hat{\sigma}_{ij}(r,\omega)$ and $\hat{u}_i(r,\omega)$ and hence may be put equal to them, giving

$$\hat{u}_i(r,\omega) = \int\limits_{-\infty}^{\infty} dt' u_i^{(e)}(r,\omega,t')e^{-i\omega t'} \,,$$

$$\hat{\sigma}_{ij}(r,\omega) = \int\limits_{-\infty}^{\infty} dt' \sigma_{ij}^{(e)}(r,\omega,t')e^{-i\omega t'} \,, \qquad (2.1.5)$$

so that the viscoelastic solutions are given by

$$u_i(r,t) = \frac{1}{2\pi} \int\limits_{-\infty}^{\infty} d\omega\, e^{i\omega t} \int\limits_{-\infty}^{\infty} dt'\, u_i^{(e)}(r,\omega,t')e^{-i\omega t'} \quad \text{and} \qquad (2.1.6)$$

$$\sigma_{ij}(r,t) = \frac{1}{2\pi} \int\limits_{-\infty}^{\infty} d\omega\, e^{i\omega t} \int\limits_{-\infty}^{\infty} dt'\, \sigma_{ij}^{(e)}(r,\omega,t')e^{-i\omega t'} \qquad (2.1.7)$$

in terms of elastic solutions obeying the same boundary conditions. Note that on interchanging the order of integration we can write (2.1.6), for example, in the form

$$u_i(r,t) = \frac{1}{2\pi} \int\limits_{-\infty}^{\infty} dt' \int\limits_{-\infty}^{\infty} d\omega\, e^{i\omega(t-t')} u_i^{(e)}(r,\omega,t') \,. \qquad (2.1.8)$$

It follows immediately from Fourier's Integral Theorem (Sect. A3.1) that if $u_i^{(e)}(r,\omega,t')$ is independent of ω at any point r, then $u_i(r,t)$ and $u_i^{(e)}(r,t)$ are equal at that point. In particular, this gives that the boundary values for the elastic solution are the same as those for the viscoelastic solution, as has been assumed. If any elastic solution is independent of the moduli at any point, then the elastic and viscoelastic solutions will be the same.

Note that the Correspondence Principle does not apply if the material is aging. This includes the case where temperature variation destroys the convolution form of the hereditary integrals. However, an extension of the proportionality assumption, described in Sects. 1.8, 1.9, provides for aging materials with only one independent relaxation function. For such materials there exist useful analogies between viscoelastic and appropriate elastic solutions. These are referred to in Sect. 2.12.

If an elastic solution to a given problem is known, the main task in applying the Classical Correspondence Principle is that of finding the inverse Fourier transform indicated in (2.1.6, 7). In non-inertial problems, this is frequently fairly straightforward. However, if inertial effects are included, it is generally very difficult. In fact, perhaps the most powerful approach to solving elastic inertial problems is by means of Fourier transforms, and in this case the greatest difficulty usually lies in trying to calculate inverse transforms. The difficulty is compounded in the viscoelastic case by the dependence of the complex moduli on frequency.

This approach is therefore most useful in the non-inertial approximation. The discussion will be confined to this case for the remainder of this section, and in fact on through later sections, with minor exceptions.

Note that the non-inertial elastic problem is the same in terms of FT quantities as in the untransformed case, apart from the fact that the specific form of the functions prescribed on the boundary is different, which does not affect the method of solution or the Green's functions of the problem. This is essentially a manifestation of the fact that time is merely an added parameter in the problem, playing no part in the structure of the equations. An implication of this observation is that in the non-inertial case, there is a correspondence between the FT form of the viscoelastic equations and the untransformed form of the elastic equations, where however, the correspondence does not extend to the specific form of the boundary functions.

The Classical Correspondence Principle was enunciated in reasonably general form by Read (1950) and Lee (1955) among others and discussed rigorously by Sternberg (1964) for the more general non-isothermal case. Sternberg (1964) reviews the older literature in some detail. Tao (1966) discusses correspondences between elastic and viscoelastic inertial problems in terms of Laplace transforms, essentially generalizing the work of Lee (1955).

The quantity $u_i^{(e)}(r, \omega, t)$ at any point is given by a Green's function representation, which we shall discuss in Sect. 2.3, involving integrals over the boundary functions at a given time. The time variable comes in only through these boundary functions. Let us write (2.1.8) in the form

$$u_i(r, t) = \int_{-\infty}^{\infty} dt' \, U_i^{(e)}(r, t - t', t') \ ,$$

$$U_i^{(e)}(r, t, t') = \frac{1}{2\pi} \int_{-\infty}^{\infty} d\omega \, e^{i\omega t} u_i^{(e)}(r, \omega, t') \ . \tag{2.1.9}$$

In the first equation, the $t - t'$ occurs in inverse transformed functions of $\hat{\mu}_{ijkl}(\omega)$, while the t' occurs in the boundary functions. Now, the state of the material at time t cannot depend upon the values of the driving quantities, namely the boundary functions, at a later time. This implies that the integral over t' must stop at t or, equivalently, that $U_i^{(e)}(r, t, t')$ is zero for $t < 0$, which is a causality condition imposing constraints on the elastic Green's functions as discussed in Sect. 1.10. By virtue of this property, we write

$$u_i(r, t) = \int\limits_{-\infty}^{t} dt'\, U_i^{(e)}(r, t - t', t')$$

$$= \frac{1}{2\pi} \int\limits_{-\infty}^{\infty} d\omega\, e^{i\omega t} \int\limits_{-\infty}^{t} dt'\, u_i^{(e)}(r, \omega, t') e^{-i\omega t'} \quad \text{and} \qquad (2.1.10)$$

$$\sigma_{ij}(r, t) = \int\limits_{-\infty}^{t} dt'\, \Sigma_{ij}^{(e)}(r, t - t', t')$$

$$= \frac{1}{2\pi} \int\limits_{-\infty}^{\infty} d\omega\, e^{i\omega t} \int\limits_{-\infty}^{t} dt'\, \sigma_{ij}^{(e)}(r, \omega, t') e^{-i\omega t'} \;, \qquad (2.1.11)$$

where

$$\Sigma_{ij}^{(e)}(r, t, t') = \frac{1}{2\pi} \int\limits_{-\infty}^{\infty} d\omega\, e^{i\omega t} \sigma_{ij}^{(e)}(r, \omega, t') \;. \qquad (2.1.12)$$

2.1.1 Separation of Space and Time Variables

In certain problems, the nature of the boundary conditions [Christensen (1982)] allows one to factor the solution as follows:

$$u_i(r, t) = u_i(r) f(t) \;, \qquad \sigma_{ij}(r, t) = \sigma_{ij}(r) g(t) \;. \qquad (2.1.13)$$

This will only be possible under the proportionality assumption discussed in Sect. 1.8. Applying this assumption, it is easy to show that $f(t)$, $g(t)$ are connected by a simple formula. One can show furthermore that $u_i(r)$, $\sigma_{ij}(r)$ are given by their static elasticity form while the time functions are easily determined from the boundary conditions.

2.2 Time-dependent Boundary Regions

The considerations of the last section apply to the case where the regions $B_u^{(i)}$, $B_\sigma^{(i)}$ are time-independent. This restriction excludes a number of problem categories of great interest, for example those involving moving boundary loads or expanding cracks. It is therefore desirable to relax this assumption. In fact, the main purpose of the present work is to present a coherent coverage of those problem categories not covered by the Classical Correspondence Principle. The discussion here is confined to the non-inertial case.

Consider relations of the form (2.1.6, 7) for the viscoelastic displacements and stresses at any point in the medium, where $\{u_i^{(e)}(r, \omega, t), \; \sigma_{ij}^{(e)}(r, \omega, t)\}$ denotes a solution of some boundary value problem for an elastic material occupying the same region as the viscoelastic medium, but where moduli have been replaced by complex moduli. From (1.8.19), we deduce that (2.1.6) obeys the dynamical equations. The boundary conditions (1.8.15) must also be imposed, which presents difficulties, if the boundary regions are time-dependent.

The observations of the previous paragraph suggest a general strategy which we now state. It is to look for solutions of the form (2.1.6, 7), in other words, to treat these as Ansätze. One must then explore what boundary conditions have to be imposed on the elastic solutions in order that the viscoelastic solutions have the correct boundary behaviour. The question arises whether this procedure is guaranteed to work, in other words, whether all possible solutions to viscoelastic boundary value problems can be expressed in the form (2.1.6, 7). We will not attempt to answer this question by means of general arguments, but instead will show that, in the problem categories of interest, the method provides solutions. Observe that even at this stage, the problem has been reduced to one involving only boundary quantities, which constitutes a considerable simplification.

The next four sections will be devoted to the application of this strategy, in fairly general terms, to certain problem categories. We will make the assumption that the elastic problem, or class of problems, corresponding to the viscoelastic configuration of interest, can be solved, in the sense that the Green's functions for the problem are known.

Again we point out that the integral over time in (2.1.9) is an integral over the input quantities, which are the specified boundary functions, and therefore, by Causality, must stop at t. Therefore (2.1.6, 7) can be replaced by (2.1.10, 11). This somewhat abstract argument will be rendered more immediate by specific examples in later sections.

The conditions that must be satisfied by the elastic solutions on the boundary follow from (1.8.15) and (2.1.10, 11). These are

$$d_i(r, t) = \frac{1}{2\pi} \int_{-\infty}^{\infty} d\omega \, e^{i\omega t} \int_{-\infty}^{t} dt' \, u_i^{(e)}(r, \omega, t') e^{-i\omega t'} \, , \qquad r \in B_u^{(i)}(t) \, , \qquad (2.2.1a)$$

$$c_i(r, t) = \frac{1}{2\pi} \int_{-\infty}^{\infty} d\omega \, e^{i\omega t} \int_{-\infty}^{t} dt' \, s_i^{(e)}(r, \omega, t') e^{-i\omega t'} \, , \qquad r \in B_\sigma^{(i)}(t) \qquad (2.2.1b)$$

where

$$s_i^{(e)}(r, \omega, t) = \sigma_{ij}^e(r, \omega, t) n_j(r) \, . \tag{2.2.2}$$

The functions $d_i(r, t)$, $c_i(r, t)$ are known for $r \in B_u^{(i)}(t)$, $B_\sigma^{(i)}(t)$, respectively. However, they will not in general be known at the same point for all earlier times, since the boundary regions are time-dependent. This is the essential complicating feature which renders the determination of $u_i^{(e)}(r, \omega, t)$, $s_i^{(e)}(r, \omega, t)$ non-trivial.

For those times t' at which $r \in B_\sigma^{(i)}(t')$ in (2.2.1a), it should be possible to re-express $u_i^{(e)}(r, \omega, t')$ in terms of $u_i^{(e)}(r', \omega, t')$, $r' \in B_u^{(i)}(t')$ and $s_i^{(e)}(r', \omega, t')$,

$r' \in B_\sigma^{(i)}(t')$, using the Green's function formalism (see Sect. 2.3); similary for those times t' at which $r \in B_u^{(i)}(t')$ in (2.2.1b). We see, in rough terms at least, how (2.2.1) may be transformed into a system of integral equations for $u_i^{(e)}(r, \omega, t')$, $r \in B_u^{(i)}(t')$ and $s_i^{(e)}(r, \omega, t')$, $r \in B_\sigma^{(i)}(t')$. A procedure related to the one just outlined will in fact be implemented in the next three sections, for certain problem classes.

What will determine the complexity of (2.2.1) is the nature of the ω dependence of the elastic solutions. The simpler this dependence is, the easier it will be to determine $u_i^{(e)}(r, \omega, t)$ and $s_i^{(e)}(r, \omega, t)$. The question is, what determines the ω dependence of these functions? In the next section, the Green's Function formalism is introduced. It emerges that, at least in special cases (of great interest), this formalism points to the dependence on ω of $u_i^{(e)}(r, \omega, t)$ and $s_i^{(e)}(r, \omega, t)$.

The basic idea behind the approach outlined here was proposed in a general form by Graham and Sabin (1973), though earlier papers, for example Graham (1968) and Ting (1969), contained similar concepts, in less general form.

An ablating body may be characterized by the observation that any material point that belongs to the body at time t has been a part of the body for all times prior to t. Since the value at time t of the viscoelastic state generated through (2.1.10, 11) is expressed in terms of elastic states at the same material point for times up to t, it is clear that (2.1.10, 11) are valid for ablating bodies. This observation has been used by Graham and Sabin (1973) to recover viscoelastic stress and displacement fields in various hollow ablating right circular cylinders. Other work on this problem may be traced by referring to the above paper and Christensen (1982).

2.3 Elastic Solutions in Terms of Green's Functions

The Green's function formalism for non-inertial elastic problems will now be used to express $u_i^{(e)}(r, \omega, t)$, $s_i^{(e)}(r, \omega, t)$ at any point in the body, in terms of specified surface quantities. The development of this formalism is standard and will not be discussed here. We refer to the treatment by Lardner (1974) for example. Really, all that is required in the present context is that the displacements and stresses can be expressed as space integrals of the boundary functions.

The Green's functions will be attributed a dependence upon ω to indicate that the moduli have been replaced by the complex moduli. We will omit subscripts, in this and the next two sections, which amounts to developing a one-dimensional rather than a three-dimensional theory. This results in a considerable tidying of the equations and the loss of generality is irrelevant in the present context, because only the one-dimensional theory is required, in any case, in later chapters. This is because attention is focussed on crack and half-space problems of such a nature that only the normal displacement and pressure are relevant to the solution of the problem.

It is not difficult to generalize the discussion, in a formal sense, to three dimensions. The Green's functions become rank two tensors. However, if the boundary regions depend upon the coordinate subscript, certain complications arise, particularly in the discussion of the integral equation introduced in Sect. 2.5.

Let $u^{(e)}(r, \omega, t)$, $s^{(e)}(r, \omega, t)$ be the solutions of the elastic problem. For r anywhere in the material, but usually for purposes of the present discussion, on the boundary, we write

$$u^{(e)}(r, \omega, t) = \int_{B_u(t)} ds'\, H(r, r', \omega, t) u^{(e)}(r', \omega, t)$$

$$+ \int_{B_\sigma(t)} ds'\, G(r, r', \omega, t) s^{(e)}(r', \omega, t) \ ,$$

$$\tag{2.3.1}$$

$$s^{(e)}(r, \omega, t) = \int_{B_u(t)} ds'\, L(r, r', \omega, t) u^{(e)}(r', \omega, t)$$

$$+ \int_{B_\sigma(t)} ds'\, Q(r, r', \omega, t) s^{(e)}(r', \omega, t)$$

where ds' is the surface element. The functions H, G, L, Q are elastic Green's functions. They are time-dependent in the non-inertial approximation if B_u, B_σ vary with time. It must be re-emphasized that $u^{(e)}(r, \omega, t)$, $r \in B_u(t)$ and $s^{(e)}(r, \omega, t)$, $r \in B_\sigma(t)$ are not in fact known. Our object is to determine them in such a manner as to ensure that (2.2.1) is satisfied. Our immediate aim is to decide what the ω dependence of these two quantities should be. We begin by making assumptions on the ω dependence of the Green's functions. The quantities H, Q are taken to be independent of ω, while G, L are assumed to have the form

$$G(r, r', \omega, t) = \hat{k}(\omega) \Delta(r, r', t) \ , \qquad L(r, r', \omega, t) = \hat{l}(\omega) T(r, r', t) \ , \quad (2.3.2)$$

where $\hat{k}(\omega)$, $\hat{l}(\omega)$ are functions of the complex moduli such that it can be arranged that

$$\hat{k}(\omega) = 1/\hat{l}(\omega) \ . \tag{2.3.3}$$

On the basis of a simple dimensional argument, these assumptions are certainly valid if the proportionality of the complex moduli, as given by (1.8.20) or (1.9.20), applies. In fact $\hat{l}(\omega)$ will be proportional to $\hat{\mu}(\omega)$ and $\hat{k}(\omega)$ to $\hat{y}(\omega)$ $[= 1/\hat{\mu}(\omega)]$, so that by trivially scaling the functions G, L, (2.3.3) can be rendered correct. It will emerge also that for certain restricted problem categories, notably crack problems and half-space contact problems with no surface shear, it is not necessary to make the proportionality assumption in order to achieve (2.3.2) and (2.3.3). Some progress at characterising this class of problems is reported by Graham and Golden (1988). Anyway, only those problems for which (2.3.2) is true will be considered in later chapters. These remarks apply to non-inertial problems.

Incorporating (2.3.2) into (2.3.1) gives that

$$u^{(e)}(r, \omega, t) = \int\limits_{B_u(t)} ds' \, H(r, r', t) u^{(e)}(r', \omega, t)$$

$$+ \hat{k}(\omega) \int\limits_{B_\sigma(t)} ds' \, \Delta(r, r', t) s^{(e)}(r', \omega, t)$$

$$(2.3.4)$$

$$s^{(e)}(r, \omega, t) = \hat{l}(\omega) \int\limits_{B_u(t)} ds' \, T(r, r', t) u^{(e)}(r', \omega, t)$$

$$+ \int\limits_{B_\sigma(t)} ds' \, Q(r, r', t) s^{(e)}(r', \omega, t) \, .$$

Let us adopt a criterion for determining the ω dependence of $u^{(e)}$, $s^{(e)}$ as follows. We choose it so as to render (2.3.4) free of frequency dependence. This will be justified later. There are two ways of rendering (2.3.4) independent of frequency. In the first place, we could assume that

$$u^{(e)}(r, \omega, t) = u^{(e)}(r, t), \qquad r \in B_u(t)$$
$$s^{(e)}(r, \omega, t) = \hat{l}(\omega) q(r, t), \qquad r \in B_\sigma(t)$$

$$(2.3.5)$$

where $u^{(e)}(r, t)$, $q(r, t)$ are independent of ω. In general, they will depend upon both the boundary conditions and the viscoelastic material parameters, so that the remark in Sect. 2.1, that the ω dependence comes in through the complex moduli and the time dependence from the boundary conditions, no longer applies. Alternatively, we could put

$$u^{(e)}(r, \omega, t) = \hat{k}(\omega) v(r, t), \qquad r \in B_u(t)$$
$$s^{(e)}(r, \omega, t) = s^{(e)}(r, t), \qquad r \in B_\sigma(t)$$

$$(2.3.6)$$

where $v(r, t)$, $s^{(e)}(r, t)$ are independent of ω. Note that if we substitute either (2.3.5) or (2.3.6) into (2.3.4), similar relations can be deduced for all r, which gives the required ω dependence. So we can regard the alternatives given by (2.3.5) and (2.3.6) as true for all r. Let us consider (2.3.6) in more detail. From the second relation and (2.2.1), we deduce that

$$s^{(e)}(r, t) = c(r, t), \qquad r \in B_\sigma(t) \, , \tag{2.3.7}$$

so that the boundary value of $s^{(e)}(r, t)$ is the same as the specified value of $s(r, t)$ for the viscoelastic problem. In fact, $s^{(e)}(r, t)$ will be the same as the viscoelastic solution $s(r, t)$ all over the material. Also, from (2.1.10):

$$u(r, t) = \int\limits_{-\infty}^{t} dt' \, k(t - t') v(r, t') \, . \tag{2.3.8}$$

Observe that the causal structure of (2.1.10) is equivalent to requiring that $\hat{k}(\omega)$ be causal, which is precisely the type of constraint explored in Sect. 1.10.

If $v(r, t)$ can be determined for $r \in B_u(t)$, then it is given for $r \in B_\sigma(t)$ by (2.3.4), which may be written as

$$v(r, t) = \int\limits_{B_u(t)} ds'\, H(r, r', t)v(r', t) + \int\limits_{B_\sigma(t)} ds'\, \Delta(r, r', t)c(r', t) \qquad (2.3.9\text{a})$$

$$s(r, t) = \int\limits_{B_u(t)} ds'\, T(r, r', t)v(r', t) + \int\limits_{B_\sigma(t)} ds'\, Q(r, r', t)c(r', t)\ , \qquad (2.3.9\text{b})$$

by virtue of (2.3.7). These relations, apart from certain coefficients which can easily be readjusted, have the form of the Green's function solution of the elastic problem with $v(r, t)$ playing the role of displacement. They play a central role in Sect. 2.5. So the crux of the matter is to determine $v(r, t)$ on $B_u(t)$, subject to the requirement that

$$d(r, t) = \int\limits_{-\infty}^{t} dt'\, k(t - t')v(r, t')\ , \qquad r \in B_u(t)\ . \qquad (2.3.10)$$

This is essentially condition (2.2.1a).

Similarly, from (2.3.5) and (2.2.1) it follows that the boundary value of $u^{(e)}(r, t)$ is

$$u^{(e)}(r, t) = d(r, t)\ , \qquad r \in B_u(t)\ , \qquad (2.3.11)$$

which is the same as the viscoelastic value. Indeed, $u^{(e)}(r, t)$ will be the same as the viscoelastic solution for all r. From (2.1.11) and (2.3.5) we have

$$s(r, t) = \int\limits_{-\infty}^{t} dt'\, l(t - t')q(r, t') \qquad (2.3.12)$$

where $s(r, t)$ is the viscoelastic boundary traction [specified on $B_\sigma(t)$]. In this case, (2.3.4) may be written as

$$u(r, t) = \int\limits_{B_u(t)} ds'\, H(r, r', t)d(r', t) + \int\limits_{B_\sigma(t)} ds'\, \Delta(r, r', t)q(r', t)$$

$$q(r, t) = \int\limits_{B_u(t)} ds'\, T(r, r', t)d(r', t) + \int\limits_{B_\sigma(t)} ds'\, Q(r, r', t)q(r', t)\ . \qquad (2.3.13)$$

Therefore, if $q(r, t)$ can be determined on $B_\sigma(t)$, it is known all over the boundary. So the crucial matter is to determine $q(r, t)$ on $B_\sigma(t)$, subject to the requirement that

$$c(r, t) = \int\limits_{-\infty}^{t} dt'\, l(t - t')q(r, t')\ , \qquad r \in B_\sigma(t)\ , \qquad (2.3.14)$$

which is essentially condition (2.2.1b).

From (2.3.12), we see that, as before, the causal structure of (2.1.11) requires that $l(t)$ be causal. Equation (2.3.3) and the causal property of both $l(t)$ and $k(t)$ imply that they are related in a manner similar to $\mu(t)$ and $\gamma(t)$, as given by (1.2.36), namely:

$$\int\limits_{0}^{t} dt'\, k(t - t')l(t') = \int\limits_{0}^{t} dt'\, l(t - t')k(t') = \delta(t)\ . \qquad (2.3.15)$$

Inverting (2.3.8, 12) gives

$$v(r, t) = \int\limits_{-\infty}^{t} dt' l(t - t') u(r, t') \quad \text{and} \tag{2.3.16}$$

$$q(r, t) = \int\limits_{-\infty}^{t} dt' k(t - t') s(r, t') . \tag{2.3.17}$$

It is on these, rather than (2.3.8, 12), that the approach developed in the following sections will rest; on these equations and (2.3.9) or (2.3.13). We remark that relations (2.3.9) and (2.3.13) could have been written down directly, at least if the proportionality assumption is made. Essentially, the point is that made in the context of (1.8.23) which we restate here within the present simpler, more concrete framework. Consider (2.3.9) for example. The proportionality assumption means that there is only one hereditary integral in the theory, and the equations of the theory are identical to the elastic equations if displacements are replaced by quantities of the form of $v(r, t)$ where $l(t)$ is proportional to $\mu(t)$, for example. It follows that elastic solutions are applicable if displacements are replaced by $v(r, t)$ and corresponding quantities for the other components. This is precisely the content of equation (2.3.9). A similar argument applies to (2.3.13). It is not necessary even to assume proportionality for certain special problems, though these problems are difficult to characterize in fundamental terms. They are mainly problems where all the dependence on material properties can be grouped into one function.

The fact that (2.3.9) and (2.3.13) are manifestly correct is the essential justification of (2.3.5) and (2.3.6). The arguments leading up to (2.3.9) and (2.3.13) could strictly have been omitted. However, they provide a link between reasoning based on formal elastic-viscoelastic correspondence as incorporated in (2.1.10), (2.1.11) and (2.3.1) and the methodology which is developed below.

We now focus on the option characterized by (2.3.7 − 10) and (2.3.16), though the developments are equally applicable to the second option. An integral equation will be derived for $v(r, t)$, $r \in B_u(t)$, once we have shown how this quantity can be decomposed in a convenient manner.

2.4 Decomposition of Hereditary Integrals

We consider a problem in this section which arises under several guises through the book, in particular, in the context of deriving an integral equation for the quantity $v(r, t)$, $r \in B_u(t)$. Let us first phrase the problem in fairly abstract terms. Let $u(t)$, $v(t)$ be two functions related by

$$v(t) = \int\limits_{-\infty}^{t} dt' l(t - t') u(t') , \quad u(t) = \int\limits_{-\infty}^{t} dt' k(t - t') v(t') , \tag{2.4.1}$$

where $k(t)$, $l(t)$ are two causal functions, related by (2.3.15). Let $\theta(t)$ be the set of the present and all past times $(-\infty, t]$, which we decompose into two disjoint sets $W_u(t)$, $W_v(t)$, i.e.

$$\theta(t) = W_u(t) \cup W_v(t) , \tag{2.4.2}$$

where $u(t')$ is given for $t' \in W_u(t)$ and $v(t')$ is given (or can be usefully represented; we shall see an example of what is meant by this phrase in the next section) for $t' \in W_v(t)$. If we could decompose $v(t)$, for example, as follows:

$$v(t) = \int_{W_u(t)} dt' \Pi_u(t, t') u(t') + \int_{W_v(t)} dt' \Pi_v(t, t') v(t') , \qquad (2.4.3)$$

then everything on the right-hand side is known, or can be usefully represented, thus giving an expression for $v(t)$. The problem is therefore to decompose a hereditary integral in this manner.

Let us begin by considering the decomposition of $v(t)$. We first define the sets $W_u(t)$, $W_v(t)$. Let $t_1, t_2, t_3, \ldots$ be the monotone decreasing sequence of times marking when t' changes from $W_u(t)$ to $W_v(t)$ or vice versa. Initially, this will be taken to be a finite sequence so that if t_n is the last member, no change from $W_u(t)$ to $W_v(t)$, or vice versa, occurs in $(-\infty, t_n]$. There are two possibilities:

(a) $[t_1, t] \in W_u(t) \Rightarrow W_u(t) = (t_1, t] \cup (t_3, t_2] \cup \ldots$

$$\qquad (2.4.4)$$

(b) $[t_1, t] \in W_v(t) \Rightarrow W_v(t) = (t_1, t] \cup (t_3, t_2] \cup \ldots$.

Case (b) is trivial; there is no problem, since $v(t)$ is known at time t. Consider case (a). Let us write $v(t)$ as

$$v(t) = \int_{t_1}^{t} dt' l(t - t') u(t') + \int_{-\infty}^{t_1} dt' l(t - t') u(t')$$

$$\qquad (2.4.5)$$

$$= \int_{t_1}^{t} dt' l(t - t') u(t') + \int_{-\infty}^{t_1} dt' T_1(t, t') v(t') \quad \text{where}$$

$$T_1(t, t') = \int_{t'}^{t_1} dt'' l(t - t'') k(t'' - t') . \qquad (2.4.6)$$

This last step involved substituting for $u(t')$ from (2.4.1) and an interchange of integrations. The procedure can be repeated. We obtain

$$v(t) = \int_{t_1}^{t} dt' l(t - t') u(t') + \int_{t_2}^{t_1} dt' T_1(t, t') v(t') + \int_{-\infty}^{t_2} dt' T_2(t, t') u(t') \quad (2.4.7)$$

where

$$T_2(t, t') = \int_{t'}^{t_2} dt'' T_1(t, t'') l(t'' - t') . \qquad (2.4.8)$$

Continuing in this way, we obtain a decomposition as in (2.4.3) where

$$\Pi_u(t, t') = T_0(t, t') R(t'; t_1, t) + T_2(t, t') R(t'; t_3, t_2)$$

$$+ T_4(t, t') R(t'; t_5, t_4) + \ldots$$

$$\qquad (2.4.9)$$

$$\Pi_v(t, t') = T_1(t, t') R(t'; t_2, t_1) + T_3(t, t') R(t'; t_4, t_3) + \ldots$$

where the function $R(t; t_2, t_1)$ projects onto the interval $[t_2, t_1]$ in the sense that

$$R(t; t_2, t_1) = \begin{cases} 1, & t \in [t_2, t_1] \\ 0, & t \notin [t_2, t_1] \end{cases} \tag{2.4.10}$$

for all t_2, t_1, t. Also:

$$T_0(t, t') = l(t - t')$$

$$T_l(t, t') \begin{cases} = \int\limits_{t'}^{t_l} dt'' T_{l-1}(t, t'') l(t'' - t'), & l \text{ even} \\ \\ = \int\limits_{t'}^{t_l} dt'' T_{l-1}(t, t'') k(t'' - t'), & l \text{ odd} . \end{cases} \tag{2.4.11}$$

Recall that n is the total number of transition times t_r. If n is odd, then $t' \in W_v(t)$ initially ($t' < t_n$) and the number of terms in Π_u, Π_v is $(n + 1)/2$ in both cases, while if n is even, $t' \in W_u(t)$ initially and there are $(n/2) + 1$ in Π_u and $n/2$ in Π_v.

Let us now consider the decomposition of $u(t)$. The non-trivial case here is (b) where $(t_1, t] \in W_v(t)$, since if $t \in W_u(t)$, $u(t)$ is known without further ado. Following an identical argument, we find that $u(t)$ can be decomposed into

$$u(t) = \int\limits_{W_u(t)} dt' \Gamma_u(t, t') u(t') + \int\limits_{W_v(t)} dt' \Gamma_v(t, t') v(t') , \tag{2.4.12}$$

where

$$\Gamma_v(t, t') = N_0(t, t') R(t'; t_1, t) + N_2(t, t') R(t'; t_3, t_2) + \cdots$$
$$\tag{2.4.13}$$

$$\Gamma_u(t, t') = N_1(t, t') R(t'; t_2, t_1) + N_3(t, t') R(t'; t_4, t_3) + \cdots ,$$

where the functions $N_l(t, t')$ are given by

$$N_0(t, t') = k(t - t')$$

$$N_l(t, t') \begin{cases} = \int\limits_{t'}^{t_l} dt'' N_{l-1}(t, t'') l(t'' - t'), & l \text{ odd} \\ \\ = \int\limits_{t'}^{t_l} dt'' N_{l-1}(t, t'') k(t'' - t'), & l \text{ even}. \end{cases} \tag{2.4.14}$$

In this case, if n is odd, then $t' \in W_u(t)$ initially and the number of terms in Γ_u, Γ_v is $(n + 1)/2$ in both cases, while if n is even, $t' \in W_v(t)$ initially and there are $(n/2) + 1$ terms in Γ_v and $n/2$ in Γ_u.

It is clear that the difference between Γ_u, Γ_v, and Π_v, Π_u amounts to an interchange of the role of the functions $l(t)$, $k(t)$.

These formulae will find application essentially unchanged in form, in the context of expanding and contracting contact regions as discussed in Sects. 2.6, 3.10 and 5.2; and to problems involving closing cracks in Sect. 4.4. In order to derive a useful decomposition of $v(r, t)$, given by (2.3.16), for use in a general context discussed in Sect. 2.5, it is necessary to generalize these formulae trivially to include a dependence on r in the time regions.

Let the boundary B be decomposed into

$$B = B_u(t) \cup B_\sigma(t) , \tag{2.4.15}$$

where $u(r, t)$ is known on $B_u(t)$ but not on $B_\sigma(t)$. In this case, it is not in general true that $v(r, t)$ is known on $B_\sigma(t)$. However, we shall see that it can be represented in a useful and convenient manner [actually by (2.3.9)], in this region. We use the notation W_σ instead of W_v and include dependence on r, so that

$$\theta(t) = W_u(r, t) \cup W_\sigma(r, t) , \tag{2.4.16}$$

where $W_u(r, t)$ is all those times $t' \leq t$ such that $r \in B_u(t')$ and $W_\sigma(r, t)$ is all those times such that $r \in B_\sigma(t')$. Applying exactly the same technique as above, we obtain the decomposition [for $t \in W_u(r, t)$ or $r \in B_u(t)$]

$$v(r, t) = \int_{W_u(r, t)} dt' \, \Pi_u(t, t'; r) u(r, t') + \int_{W_\sigma(r, t)} dt' \, \Pi_\sigma(t, t'; r) v(r, t'), \tag{2.4.17}$$

where

$$\Pi_u(t, t'; r) = T_0(t, t') R(t'; t_1(r), t) + T_2(t, t'; r) R(t'; t_3(r), t_2(r))$$
$$+ T_4(t, t'; r) R(t'; t_5(r), t_4(r)) + \dots \tag{2.4.18}$$

$$\Pi_\sigma(t, t', r) = T_1(t, t'; r) R(t'; t_2(r), t_1(r))$$
$$+ T_3(t, t'; r) R(t'; t_4(r), t_3(r)) + \dots$$

and

$$T_0(t, t') = l(t - t')$$

$$T_l(t, t'; r) \begin{cases} = \displaystyle\int_{t'}^{t_l(r)} dt'' \, T_{l-1}(t, t''; r) l(t'' - t'), & l \text{ even} \\[4mm] = \displaystyle\int_{t'}^{t_l(r)} dt'' \, T_{l-1}(t, t''; r) k(t'' - t'), & l \text{ odd} , \end{cases} \tag{2.4.19}$$

the times $t_l(r)$ being times of transition of r between $B_u(t)$, $B_\sigma(t)$. It should be noted that the transition times $t_l(r)$ (and the t_l considered earlier) depend, in a discontinuous manner, on the current value of t. However, to simplify the notation, this has not been shown explicitly.

A similar decomposition is possible for $u(r, t)$, $r \in B_\sigma(t)$, obtained from (2.5.12 – 14) by simply including a space variable dependence. Furthermore, the quantities $s(r, t)$, $q(r, t)$, related by (2.3.12, 17), may be decomposed in a manner entirely analogous to $u(r, t)$, $v(r, t)$. We shall not give the formulae explicitly as they will not be needed.

The fundamental decomposition developed in this section is an alternative form of that given by Graham (1967), Ting (1968) and later, for the aging case, by Sabin and Graham (1980), in the context of normal contact problems. The form developed here is more convenient in some respects than those given previously. It can also be given for the Stieltjes form of the hereditary integral. Since this is not required later, it is left as an exercise to the reader.

2.5 The Integral Equation

The decomposition (2.4.17) of (2.3.16), combined with (2.3.9a) immediately gives an integral equation for $v(r, t)$, $r \in B_u(t)$. Let us substitute from (2.3.9a) into the right-hand side of (2.4.17) to obtain

$$v(r, t) = \int_{W_\sigma(r, t)} dt' \int_{B_u(t')} ds' \, K(r, r'; t, t') v(r', t') + I(r, t), \quad r \in B_u(t) \quad (2.5.1)$$

where

$$K(r, r'; t, t') = \Pi_\sigma(t, t'; r) H(r, r'; t') \quad (2.5.2)$$

and

$$I(r, t) = \int_{W_u(r, t)} dt' \, \Pi_u(t, t'; r) u(r, t')$$

$$+ \int_{W_\sigma(r, t)} dt' \, \Pi_\sigma(t, t'; r) \int_{B_\sigma(t')} ds' \Delta(r, r'; t') c(r', t') . \quad (2.5.3)$$

If the sets $W_u(r, t)$, $W_\sigma(r, t)$ are known, then the inhomogeneous term $I(r, t)$ is known.

We expect a singularity in $K(r, r'; t, t')$ where $r = r'$, due to the Green's function factor $H(r, r'; t')$ in its definition. However, the regions of integration over space and time in (2.5.1) are such that r, r' cannot in general become equal. This follows from the definition of $W_\sigma(r, t)$, given by (2.4.16). For certain times in the integration range, it may happen that they can become equal however. We refer to the transition times $t_l(r)$. This does not necessarily render the equation singular however, as we shall see in Sect. 3.4., and for a worked-out example, in Sect. 3.6.

Let us denote by P the class of boundary value problems for which a unique solution to (2.5.1) exists. We shall not seek to determine the size of this class. However, in Chap. 3, it will emerge that simple contact problems involving transverse motion are members of P, so that at least it is non-empty. It is likely in fact that P is quite extensive, in which case (2.5.1) provides a general method for solving non-inertial boundary value problems. For certain problem categories, the question of whether or not they belong to P is irrelevant because easier methods exist for solving them, as we shall see in Sect. 2.6.

For contact problems involving transverse motion, it would seem that no simpler method than solving (2.5.1) is available, without special assumptions on the form of the viscoelastic functions.

Equation (2.5.1) involves both space and time variables occurring in an interdependent manner, which renders it unlikely that exact solutions will be available, even for simple problems. The two methods of solution, with any wide degree of applicability, are (a) numerical and (b) iterative. In the former case, one starts at the initial time and progresses through the motion, solving at each stage an integral equation in the space variable.

The standard iteration procedure is to take $v_0(r, t) = I(r, t)$ and obtain $v_1(r, t)$ by substituting $v_0(r, t)$ on the right of (2.5.1), repeating the procedure to achieve higher order approximations. This should lead to a convergent series, subject to some condition analogous to, and generalizing, the bound on the square integral of the kernel in Fredholm theory (Sect. A4.1). There is an alternative procedure, however, closer to the physics of the problem. If the material is such that viscoelastic effects are smaller than elastic effects, a typical ratio of the two being say $r < 1$, then it should be possible to obtain solutions to (2.5.1), systematically, to any desired order, in r. Briefly, the procedure would be to calculate $l(t)$, $k(t)$ to the desired order, and hence Π_u, Π_σ. A series of iterations should then give the desired approximate solution. The algebra involved in carrying through such a program becomes quickly unmanageable beyond first order calculations. Some examples of first order calculations are given in later chapters.

Considerable simplification is achieved if only steady-state solutions, involving uniform linear motion, are sought. We shall see, in Chap. 3, that the time variable is essentially eliminated, leaving an integral equation in the space variable. At least for simple problems, it is possible to make considerable analytical progress before resorting to numerical methods.

2.6 Expanding and Contracting Boundary Regions

As mentioned in the previous section, we shall now isolate a class of problems which can be treated by simpler methods than that based on the solution of (2.5.1). We start by considering monotonically expanding or contracting boundary regions.

2.6.1 The Extended Correspondence Principle

Consider the case where the boundary region $B_u(t)$ is non-expanding, or in other words, is stationary or contracting. By this we mean that $B_u(t_2) \subseteq B_u(t_1)$ for all t_1, t_2 such that $t_2 \geq t_1$. Then $v(r, t)$, given by (2.3.16), is known on $B_u(t)$, since $u(r, t')$ is specified on $B_u(t')$ for $t' \leq t$. Relations (2.3.9) give $v(r, t)$ on $B_\sigma(t)$ and $s(r, t)$ on $B_u(t)$, so the problem is solved.

Recall that the kernel functions in (2.3.9) are the elastic Green's functions for the same problem, up to a multiplying factor, depending on the moduli. Therefore, $v(r, t)$ and $s(r, t)$ are given by the elastic solution, up to trivial coefficients which can be easily adjusted to unity, of the corresponding elastic boundary value problem, where $v(r, t)$ plays the part of the specified displacement on $B_u(t)$. In particular, if $u(r, t')$ is zero on $B_u(t')$ for all $t' < t$, then $v(r, t)$ is zero also, and (2.3.9b) gives that the stress is identical to the elastic stress — since $Q(r, r', t)$ is independent of the moduli. On $B_\sigma(t)$, $v(r, t)$ is proportional to the expression for the elastic displacement, so that the viscoelastic displacement can be calculated by using (2.3.8).

If, on the other hand, $B_u(t)$ is non-contracting so that $B_\sigma(t)$ is contracting or stationary, the natural approach is to use $q(r, t)$, given by (2.3.17). Here, $q(r, t)$ is known in $B_\sigma(t)$, and (2.3.13) gives its value at any point in $B_u(t)$. Hence, the problem is solved. In this case, $u(r, t)$ and $q(r, t)$ are given by the elastic form, up to constant coefficients, where $q(r, t')$ is specified on $B_\sigma(t)$, and thus plays the role of the stress. Again, if the stress $s(r, t)$ is specified to be zero, for all $t' \leq t$, on $B_\sigma(t)$, then $q(r, t)$ is zero on $B_\sigma(t)$, and the displacement is given by an expression identical to the elastic solution. Also, $q(r, t)$ is proportional to the elastic stress, and the viscoelastic stress is given by (2.3.12).

These observations express the main content of theorems extending the Correspondence Principle by Graham (1968), generalized by Ting (1969) and further discussed by Graham and Sabin (1973), though the treatment by these authors is in fact somewhat more general than that given here, insofar as it is three-dimensional and avoids the special assumptions contained in (2.3.9). Graham (1969) considers the non-isothermal case for a material to which the proportionality assumption applies. The results just outlined will be referred to as the Extended Correspondence Principle.

It follows from these observations on expanding and contracting boundary regions that a history consisting of successive expansions and contractions is also susceptible to solution by starting when the deformation begins, and systematically constructing the solution for each phase. One may see this by noting that a hereditary integral, say $v(r, t)$, may be written as

$$v(r, t) = \int_{-\infty}^{t_m} dt'\, l(t-t')u(r, t') + \int_{t_m}^{t} dt'\, l(t-t')u(r, t') \,, \qquad (2.6.1)$$

where t_m is the time of the last maximum in for example $B_u(t)$. The first term on the right is assumed to be known. The remaining integral is also known by virtue of the argument applied earlier to $v(r, t)$.

2.6.2 The Generalized Partial Correspondence Principle

A further generalization of the Correspondence Principle is possible. Let us assume that, for all $t' \leq t$, $B_u(t') \subseteq B_u(t)$. Note that this is not saying that $B_u(t)$ is non-contracting. Any history is allowed. All that we insist upon is that, at time t, $B_u(t)$ contains all earlier regions $B_u(t')$, $t' < t$. Now consider (2.3.9a) and visualize a somewhat different boundary value problem, where the region of specification of $v(r, t')$, for all $t' < t$, is the *fixed* region $B_u(t)$. In general, we do not in fact know $v(r, t')$ on $B_u(t)$. However, as we shall observe in a moment, this does not matter. The forms of the Green's functions for this new problem are precisely those given by (2.3.9). To see this, recall that these are elastic Green's functions and are determined only by $B_u(t)$, $B_\sigma(t)$, which are the (fixed) boundary regions for the problem. Therefore, we write the solution to this other boundary value problem as

$$v(r, t') = \int_{B_u(t)} ds' H(r, r', t) v(r', t') + \int_{B_\sigma(t)} ds' \Delta(r, r', t) c(r', t'), \quad t' \le t .$$
$$(2.6.2)$$

Note that $c(r', t')$ is known for all $t' < t$. This is a consequence of the crucial assumption that $B_u(t)$ contains all previous $B_u(t')$, which implies that $B_\sigma(t)$ contains only those points that were always in $B_\sigma(t')$, $t' \le t$. Applying (2.3.8) to (2.6.2) gives

$$u(r, t) = \int_{B_u(t)} ds' H(r, r', t) u(r', t) + \int_{B_\sigma(t)} ds' \Delta(r, r', t) q(r', t) \qquad (2.6.3)$$

where

$$q(r, t) = \int_{-\infty}^{t} dt' k(t - t') c(r, t') \qquad (2.6.4)$$

is known at time t for $r \in B_\sigma(t)$. By definition of $B_u(t)$, $u(r', t)$ is known on it. Therefore, the displacement is given by the elastic form, but with $q(r, t)$, a known quantity, substituted for the specified stress (up to a multiplying constant). If the specified stress is always zero, as happens in certain contact problems for example, $q(r, t)$ will also be zero, and the normal boundary displacement will be given precisely by the elastic form. In contrast to the special case of expanding or stationary $B_u(t)$, we can make no useful statement about the stress, in the general case.

If, on the other hand, we assumed that, at time t, $B_\sigma(t)$ has the property that $B_\sigma(t') \subseteq B_\sigma(t)$ for all $t' \le t$, then the roles of displacement and stress are interchanged. By means of an identical argument, we find that the stresses are given by the elastic stresses but where $v(r, t)$ plays the role of specified displacement, up to a multiplying constant. If the displacement is always zero in $B_u(t')$, $t' \le t$, as happens in certain crack problems, then the boundary stresses are identical to the elastic boundary stresses.

Again, in contrast to the special case of non-contracting $B_\sigma(t')$, we can make no useful statement about the displacement in this general case.

These results will be referred to as the Generalized Partial Correspondence Principle, to distinguish it from the more detailed, and specialized, Extended Correspondence Principle. A more general derivation of this result, which does not rely on the Green's function representation of the solution, has been given by Graham and Golden (1988).

2.6.3 Repetitive Expansion and Contraction

We remarked above that histories where the regions $B_u(t)$, $B_\sigma(t)$ are alternatively expanding and contracting may be handled by consecutive applications of the Extended Correspondence Principle. There is a special category of such histories for which one can go further and give more explicit solutions. Consider a configuration where, in the area that the surface stresses are known, they are zero. This is the case, usually, for contact problems. Let the region $B_u(t)$ be expanding and contracting with time in such a manner that it is continually passing through a fixed one-parameter series of states. This will be referred to as repetitive expansion and contraction. The most immediate example is that of contact under a load that is consecutively increasing and decreasing [Graham (1967), Ting (1968), Sabin and Graham (1980)].

Instead of (2.3.9a) let us use the Green's function relation for a boundary value problem of the first kind, that is, where the stresses are assumed to be given over the entire boundary. In (2.3.9a), we replace $B_u(t)$ by the empty set, and $B_\sigma(t)$ by B, to obtain the desired result. The Green's function will be time-independent. The relation will have the form

$$v(r, t) = \int_{B_u(t)} ds' \, \Lambda(r, r') s(r', t) \tag{2.6.5}$$

under the assumption that the stresses are zero on $B_\sigma(t)$. The quantity $\Lambda(r, r')$ is the appropriate (typically relatively simple) Green's function for elastic boundary value problems of the first kind, up to a multiplying factor.

We now require a decomposition of $v(r, t)$ of the type discussed in Sect. 2.4, though somewhat different to that used in Sect. 2.5. In this case, there is no space dependence. We define the times $\theta_l(t)$, $l = 1, 2, \ldots$, by the requirements that

$$B_u(\theta_l(t)) = B_u(t) . \tag{2.6.6}$$

In other words, these are the times in the past at which B_u has had the same extent as it occupies currently. Note that such times can be defined only because of the repetitive nature of the expansion and contraction, which we have assumed. If $B_u(t)$ is expanding at time t, it is contracting at time $\theta_1(t)$, expanding at time $\theta_2(t)$ and so on. The relationship between the $t_l(r)$ introduced in Sect. 2.4 and the $\theta_l(t)$, defined by (2.6.6), may be expressed as follows. If $r_b(t)$ is any point on the boundary of $B_u(t)$, then

$$\theta_l(t) = t_l(r_b(t)) . \tag{2.6.7}$$

For a contraction phase, $\theta_l(t) \le t_l(r)$, $r \in B_u(t)$, while, for an expansion phase, $\theta_l(t) \ge t_l(r)$. We can now decompose $v(r, t)$, $u(r, t)$ in the manner described in Sect. 2.4, as given by (2.4.3) and (2.4.12). The times $\theta_l(t)$ here replace the t_l used in that section. We again substitute the subscript "σ" for "v". The set $W_u(t)$ consists of those times t' for which $B_u(t') \supseteq B_u(t)$, while $W_\sigma(t)$ is the set of times t' for which $B_u(t') \subset B_u(t)$.

If $B_u(t)$ is in a contraction phase, then $[\theta_1(t), t]$ belongs to $W_u(t)$ and we put

$$v(r, t) = \int_{W_u(t)} dt' \, \Pi_u(t, t') u(r, t') + \int_{W_\sigma(t)} dt' \, \Pi_\sigma(t, t') v(r, t') \tag{2.6.8}$$

after (2.4.3), where Π_u, Π_σ are given, with appropriate alterations, by (2.4.9) to (2.4.11). If, on the other hand, $B_u(t)$ is expanding at time t, then $[\theta_1(t), t]$ belongs to $W_\sigma(t)$ and we put

$$u(r, t) = \int_{W_u(t)} dt' \, \Gamma_u(t, t') u(r, t') + \int_{W_\sigma(t)} dt' \, \Gamma_\sigma(t, t') v(r, t') \tag{2.6.9}$$

after (2.4.12), where Γ_u, Γ_σ are given by (2.4.13), again with the noted alterations.

First consider the case where $B_u(t)$ is contracting. Let us substitute (2.6.5) for $v(r, t)$, where it occurs in (2.6.8), to obtain

$$\int\limits_{W_u(t)} dt' \, \Pi_u(t, t') u(r, t') = \int\limits_{B_u(t)} ds' \, \Lambda(r, r') s^{(1)}(r', t) \tag{2.6.10}$$

where

$$s^{(1)}(r', t) = s(r', t) - \int\limits_{W_\sigma(t)} dt' \, \Pi_\sigma(t, t') s(r', t') . \tag{2.6.11}$$

The crucial step in writing down (2.6.10) is the interchange of the space and time integration, which is possible because

$$\int\limits_{W_\sigma(t)} dt' \, \Pi_\sigma(t, t') \int\limits_{B_u(t')} ds' \, \Lambda(r, r') s(r', t')$$

$$= \int\limits_{W_\sigma(t)} dt' \, \Pi_\sigma(t, t') \int\limits_{B_u(t)} ds' \, \Lambda(r, r') s(r', t') . \tag{2.6.12}$$

This follows from the assumed property that $s(r', t')$ is zero for $r' \notin B_u(t')$ and from the fact that for $t' \in W_\sigma(t)$, $B_u(t') \subseteq B_u(t)$. The interchange of integrations is now trivial.

The left-hand side of (2.6.10) is known for $r \in B_u(t)$, because of the definition of $W_u(t)$. Therefore, this equation is formally the same as the corresponding elastic equations and may be solved by the same techniques.

The solution will give us the form of $s^{(1)}(r, t)$ on $B_u(t)$. What is finally of interest, however, is the form of $s(r, t)$. For a finite number of maxima and minima, (2.6.11) provides a series of equations relating $s^{(1)}(r, t)$ and $s(r, t)$, which it should be possible, in principle, to solve for $s(r, t)$ in the various phases by starting at the earliest, and working systematically from there, provided the expanding phases, which we discuss below, are also incorporated. We will consider this formalism in a somewhat more explicit manner in Sects. 3.10 and 3.11. Practically speaking, the results will be sufficiently compact to be of interest only for the first cycle or two, and also in the steady state limit, which is discussed in Sect. 3.11.

For the expanding phase we substitute for $v(r, t')$ on the right of (2.6.9) from (2.6.5) to obtain

$$u(r, t) - \int\limits_{W_u(t)} dt' \, \Gamma_u(t, t') u(r, t') = \int\limits_{B_u(t)} ds' \, \Lambda(r, r') q^{(1)}(r', t) \tag{2.6.13}$$

where

$$q^{(1)}(r', t) = \int\limits_{W_\sigma(t)} dt' \, \Gamma_\sigma(t, t') s(r', t') \tag{2.6.14}$$

after interchanging space and time integration by the same argument as before. Again, the left-hand side is known for $r \in B_u(t)$, so that this has the same form as the elastic equation, and it may be possible, by analogy, to deduce the form of $q^{(1)}(r, t)$ on $B_u(t)$, and hence, by induction through the various phases, $s(r, t)$. In fact, (2.6.11) and (2.6.14) must be considered together.

2.7 Viscoelastic Papkovich-Neuber Solution

For isotropic materials, in the non-inertial approximation, we can write down the solution

$$2 \int_{-\infty}^{t} dt' \mu(t-t') u_i(r, t') = \int_{-\infty}^{t} dt' \kappa(t-t') \psi_i(r, t') - x_j \psi_{j,i}(r, t) - \psi_{0,i}(r, t) \tag{2.7.1}$$

where, in the case of zero body force, $\psi_i(r, t)$ $i = 0, 1, 2, 3$ are all harmonic functions over the region occupied by the material. This relation generalizes the well-known Papkovich-Neuber solution in elastostatics [Sokolnikoff (1956), Chap. 6]. The quantity $\kappa(t)$ is related to $v(t)$, the generalized Poisson's ratio for the material by

$$\hat{\kappa}(\omega) = 3 - 4\hat{v}(\omega) , \tag{2.7.2}$$

where $\hat{v}(\omega)$ is given by (1.9.19). The function $\hat{\kappa}(\omega)$ is causal, in the sense of Sect. 1.10, by virtue of the same property in $\hat{v}(\omega)$; $\kappa(t)$ is also real; see Problem (1.10.1).

An intuitive derivation of (2.7.1) can be given immediately by simply recalling the formal identity between viscoelastic and elastic equations, after Fourier transformation, exemplified by (1.9.18). A rigorous derivation, using a somewhat different notation, has been given by Gurtin and Sternberg (1962), who demonstrate the completeness of the representation − in the sense that any displacement field can be represented in this manner. While on this point, note that by operating on (2.7.1) with $\gamma(t)$, we can eliminate the hereditary integral on the left, giving an explicit expression for the displacement.

The use of this representation for solving viscoelastic boundary value problems for ablating bodies is discussed in the context of a specific problem by Graham (1965b). While it may provide solutions to particular problems, there is no well-defined procedure for obtaining solutions to general categories of problems in three dimensions, using this representation. In its three-dimensional form, it will not be used in later chapters. However, a special two-dimensional form is given in the next section, which is extremely useful.

Sternberg (1964) extended this representation to non-isothermal conditions while Efimov (1966b) and Graham and Williams (1972) generalized it to the case of an aging material.

Problem 2.7.1: Show that (2.7.1) is unaltered if $\psi_i(r, t)$, $i = 1, 2, 3$ and $\psi_0(r, t)$ are replaced by

$$\psi_{i1}(r, t) = \psi_i(r, t) + f_{,i}(r, t) \tag{2.7.1p}$$

$$\psi_{01}(r, t) = \psi_0(r, t) + f(r,t) + \int_{-\infty}^{t} dt' \kappa(t-t') f(r, t') - x_i f_{,i}(r, t) ,$$

where $f(r, t)$ is a harmonic function.

2.8 Plane Strain in Linear Viscoelasticity

Considerable simplification is obtained if the number of dimensions in a problem reduces to two, just as in the elastic case. One specific way that this happens, which will be of interest in later chapters, is when plane strain conditions prevail. We will discuss briefly the characteristics of plane strain, omitting mention of the closely related topic of plane stress.

Under plane strain conditions parallel to the xy plane, the displacement $u_z(r, t)$ is constrained to be zero, while the other components are independent of z. This would occur for example if all the bodies in the problem were uniform and infinitely extensive in the z direction, and are acted upon by external stresses independent of z. It follows that

$$\varepsilon_{xz} = \varepsilon_{yz} = \varepsilon_{zz} = 0 \ . \tag{2.8.1}$$

We confine the discussion to the case of an isotropic material. The constitutive equations (1.9.6) give that

$$\sigma_{xz} = \sigma_{yz} = 0 \ . \tag{2.8.2}$$

However, σ_{zz} is not zero. It is given by

$$\sigma_{zz}(r, t) = \int_{-\infty}^{t} dt' \lambda (t - t') \varepsilon(r, t')$$

$$\varepsilon(r, t') = \varepsilon_{xx}(r, t') + \varepsilon_{yy}(r, t') \tag{2.8.3}$$

$$r = (x, y) \ .$$

This normal stress must operate in order to maintain plain strain conditions.

Problem 2.8.1: Show that

$$\sigma_{zz}(r, t) = \int_{-\infty}^{t} dt' v(t - t') [\sigma_{xx}(r, t') + \sigma_{yy}(r, t')] \ , \tag{2.8.1p}$$

where $v(t)$ is defined by (1.9.19). What happens under the proportionality assumption? See (1.9.20).

Consider the Papkovich-Neuber form (2.7.1) in the two-dimensional case. All dependence on $z = x_3$ must drop out, so we put $\psi_3 = 0$. There are two equations. It is our object to combine them into one complex equation. Let us for a moment assume that [Gladwell (1980)]

$$\Phi(z, t) = \psi_1(r, t) + i \psi_2(r, t)$$

$$\tag{2.8.4}$$

$$\Psi(z, t) = \frac{\partial}{\partial x} \psi_0(r, t) - i \frac{\partial}{\partial y} \psi_0(r, t)$$

$$z = x + iy$$

are analytic complex functions over the extent of the material. Then, it is a simple exercise to show that (2.7.1) takes the form

$$2 \int_{-\infty}^{t} dt' \mu(t-t') D(r,t') = \int_{-\infty}^{t} dt' \kappa(t-t') \Phi(z,t') - z \bar{\Phi}'(\bar{z},t) - \bar{\Psi}(\bar{z},t), \quad (2.8.5)$$

where

$$D(r,t) = u_1(r,t) + i u_2(r,t) \quad (2.8.6)$$

and where the bar indicates complex conjugation. Over the function alone, it indicates complex conjugation, leaving z untouched, while over z it affects only that quantity. The derivation of (2.8.5) requires (A2.1.2, 3).

It must however be shown that $\Phi(z,t)$, $\Psi(z,t)$ are analytic. The argument is not essentially different from the elastic case and will not be repeated in detail. We refer instead to Gladwell (1980), for example who proves this result with the aid of the elastic limit of (2.7.1 p).

The stresses may be calculated by differentiating (2.8.5). One finally obtains the viscoelastic generalization of the Kolosov-Muskhelishvili equations which we collect together as follows:

$$\sigma_{11}(r,t) + \sigma_{22}(r,t) = 2[\Phi'(z,t) + \bar{\Phi}'(\bar{z},t)] = 4\,\mathrm{Re}\{\Phi'(z,t)\}$$

$$\sigma_{22}(r,t) - i\sigma_{12}(r,t) = \Phi'(z,t) + \bar{\Phi}'(\bar{z},t) + z\bar{\Phi}''(\bar{z},t) + \bar{\Psi}'(\bar{z},t) \quad (2.8.7)$$

$$2 \int_{-\infty}^{t} dt' \mu(t-t') D(r,t') = \int_{-\infty}^{t} dt' \kappa(t-t') \Phi(z,t') - z\bar{\Phi}'(\bar{z},t) - \bar{\Psi}(\bar{z},t) \;.$$

Problem 2.8.2: Let

$$R(z,\bar{z},t) = \Phi(z,t) + z\bar{\Phi}'(\bar{z},t) + \bar{\Psi}(\bar{z},t) \;. \quad (2.8.2\mathrm{p})$$

Show that

$$\frac{d}{dx} R(z,\bar{z},t) = \sigma_{22} - i\sigma_{12} \;, \qquad \frac{d}{dy} R(z,\bar{z},t) = i(\sigma_{11} + i\sigma_{12}) \;, \quad (2.8.3\mathrm{p})$$

using the first two relations of (2.8.7), and (A2.1.2). Hence show that along an arbitrary curve

$$\frac{dR(z,\bar{z},t)}{ds} = \sin\theta \frac{dR}{dx} - \cos\theta \frac{dR}{dy} = \tau_y - i\tau_x \;, \quad (2.8.4\mathrm{p})$$

where the normal on the left as one travels along the curve has direction $(\cos\theta, \sin\theta)$. Show, using (1.1.1), that (τ_x, τ_y) are the stresses acting, from the left, on the curve.

Problem 2.8.3: Show that the displacement and stresses are unaffected if an arbitrary function $c(t)$ is added to $\Phi(z,t)$ and

$$d(t) = \int_{-\infty}^{t} dt' \kappa(t-t') \bar{c}(t') \quad (2.8.5\mathrm{p})$$

is added to $\Psi(z,t)$.

In many cases considered in later chapters, $D(r, t)$ contains a logarithmic divergence. This point is noted below, in the present section, in a general context. It is therefore necessary, and in fact convenient, to differentiate the displacement equation in (2.8.7) with respect to x. On doing so, we can present the Kolosov-Muskhelishivili equations in an alternative form which is more convenient for practical applications. Let

$$\phi(z, t) = \Phi'(z, t) , \quad \psi(z, t) = \Psi'(z, t) . \tag{2.8.8}$$

Then, we have

$$\sigma_{11}(r, t) + \sigma_{22}(r, t) = 2[\phi(z, t) + \bar{\phi}(\bar{z}, t)] = 4\,\mathrm{Re}\{\phi(z, t)\} \tag{2.8.9}$$

$$\Sigma(r, t) = \sigma_{22}(r, t) - i\sigma_{12}(r, t) = \phi(z, t) + \bar{\phi}(\bar{z}, t) + z\bar{\phi}'(\bar{z}, t) + \bar{\psi}(\bar{z}, t)$$

$$2\int_{-\infty}^{t} dt'\mu(t-t')D'(r, t') = \int_{-\infty}^{t} dt'\kappa(t-t')\phi(z, t') - \bar{\phi}(\bar{z}, t)$$

$$-z\bar{\phi}'(\bar{z}, t) - \bar{\psi}(\bar{z}, t) ,$$

where

$$D'(r, t) = \frac{\partial}{\partial x} [u_1(r, t) + iu_2(r, t)] . \tag{2.8.10}$$

These equations are central to the considerations of the next two chapters. We shall sometimes refer to $\phi(z, t)$, $\psi(z, t)$ as the complex potentials, though this name is often used to refer to $\Phi(z, t)$, $\Psi(z,t)$.

The conditions that must be obeyed by the complex potentials at infinity and elsewhere under different physical configurations are discussed extensively by Muskhelishvili (1963) and Green and Zerna (1968); see also Sokolnikoff (1956), England (1971) and Gladwell (1980). These arguments go over, with minor modification, to the viscoelastic case, and will not be discussed here. For later reference, however, we will need certain results. The two configurations which will be of interest later are the infinite half-plane under load and an infinite plane containing a crack. In the case of a half-plane under load, we assume that stresses and rotations vanish at infinity. Also, let the boundary stresses along the x-axis fall off as $1/x^2$ or faster, at large distances from the origin. Then [Green and Zerna (1968), Chap. 8, for example], at large $|z|$,

$$\phi(z, t) \sim -\frac{Q}{2\pi z} , \quad \psi(z, t) \sim \frac{\bar{Q}}{2\pi z} \tag{2.8.11}$$

where $Q = X + iY$ is the resultant of the external forces acting on the x-axis.

For the problem of the crack in an infinite plane, we again assume that stresses and rotations vanish at infinity. Also, the resultant of all the forces acting on the crack face is assumed to cancel to zero. This is the standard assumption. Then, it can be shown that

$$\phi(z, t) \sim 0(1/z^2) , \qquad \psi(z, t) \sim 0(1/z^2) \tag{2.8.12}$$

at large z.

Observe that if the complex potentials $\phi(z, t)$, $\psi(z, t)$ behave as $1/z$, it implies that $\Phi(z, t)$, $\Psi(z, t)$ have logarithmic singularities at infinity. This means that the displacements, given by (2.8.7), also diverge logarithmically at large z. This difficulty will exist unless stresses and rotations at infinity vanish and the resultant of all stresses acting on finite boundaries is also zero. It is an underlying flaw in the theory, but its effects can be hidden, at least mathematically, by operating with displacement derivatives, rather than with the displacements themselves, as previously noted. However, physically, it means that absolute (as opposed to relative) displacements cannot be calculated.

It is clear from (2.8.11, 12) that this problem will be present in contact problems but not in crack problems.

Our method of attacking plane, non-inertial problems will be, in the first instance, to reduce (2.8.9) to a Hilbert problem, in precisely the manner developed by Muskhelishvili (1963), and then to handle the specifically viscoelastic aspects, essentially by the methods outlined in Sects. 2.4 – 6. We remark that an alternative way of approaching the first stage is the dual integral equation method originally used in this context by Sneddon (1951) but with a long history of mathematical development summarized by Gladwell (1980).

In some circumstances, it will be convenient to move the hereditary integral from the displacements onto the stresses. This may be achieved as follows. We define

$$\chi(z, t) = \int_{-\infty}^{t} dt' \gamma(t-t')\phi(z, t') , \qquad \Lambda(z, t) = \int_{-\infty}^{t} dt' \gamma(t-t')\psi(z, t') , \tag{2.8.13}$$

where $\gamma(t)$ is the inverse of $\mu(t)$ in the sense of (1.2.35). The quantities $\chi(z, t)$, $\Lambda(z, t)$ will have the same analyticity properties as $\phi(z, t)$, $\psi(z, t)$, under fairly general conditions, including the asymptotic behaviour (2.8.11) or (2.8.12). In terms of the new functions, we write (2.8.9) as

$$\int_{-\infty}^{t} dt' \gamma(t-t') [\sigma_{11}(r, t') + \sigma_{22}(r, t')] = 4\operatorname{Re}\{\chi(z, t)\} , \tag{2.8.14a}$$

$$\int_{-\infty}^{t} dt' \gamma(t-t')\Sigma(r, t') = \chi(z, t) + \bar{\chi}(\bar{z}, t) + z\bar{\chi}'(\bar{z}, t) + \bar{\Lambda}(\bar{z}, t) , \tag{2.8.14b}$$

$$2D'(r, t) = \int_{-\infty}^{t} dt' \kappa(t-t')\chi(z, t') - \bar{\chi}(\bar{z}, t) - z\bar{\chi}'(\bar{z}, t) - \bar{\Lambda}(\bar{z}, t) . \tag{2.8.14c}$$

Note that, in writing the last equation, the commutativity of convolution integration has been used (Problem 1.2.2).

2.9 Contact between Viscoelastic Media

All the boundary value problems considered in later chapters may be regarded as problems involving contact between similar or dissimilar viscoelastic and elastic (including rigid) bodies. General theorems concerning the existence and uniqueness of solutions to such problems have been given by Signorini (1959) and Fichera (1972) for the elastic case and Boucher (1975) for the viscoelastic case.

Here we will focus upon a different type of general result, given, for elastic materials, by Dundurs and Stippes (1970) and Dundurs (1975); and for viscoelastic materials by Comninou (1976). This topic, in the three-dimensional case, will be discussed in this section. Stronger results for the plane case, also due to these authors, are presented in Sect. 2.10.

A fundamental distinction central to the topic, is that between receding and advancing contact. Problems involving receding contact are defined as follows. Let C_0 be the contact region before deformation, that is, before application of the loads. At time $t = 0$, the loads are applied and $C(t)$, $t \geq 0$, is the contact region. If $C(t) \subseteq C_0$ for all $t \geq 0$, the contact is receding. If the contact is not receding, it is advancing. Let C_0' be the complement of C_0 on the boundary. If $C(t) \cap C_0'$ is not empty for some $t \geq 0$, the contact is advancing.

If $C(t)$ does not vary in time, apart from an initial, instantaneous change, the contact is said to be stationary.

It is important to distinguish between receding and advancing contact in the sense just defined, and contracting and expanding contact in the sense used in Sect. 2.6, and in various contexts throughout this volume. The former is a comparison of two situations: before load application and after, while the latter refers to how $C(t)$ changes with time. Note that a receding contact could be expanding, after an initial contraction. Also, if the contact is advancing, it does not follow that $C(t) \supseteq C_0$. (This would follow if, for all $t \geq 0$, the contact were expanding.)

The basic assumption underlying the results proved in this and the following section is that the contact be receding. This excludes most contact problems in the ordinary usage of the term. Examples of contact problems which have this property are given by Dundurs (1975). The only configuration relevant to later chapters for which the contact is receding is a crack, initially closed, which opens under a tensile stress. In this example, C_0 is the entire crack face and $C(t)$ is the closed portion of it, if any.

The region $C(t)$ and its complement on the boundary $C'(t)$ correspond respectively to $B_u(t)$, $B_\sigma(t)$, used earlier.

Consider the frictionless, isothermal contact between two viscoelastic bodies, where the contact is receding. The constitutive relations for the two bodies have the form (1.8.8):

$$\sigma_{ij}^a(r, t) = [G_{ijkl}^a * d\varepsilon_{kl}^a](r, t) \ , \tag{2.9.1}$$

where the superscript a distinguishes the two bodies. Recall that the strains $\varepsilon_{ij}^a(r, t)$, given by (1.1.4), are linear in the displacements $u_i^a(r, t)$. We now write down the dynamical equations and boundary conditions for the contact problem.

For the moment, let us confine ourselves to the non-inertial approximation. The dynamical equations (1.1.3) read:

$$\sigma_{ij,j}^a(r, t) = -b_i^a(r, t) \ ,\tag{2.9.2}$$

where the σ_{ij}^a are given by (2.9.1) and the b_i^a are body forces at a given point. Let the stresses be prescribed on the boundaries, outside of C_0, giving

$$\sigma_{ij}^a(r, t)n_j^a(r) = c_i^a(r, t) \ , \quad r \in C_0'\tag{2.9.3}$$

where $n_j^a(r)$, $a = 1, 2$, are the normal vectors to the boundaries. In a linear theory, these can be determined from the undeformed boundaries. The $c_i^a(r, t)$ are the prescribed stresses. Consider now the boundary conditions within C_0. On $C(t)$ itself, normal displacement must be equal and opposite, giving

$$n_i^1 u_i^1 + n_i^2 u_i^2 = 0 \ , \quad r \in C(t) \ .\tag{2.9.4}$$

We note in passing that $n_i^1 = -n_i^2$ in $C(t)$. Also, wherever contact is possible (C_0 in the receding case) the two normals will be nearly equal, differing only by second order terms, since in a linear theory any gap must be first order.

If the contact were advancing, (2.9.4) would not be valid. On those parts of $C(t)$ outside C_0, a separation existed initially which must be eliminated before contact begins, so instead of (2.9.4), one has

$$n_i^1 u_i^1 + n_i^2 u_i^2 = d_0 \ ,\tag{2.9.5}$$

where d_0 is the initial separation at the point of interest. It is the fact that (2.9.4) is homogeneous, in contrast to (2.9.5), that gives rise to the simple results, presented below, for receding contacts.

Returning to the receding contact case, we have also that normal tractions must be equal and non-tensile, giving [see (1.1.1)]:

$$\sigma_{ij}^1 n_i^1 n_j^1 = \sigma_{ij}^2 n_i^2 n_j^2 \le 0 \ , \quad r \in C(t) \ ,\tag{2.9.6}$$

while the frictionless condition gives that (no summation over a):

$$\sigma_{ij}^a n_j^a m_i^a = 0 \ , \quad a = 1, 2 \ , \quad r \in C(t)\tag{2.9.7}$$

where $m_i^a(r, t)$ is any unit tangent at r. For points on $C_0 - C(t)$, from which contact has receded, we have the requirement of no penetration

$$n_i^1 u_i^1 + n_i^2 u_i^2 < 0 \ , \quad r \in C_0 - C(t)\tag{2.9.8}$$

and vanishing tractions

$$\sigma_{ij}^a n_j^a = 0 \ , \quad r \in C_0 - C(t) \ .\tag{2.9.9}$$

For crack problems, where the load is viewed as being applied to the crack surface (see Sect. 4.1), (2.9.9) must be replaced by (2.9.3), extended to this region. The argument given below is not affected, provided the stresses in $C_0 - C(t)$ are scaled in the same manner as on C_0'.

We note that it may not be necessary to impose some of the conditions, specifically the inequalities in (2.9.6, 8) in order to obtain a solution. However, these must be confirmed afterwards. Also, the region $C(t)$ is in many cases not known in advance, so that (2.9.4) must be regarded as the condition defining $C(t)$. The external loads $b_i^a(r, t)$, $c_i^a(r, t)$ may have to satisfy some global equilibrium conditions.

Of (2.9.2 – 9), the only inhomogeneous relations, in terms of the displacements, are (2.9.2, 3) and, as mentioned, possibly (2.9.9). It is clear, therefore, that if the displacements are scaled by a factor k, the equations remain true provided the external stresses $b_i^a(r, t)$, $c_i^a(r, t)$ are similarly scaled. The scaling factor k must be positive. This is apparent from physical intuition and also from the fact that a negative k would change the sense of the inequalities in (2.9.6) and (2.9.8). It is further to be observed that such scaling would leave the contact area unchanged. This is apparent from (2.9.4), which determines the extent of $C(t)$. We conclude that the solution of the problem has the following properties:

(1) The displacements and hence the stresses and strains are linear in the applied loads $b_i^a(r, t)$, $c_i^a(r, t)$.
(2) The contact area is independent of the load.

We will see examples in Chap. 3 of advancing contact problems where these properties manifestly do not hold. However Sect. 4.8 contains an interesting example of a receding contact problem that may be used to illustrate the results given in this section (see Problem 4.8.1).

Consider the elastic problem for a moment. Since the contact C does not depend on load, the moment any load is applied, there is a discontinuous change from C_0 to C, where C depends only on the geometry of the problem and the moduli of the bodies. This is an interesting phenomenon in that it is truly discontinuous; if even an infinitesimal load is applied, the system "clicks" from one configuration into the other. The exception is where C happens to be the same as C_0.

In the viscoelastic case, $C(t)$ may be time-dependent since the moduli depend on time. Initially, however, we recall (Sect. 1.4) that viscoelastic materials generally behave as elastic materials with moduli equal to the instantaneous moduli. Therefore:

(3) Initially, there will be a discontinuous alteration from C_0 to $C(0^+)$ on application of any load, unless these two regions happen to be the same.

Observe that $C(t)$ may depend on the distribution of external loads. It is independent only of their overall magnitude.

As $t \to \infty$, the viscoelastic functions approach the long-time (elastic) moduli (Sect. 1.4) and the contact area takes up its final form which will be the elastic contact area corresponding to these moduli. This applies only if final equilibrium moduli exist, which will not be true for viscoelastic liquids.

These considerations also apply to multiple contacts and inhomogeneous bodies, since the argument hinges totally on the fact that (2.9.4) has a zero on the right-hand side.

The following statements may also be checked trivially. If:

(4) $G_{ijkl}^a(t)$ is replaced by $kG_{ijkl}^a(t)$, and the external stresses $b_i^a(r, t)$, $c_i^a(r, t)$ left unaltered, then $C(t)$ and the stress tensors are not affected and the displacements are scaled according to u_i^a / k.

(5) If $G_{ijkl}^a(t) \to k G_{ijkl}^a(t)$, and $b_i^a(r, t)$, $c_i^a(r, t)$ are scaled similarly, then the displacements are not affected and the stresses are scaled according to $\sigma_{ij}^a \to k \sigma_{ij}^a$.

These last two observations were made for Elasticity by Dundurs (1975).

(6) If the contact is stationary, having instantaneously receded, in other words, if $C(t) = C(0^+) \subseteq C_0$, then stronger statements can be made. All the above observations hold except that $k(t)$ is now a time-dependent, causal function and multiplication is replaced by the Stieltjes product.

To show this is left as an exercise. The essential point is contained in Problem 1.2.2, which tells us that the Stieltjes product (1.2.29), in the non-aging case, is commutative, so that $k(t)$ can be moved across the viscoelastic functions. It must be positive in the sense that $[k*du] > 0$ for all positive functions $u(t)$, if the inequalities are to remain unaffected.

2.9.1 Inclusion of Inertial Affects

In this case, the dynamical equations (2.9.2) become

$$\sigma_{ij,j}^a(r, t) - \varrho \frac{\partial^2 u_i^a}{\partial t^2} = -b_i^a(r, t) \tag{2.9.10}$$

and the boundary conditions are unaffected. It follows that properties (2.9.1 − 5) hold also in this case. The only question that arises is with respect to the final property (2.9.6), and that is whether $k(t)$ commutes with the differential operator $\partial^2/\partial t^2$ in the sense that

$$\left(k * d \frac{\partial^2 u_i^a}{\partial t^2} \right)(t) = \left(\frac{\partial^2}{\partial t^2}(k * du_i) \right)(t) . \tag{2.9.11}$$

It can be shown, by means of two partial integrations, that (2.9.11) holds.

2.10 Receding Contact in Plane Viscoelasticity

More detailed results are possible if the bodies in contact are isotropic and such that plane strain, or stress, conditions apply and body forces are zero. Let us suppose, as in the last section, that outside of the contact region, only surface tractions (as opposed to displacements) are prescribed on the boundary and further that any line integral of the stresses around a closed contour T, covering any portion of either or both bodies, is zero:

$$\int_T \sigma_{ij} n_j ds = 0 . \tag{2.10.1}$$

The vector n_j is the outward normal to the contour at a given point. This integral is proportional to the vector sum of the forces acting on T. It will be zero if no net forces are acting across the contours. If the bodies are simply connected, that is without cracks or cavities, condition (2.10.1) will be valid if quasi-equilibrium prevails throughout – in the sense that acceleration terms are negligible and body forces zero. If cavities are present, then the resultant of the forces acting on the cavity faces must be zero in each case.

Michell (1899) showed that for a single elastic body, condition (2.10.1), valid for all closed contours together with the restriction to stress boundary conditions, implies that the stresses are independent of the material constants. A simple dimensional argument shows that they could at most depend upon Poisson's ratio. This theorem eliminates even that dependence. An intuitive demonstration of the result would proceed as follows. Since the first two equations of (2.8.7) have no dependence on the material parameters, the stresses will depend on these quantities if the complex potentials do. Given the nature of the boundary conditions, the only way such dependence on the complex moduli can enter is through a constraint that the displacements and stresses be single-valued. This condition forces certain properties on the complex potentials as shown in the literature quoted in Sect. 8, and can introduce a dependence on the moduli. However, an examination of how this occurs in Muskhelishvili (1963), and the other standard references, indicates that the moduli enter only if contours exist such that (2.10.1) does not hold.

The conclusion is also valid for viscoelastic bodies – if the non-inertial approximation applies. This follows immediately by invoking the Classical Correspondence Principle. Our object in this section is to generalize the result to the case of two viscoelastic bodies in contact.

Consider Problem 2.8.2. Let us denote by $R_1(z, \bar{z}, t)$, $R_2(z, \bar{z}, t)$ the function given by (2.8.2p) in the two bodies. Relations (2.8.4p), (2.9.6, 7) give that

$$\frac{dR_1}{ds} = \frac{dR_2}{ds} \tag{2.10.2}$$

along the contact region $C(t)$. Therefore R_1 and R_2 differ at most by a term that does not vary along $C(t)$. By using the arbitrariness in $\Phi(z, t)$, $\Psi(z, t)$, given by Problem 2.8.3, we can transform this difference to zero along $C(t)$, giving that

$$R_1(z, \bar{z}, t) = R_2(z, \bar{z}, t) \tag{2.10.3}$$

over the contact region. Now, the normal displacement on the boundary can be deduced from (2.8.7) to be [Dundurs (1975)]:

$$N_a(r, t) = \tfrac{1}{2}(\gamma_a * \mathrm{Re}\ [e^{-i\theta_a} S_a])(z, \bar{z}, t) \tag{2.10.4}$$

$$S_a(z, \bar{z}, t) = [\kappa_a * \Phi_a](z, t) - z\,\bar{\Phi}'_a(\bar{z}, t) - \bar{\Psi}_a(\bar{z}, t) , \qquad a = 1, 2$$

for each body, distinguished by the parameter, a. The quantity $\gamma(t)$ and the Faltung notation, both defined by (1.2.32), have been used. The quantity θ_a is the angle between the outward normal, for a given body, and the x-axis. Note that in the region of contact

$$e^{-i\theta_1} = -e^{-i\theta_2} . \tag{2.10.5}$$

Equation (2.10.4) contains hereditary integrals. More will be introduced below. We need to assume that $C(t)$ is not only receding but either contracting or stationary so that if $z \in C(t)$ it will have been in the contact region at all previous times.

Equation (2.9.4) gives that

$$[\gamma_1 * (e^{-i\theta_1} S_1)] (z, \bar{z}, t) + [\gamma_1 * (e^{i\theta_1} \bar{S}_1)] (\bar{z}, z, t)$$

$$+ [\gamma_2 * (e^{-i\theta_2} S_2)] (z, \bar{z}, t) + [\gamma_2 * (e^{i\theta_2} \bar{S}_2)] (\bar{z}, z, t) = 0 \ . \tag{2.10.6}$$

Using (2.10.5), we can also rewrite (2.10.3) as

$$e^{-i\theta_1} R_1 (z, \bar{z}, t) + e^{-i\theta_2} R_2 (z, \bar{z}, t) = 0$$

$$e^{i\theta_1} \bar{R}_1 (\bar{z}, z, t) + e^{i\theta_2} \bar{R}_2 (\bar{z}, z, t) = 0 \ . \tag{2.10.7}$$

Let us add these two equations together and convolute $\frac{1}{2} (\mu_1 + \mu_2) (t)$ with the result. We now add this to $[\mu_1 * \mu_2] (t)$ convoluted with (2.10.6) and finally convolute with the expression

$$\Gamma(t) = [\mu_2 * (\delta + \kappa_1) + \mu_1 * (\delta + \kappa_2)]^{-1}(t) \tag{2.10.8}$$

where δ is the unit operator under the convolution product normally realized by a δ-function. The inverse in (2.10.8) refers to the convolution product. One finds as a result of these operations that the condition that normal displacements are equal in the contact region contains only two material parameters $a(t)$, $\beta(t)$ given by

$$a(t) = \{\Gamma * [\mu_2 * (\delta + \kappa_1) - \mu_1 * (\delta + \kappa_2)]\}(t)$$

$$\beta(t) = \{\Gamma * [\mu_1 * (\delta - \kappa_2) - \mu_2 * (\delta - \kappa_1)]\}(t) \tag{2.10.9}$$

which are the viscoelastic generalizations of the Dundurs parameters [Dundurs (1975)].

Recall that our interest is in the dependence of the stresses on the material parameters. Potentially these could depend upon $\mu_1(t)$, $\mu_2(t)$, $\kappa_1(t)$, $\kappa_2(t)$. Let us consider the ways that such dependence could arise. We observe as before that since the first two equations of (2.8.7) have no dependence on the material parameters, the dependence of the stresses on these quantities will be a consequence of the dependence of the complex potentials on them.

The boundary conditions outside of the contact region will introduce no dependence on the moduli since these prescribe stress only. Condition (2.10.3) introduces no dependence. Condition (2.10.6) does, and in its modified form, described above, we see that it results in a dependence only on $a(t)$, $\beta(t)$.

There is the further condition, also noted previously, on the displacements which would cause the complex potentials and hence the stresses to acquire dependence on other material parameters − a requirement that the stresses and displacements be single-valued. This possibility is excluded, by virtue of (2.10.1), which follows by the same argument as sketched above, for a single body.

Therefore, the stresses depend only on the two parameters $a(t)$, $\beta(t)$. This is also true of the extent of the contact area since (2.10.6), which in modified form depends only on $a(t)$ and $\beta(t)$, is essentially condition (2.9.4) which determines the extent of the contact area.

For this result to be valid in the viscoelastic case, it is necessary that the boundary conditions outside the contact region be of the stress only type, that the contact region be receding but also either stationary or contracting and that (2.10.1) hold for all contours in the region covered by both bodies.

If the materials are identical, $a(t) = \beta(t) = 0$ and the stresses and contact region are independent of material properties, which is a statement akin to Michell's result.

A further simplification is obtained if the boundary over the contact region is a straight line. Let us take the x-axis along this boundary. Differentiating (2.9.4) gives the condition

$$\frac{\partial u_{1y}}{\partial x} = \frac{\partial u_{2y}}{\partial x} , \tag{2.10.10}$$

so that we can use the differentiated form of the Kolosov-Muskhelishvili equations (2.8.9). The absence of friction in the contact region gives that

$$\mathrm{Im}\,\{z\bar{\phi}_i'(\bar{z}, t) + \bar{\psi}_i(\bar{z}, t)\} = 0 , \quad i = 1, 2 . \tag{2.10.11}$$

Using this in the displacement equation of (2.8.9) gives that (no summation over "i"):

$$2\left[\mu_i * \frac{\partial}{\partial x}\, u_{iy}\right](r, t) = [(\delta + \kappa_i) * \mathrm{Im}\,\{\phi\}](z, t) , \quad i = 1, 2 . \tag{2.10.12}$$

Applying condition (2.10.10) to the displacements results in

$$[(1 + a) * \mathrm{Im}\,\{\phi_1\}](z, t) = [(1 - a) * \mathrm{Im}\,\{\phi_2\}](z, t) . \tag{2.10.13}$$

Therefore, the stresses and the extent of the contact interval depend only on $a(t)$.

If one of the bodies is rigid, then $a(t) \to \pm 1$ and the stresses and contact interval are independent of the viscoelastic properties of the material, just as when the two materials are identical.

The generalization of Dundurs' results to Viscoelasticity in the plane case is due to Comninou (1976) also. Her notation, however, is different from that adopted here. Dundurs (1975) discusses generalization of these results, for the elastic case, to assemblages of many bodies.

2.11 Energy Loss in Moving Contact Problems

In later chapters, we will be dealing with problems involving rigid identors pressed into a viscoelastic half-space and moving both vertically and horizontally. It is of interest to give an expression for the rate of energy input required to cause this movement. We continue to neglect inertial effects.

Firstly, it is relevant here and later to relate the stress tensor components on the surface to the applied stresses, with the aid of (1.1.1). Let the half-space occupy $y > 0$. The contact pressure $p(r, t)$ is given by

$$p(r, t) = -\sigma_{22}(r, t) \tag{2.11.1}$$

as remarked in Sect. 1.1. Let the indentors be in motion along the x-axis for simplicity, and with velocity U, not necessarily constant. Then the shear stress acting on the boundary is given by

$$s(r, t) = \mathrm{sgn}(U)|s(r, t)| = -\sigma_{12}(r, t) \ . \tag{2.11.2}$$

If Coulomb's Law applies, then

$$|s(r, t)| = fp(r, t) \quad \text{or} \tag{2.11.3}$$

$$s(r, t) = \mathrm{sgn}(U)fp(r, t) \ . \tag{2.11.4}$$

Let us denote the contact area by $C(t)$. In fact, it generally corresponds to $B_u(t)$. From (1.1.8), we see that the rate of energy input, by the moving load, is given by

$$\dot{E} = \int_{C(t)} ds \, s_i(r, t) \dot{u}_i(r, t) \tag{2.11.5}$$

where $s_i(r, t)$ are the surface tractions and $u_i(r, t)$, the surface displacements. Consider first the case of vertical movement. Surface friction will be neglected in this case. We denote the surface pressure by $p(r, t)$. Also, the normal displacement into the body will typically be of the form

$$u_2(r, t) = u_y(r, t) = D(t) - S(r) \ , \tag{2.11.6}$$

where $D(t)$ is the displacement under the lowest point of the indentor, taken to be the origin, and $S(r)$ describes its profile. It is clear that $S(r)$ will be zero at the lowest point. The form given by (2.11.6) will be valid in the contact area, which is all that is required here. From (2.11.6),

$$\dot{E} = \int_{C(t)} ds \, p(r, t) \dot{u}_y(r, t) = \dot{D}(t) W(t) \ , \tag{2.11.7}$$

where $W(t)$ is the load on the indentor causing the movement. This is of course the result that would be intuitively expected. It is relevant to the considerations of Sect. 5.3.

Consider next the case of horizontal motion, along the x-axis. Frictionless contact will be discussed first. Let the position of the lowest point be $r_0(t) = (x_0(t), 0)$, where $\dot{x}_0(t) = U$, the punch velocity. In the contact area,

$$u_y = D(t) - S(r - r_0(t)) \ , \tag{2.11.8}$$

so that

$$\dot{E}(t) = \dot{D}(t) W(t) - \int_{C(t)} ds \, p(r, t) \frac{\partial}{\partial x} u_y(r, t) U(t)$$

$$= \dot{D}(t) W(t) + F_H U(t) \mathrm{sgn}\,[U(t)] \ , \tag{2.11.9}$$

where F_H is the magnitude of the force resisting the horizontal motion. It follows that

$$F_H = -\operatorname{sgn}(U) \int_{C(t)} ds\, p(r,t)\, \frac{\partial}{\partial x} u_y(r,t) \ . \tag{2.11.10}$$

This formula was given by Hunter (1961) for the special case of plane strain conditions and a cylindrical punch. For an elastic medium, this force is zero. However, for a viscoelastic medium, we shall see that this is not the case. The deformation caused by the moving load results in mechanical energy loss, which is manifested by the presence of a resisting force. This is the well-known force of hysteretic friction, first demonstrated experimentally by Tabor (1952).

The coefficient of hysteretic friction is defined as

$$f_H = \frac{F_H}{W} = \frac{-\operatorname{sgn}(U)}{W} \int_{C(t)} ds\, p(r,t)\, \frac{\partial}{\partial x} u_y(r,t) \ . \tag{2.11.11}$$

If U has been constant for a long period, then steady-state conditions eventually prevail and all quantities (instead of just $u_y(r,t)$ in the constact region) depend on space and time co-ordinates only through the combination $r - r_0(t)$ where $r_0 = (Ut, 0)$. Transferring to coordinates moving with the punch, with origin at the lowest point of the punch $r_0(t)$, we obtain

$$f_H = \frac{-\operatorname{sgn}(U)}{W} \int_C ds\, p(r)\, \frac{\partial}{\partial x} u_y(r) = \frac{\operatorname{sgn}(U)}{W} \int_C ds\, p(r)\, \frac{\partial}{\partial x} S(r) \tag{2.11.12}$$

with the aid of (2.11.8). The contact area is no longer time-dependent.

Equation (2.11.11) must be modified if the contact is frictional. We must assume to begin with that steady-state conditions apply. The reason for this is that, otherwise, we cannot be sure that $u_x(r,t)$ depends only on $r - r_0(t)$, which is a necessary property for the argument to go through. Under steady-state conditions, $\dot{D}(t)$ is zero. The same argument which gave (2.11.11) now gives

$$f_H = \frac{-\operatorname{sgn}(U)}{W} \int_{C(t)} ds\, \left[p(r,t)\, \frac{\partial}{\partial x} u_y(r,t) + s(r,t)\, \frac{\partial}{\partial x} u_x(r,t) \right] \ . \tag{2.11.13}$$

In moving coordinates, instead of (2.11.12), we have

$$f_H = \frac{-\operatorname{sgn}(U)}{W} \int_C ds\, \left[p(r)\, \frac{\partial}{\partial x} u_y(r) + s(r)\, \frac{\partial}{\partial x} u_x(r) \right] \ . \tag{2.11.14}$$

If Coulomb's law applies, then $s(r)$, $p(r)$ are related by (2.11.4). Equation (2.11.14) gives the hysteretic contribution. There is also the ordinary frictional force

$$F_f = \int_C ds\, s(r) = \operatorname{sgn}(U) f W \tag{2.11.15}$$

if (2.11.4) is valid. The formula for the hysteretic friction coefficient was given by Golden (1986a). In an earlier paper [Golden (1979a)], only the first term was used. The relationship between hysteretic and adhesive rubber friction is discussed by Golden (1980).

2.12 Solution of Problems Involving Aging Materials or Non-isothermal Conditions

Our discussion so far has been largely confined to materials that are non-aging and whose viscoelastic functions are independent of position. As noted however in Sect. 1.7, both of these properties may be lost under non-isothermal conditions. Also there are of course materials that are aging under isothermal conditions. We therefore briefly indicate methods of solution applicable to such materials.

2.12.1 Aging Materials

The fundamental mathematical difference between the theories of aging and non-aging materials is the fact that the product in the operator algebra discussed in Sect. 1.2 is non-commutative for aging materials (see Problem 1.2.2). There is also the practical consideration that operator inverses will in some cases be relatively easy to find in the non-aging case – using Fourier transform techniques – and generally difficult to find in the aging case, where such techniques are not applicable.

However, formally speaking, non-commutative algebra presents no insuperable barriers and various standard results may readily be generalized to aging material. The oldest results provide a correspondence between non-inertial viscoelastic and appropriate elastic solutions for materials that have only one independent relaxation function. These results have been described by Arutyunyan (1952), Predeleanu (1965) and Bazant (1975) and are extensions of ideas due to McHenry (1943). They are generalizations to the aging case of results that would follow from the correspondence principle for materials with one relaxation function.

The decomposition of hereditary integrals described in Sect. 2.4 for non-aging materials and the general techniques based on that decomposition may be extended to the case of aging viscoelastic bodies. A form of this decomposition has been used by Graham (1980) to develop solutions for frictional contact problems that involve varying contact area.

Extensions of the correspondence principle, given in Sect. 2.6 for non-aging materials, may also be proved for aging viscoelastic materials and used to solve problems [see Graham and Williams (1972), Graham (1978b) and Graham and Golden (1987)].

For aging viscoelastic materials with two independent relaxation functions, a form of the Papkovich-Neuber stress function solution (see Sect. 2.9) has been developed by Efimov (1960) and Graham and Williams (1972), who used it to solve a problem for an ablating body. Also, in conjunction with the decomposition of hereditary integrals, Sabin and Graham (1980) have used this integral to obtain solutions for normal contact problems that involve variable contact area. Graham (1974) contains solutions, obtained from first principles by a semi-inverse method, to problems involving the flexure of aging viscoelastic beams that are valid for materials with two independent relaxation functions.

In conclusion, the various techniques discussed in earlier sections of this chapter can be generalized in a formal sense to handle aging materials. The practical algebraic details may however be considerably more complex than in an equivalent non-aging problem.

2.12.2 Non-isothermal Problems

Problems involving non-isothermal conditions are difficult to solve even in Elasticity Theory. Our interest lies mainly in problems where a material inhomogeneity derives from spacial temperature variations. In such cases, the instantaneous elastic portion of the viscoelastic functions remains independent of spacial co-ordinates, so that the limiting instantaneous problem is homogeneous. The inhomogeneity is only in the specifically viscoelastic portion of the equations.

We will outline below a method due to Fichera (1965), Edelstein (1969a, b) and used extensively by F. Williams (1975) which is applicable to such problems. In fact, it is applicable in principle to any problem for which the elastic Green's Functions are known, though it will not yield analytic results in any except the simplest cases.

The method involves transforming the integro-differential equations determining the behaviour of the material into an integral equation which can be solved by standard methods. Our object will be to sketch briefly the principle of the approach, which in fact is very simple. Later, in Chap. 6, it will be applied to the solution of a specific problem.

The principle of the method may be written down very simply in formal operator terms, if one remembers that the Green's function of a differential operator is simply the inverse of the operator. Let the equation to be solved have the form

$$Du = Q , \tag{2.12.1}$$

where u is the vector of displacements for example (or it could be some combination of stresses) and D is an integro-differential operator of the form

$$D = D_0 + D_1 \tag{2.12.2}$$

the operator D_0 being a differential operator, usually the instantaneous elastic limit, while D_1 involves differential operators with respect to space and integral operators with respect to time. The operator D_1 contains the specifically viscoelastic effects. The quantity Q is a given vector function. It is assumed that we know the elastic Green's function for the boundary value problem of interest. Write (2.12.1) in the form

$$D_0 u = -D_1 u + Q = g , \tag{2.12.3}$$

so that

$$u = D_0^{-1} g + S , \tag{2.12.4}$$

where D_0^{-1} is the Green's function operator and S is the surface term involving given boundary functions which is annihilated by D_0. Acting with D_1 on both sides and using (2.12.3) gives the relation

$$g + D_1 D_0^{-1} g = Q - D_1 S ,\tag{2.12.5}$$

which is an integral equation for g, that may be solved by iteration. Conditions ensuring the convergence of the generated series are discussed by Edelstein (1969) and F. Williams (1975). These authors also show that the method may be used to obtain viscoelastic solutions valid for ablating bodies. Once g is determined, the displacement vector u is given by (2.12.4).

2.13 Summary

General methods of solving viscoelastic boundary value problems are described in this chapter.

I. Viscoelastic Solutions in Terms of Elastic Solutions. The fundamental result is the Classical Correspondence Principle. It is based on the observation that the time Fourier transform (FT) of the governing equations of Linear Viscoelasticity may be obtained by replacing elastic constants by corresponding complex moduli in the FT of the elastic field equations. It follows that, whenever those regions over which different types of boundary conditions are specified do not vary with time, viscoelastic solutions may be generated in terms of elastic solutions that satisfy the same boundary conditions. In practical terms this method is largely restricted to the non-inertial case, since then a wide variety of elastic solutions are available and transform inversion is possible.

The method may be extended to problems involving time-dependent regions; however in that case, determination of boundary values for the elastic solutions becomes part of the problem. For bodies that occupy fixed regions of space, the solution of a wide class of problems has been reduced to the solution of a single space and time interdependent integral equation (2.5.1). In particular cases where the boundary regions over which different types of boundary conditions are specified vary monotonically with time, formulae may be derived for the boundary values of the elastic solutions and the problem is, at least in principle, completely solved. This result is referred to as the Extended Correspondence Principle (Sect. 2.2.6).

The Generalized Partial Correspondence Principle (Sect. 2.2.6) provides partial solutions to problems at specific instants of time when regions over which certain types of boundary conditions prevail are contained in (or contain) the regions where that type of boundary condition was given at all previous times.

If regions over which different types of boundary conditions are given successively expand and contract through the same one-parameter system of regions, then the equations governing these problems may be solved by analogy with the corresponding elastic problem (Sect. 2.2.6).

II. Stress Function Solutions. Equation (2.7.1) provides an extension of the Papkovich-Neuber stress function solution of three-dimensional elasticity to viscoelasticity.

Equations (2.8.9) extend to viscoelasticity the Kolosov-Muskhelishvili solution of plane elasticity, which is widely used in Chaps. 3, 4.

III. Contact Between Viscoelastic Bodies: General Results. (a) Sects. 2.9 and 2.10 contain various qualitative results for problems that involve receding contacts: contact problems where the contact region never contains points that were not in it before the deformation began.

(b) Sect. 2.11 contains results on the non-inertial energy loss due to the motion over a half-space of rigid indentors moving with velocity U. Formulae for the coefficient of hysteretic friction involved in this motion are given by (2.11.11, 12, 14).

IV. Aging or Non-isothermal Problems. Most of the techniques described above can be generalized to handle aging materials or non-isothermal problems, though in the latter case with some restriction on the temperature field. Sect. 2.12 outlines a method whereby problems not solvable by any of the above techniques may sometimes be solved. The method relies upon an iterative solution of the integral equation (2.12.5).

3. Plane Non-inertial Contact Problems

We consider, in the present chapter, problems involving moving or varying loads acting on the boundaries of viscoelastic materials, where plane strain (see Sect. 2.8) conditions prevail. Inertial effects are neglected.

Plane elastic contact problems have been extensively investigated, mainly using complex variable techniques, based on the Kolosov-Muskhelishvili formulae. Various problems are considered by Muskhelishvili (1963), by casting them in the form of a Hilbert problem (see Sect. A2.3). Galin (1961) also considers a wide range of problems, by casting them in the form of a Riemann-Hilbert problem, also discussed briefly in Sect. A2.3. More recent references on this topic are Kalker (1975) and Gladwell (1980) who deals extensively with plane problems. Ling (1973) discusses in summary form various contact problems.

The case of stationary, unvarying loads on a viscoelastic medium reduces to an elastic problem. Moving or changing loads cause considerable extra complexity however, though elastic solutions, and the methods used to obtain these solutions, remain very relevant. This might be expected in view of the discussion in the early stages of Chap. 2.

A general methodology will be developed, based on the viscoelastic Kolosov-Muskhelishvili equations (2.8.9). We will see that there are two stages to the approach, one analogous to the solution of the elastic problem − and here we use the techniques of Muskhelishvili (1963) − while the other deals with the specifically viscoelastic aspects. The procedure will be developed from first principles, rather than relying on the general results of Sects. 2.2, 3, both because there is no great difficulty in doing so, and because it provides an excellent example of the general arguments, of a rather abstract nature, which were developed in those sections. The correspondence between the two developments will be noted at appropriate places.

As pointed out in Sect. 2.2, the crux of the problem is the determination of certain functions on the boundary. This results in an apparent lack of emphasis on the behaviour of the interior of the material. In fact, however, once the boundary functions are known, everything else can be determined in a relatively straightforward manner, if there is an interest in doing so − which in some cases there is not.

Attention is confined to isotropic materials. Also, we deal only with half-plane problems and rigid indentors. However, the results are applicable to mildly curved surfaces and, with certain modifications, to the case of contact between two viscoelastic bodies. This is the familiar argument used in the theory of Hertzian contact. The modifications mentioned are not trivial in the viscoelastic case, as they are in the elastic case, involving as they do, the combining of viscoelastic functions of the two materials into one effective viscoelastic function, and operating with that [Graham (1965)].

3.1 Kolosov-Muskhelishvili Equations Adapted to the Half-Plane

Our object is to apply (2.8.9) to problems involving loads on viscoelastic half-spaces. For such problems, it is desirable to re-express these equations in an alternative form [Muskhelishvili (1963)] which facilitates reduction to a Hilbert problem. Let the material occupy the upper half-plane $y > 0$ so that $\phi(z, t)$ is analytic in this region. It is convenient to extend the region of analyticity of $\phi(z, t)$ to the lower half-plane also. Then, as we shall see, it is possible to explore the discontinuities in this function across the real axis, which gives a Hilbert problem. Another approach, possibly more direct, is that of Galin, mentioned previously, which leads to problems of the Riemann-Hilbert type. These however are somewhat more difficult to deal with, from a mathematical point of view.

Let us extend the region of analyticity of $\phi(z, t)$ as follows. We recall that $\bar{\phi}(z, t), \bar{\psi}(z, t)$ are analytic in the lower half-plane. Define

$$\phi(z, t) = -\bar{\phi}(z, t) - z\bar{\phi}'(z, t) - \bar{\psi}(z, t) \tag{3.1.1}$$

for $\text{Im}\{z\} < 0$. For $\text{Im}\{z\} > 0$, (3.1.1) may be written

$$\bar{\psi}(\bar{z}, t) = -\phi(\bar{z}, t) - \bar{\phi}(\bar{z}, t) - \bar{z}\bar{\phi}'(\bar{z}, t) \ , \tag{3.1.2}$$

since $\bar{z}$ is in the lower half-plane. This formula can be substituted for $\bar{\psi}(\bar{z}, t)$ in (2.8.9), to give, for $\text{Im}\{z\} > 0$:

$$\sigma_{11} + \sigma_{22} = 2[\phi(z, t) + \bar{\phi}(\bar{z}, t)] = 4\,\text{Re}\{\phi(z, t)\} \tag{3.1.3a}$$

$$\Sigma(r, t) = \sigma_{22} - i\sigma_{12} = \phi(z, t) - \phi(\bar{z}, t) + (z - \bar{z})\bar{\phi}'(\bar{z}, t) \tag{3.1.3b}$$

$$2 \int_{-\infty}^{t} dt'\mu(t - t')D'(r, t') = \int_{-\infty}^{t} dt'\kappa(t - t')\phi(z, t') + \phi(\bar{z}, t) + (\bar{z} - z)\bar{\phi}'(\bar{z}, t) \ . \tag{3.1.3c}$$

The function $\phi(\bar{z}, t)$ is evaluated in the lower half-plane so that as z approaches the real axis from within the material, $\phi(z, t)$ and $\phi(\bar{z}, t)$ approach $\phi^+(x, t)$ and $\phi^-(x, t)$, respectively, which are the limits of this complex function from above and below. We see from (3.1.3b), that, at points on the real axis where the boundary stresses are zero, $\phi(z, t)$ has no discontinuity. This is the essential reason for the choice of (3.1.1), that it gives this property. For contact problems, it means that the discontinuities in $\phi(z, t)$ are confined to the regions of contact.

From (2.8.11) it follows that, if there are no stresses or rotations at infinity, and if the stresses on the half-plane fall off as $1/x^2$ or faster at large distances from the origin, then $\phi(z, t)$ behaves as $1/z$ at large $|z|$.

We will sometimes refer to $\Sigma(r, t), D'(r, t)$ as the complex stress and the complex displacement derivative.

Most methods of solution of viscoelastic boundary value problems developed to date have relied heavily on analogy with Elasticity. The most striking formal difference between (3.1.3) and the corresponding elastic equations is the integral with kernel $\kappa(t-t')$ on the right of (3.1.3c). Unless this can be eliminated, there is no hope of applying standard methods which have been developed for elastic problems, in particular, the approach of Muskhelishvili (1963). The obvious restriction that will remove this integral is to assume that the material possesses the proportionality property discussed in Sect. 1.9 which gives that

$$\kappa(t) = \kappa\delta(t) \tag{3.1.4}$$

where $\kappa = 3 - 4\nu$, in terms of the unique Poisson's ratio ν of the material [see (1.9.19)]. Then (3.1.3c) becomes

$$2 \int_{-\infty}^{t} dt'\mu(t-t')D'(r,t') = \kappa\phi(z,t) + \phi(\bar{z},t) + (\bar{z}-z)\bar{\phi}'(\bar{z},t) \ . \tag{3.1.5}$$

Some cases where it is not necessary to impose this restriction are discussed later.

In this chapter, we will consider the first and second boundary value problems and then mixed problems, where the contacts are limited to certain, finite, time-dependent regions and the stresses are zero elsewhere on the boundary. Only limiting friction and zero friction problems will be discussed.

As z approaches the real axis, (3.1.3b) and (3.1.5) give relations between $\phi^{\pm}(x,t)$ and the boundary stresses and displacement derivatives $\Sigma(r,t)$, $D'(r,t)$. In the problems dealt with in this chapter, it is generally the case that $D'(r,t)$ is not known completely at any point, however. What is known, in the contact regions, is the imaginary part of $D'(r,t)$, which is the normal displacement derivative $u'_2(x,0,t)$. This means that we must take real and imaginary parts of (3.1.5), which introduces another function, namely $\bar{\phi}(z,t)$. If the problem is not to become too complicated, it is necessary to determine some simple relationship between $\phi(z,t), \bar{\phi}(z,t)$. This will now be done for the case of limiting friction on the boundary [Muskhelishvili (1963)]. We will take the motion to be in the negative x direction. This choice yields the minor advantage of symmetry between the signs of space and time quantities at a later stage. Equations (2.11.1 − 4) then give that

$$\sigma_{12}(r,t) = -f\sigma_{22}(r,t) \tag{3.1.6}$$

for r on the x-axis. We therefore deduce from (3.1.3b) that

$$(1+if)\sigma_{22} = \phi^{+}(x,t) - \phi^{-}(x,t)$$

$$(1-if)\sigma_{22} = \bar{\phi}^{-}(x,t) - \bar{\phi}^{+}(x,t) \ . \tag{3.1.7}$$

It follows that

$$\bar{\phi}^{+} - a\phi^{+} = \bar{\phi}^{-} - a\phi^{-} \ , \qquad a = \frac{1-if}{1+if} \tag{3.1.8}$$

at all points on the x-axis. Since $\phi(z, t)$ is analytic in both the upper and lower half-plane, except on parts of the real axis, it follows that $\bar{\phi}(z, t)$ is also. What (3.1.8) tells us is that the combination $\bar{\phi}(z, t) - a\phi(z, t)$ has no discontinuity across the real axis. Since $\phi(z, t)$, and therefore $\bar{\phi}(z, t)$ vanish as z^{-1} at infinity, it follows from Liouville's Theorem (Sect. A2.1) that

$$\bar{\phi}(z, t) = a\phi(z, t) \tag{3.1.9}$$

for all z. This is the desired relationship. Note that its derivation did not require (3.1.4). The complex conjugate of (3.1.5) therefore reads

$$2 \int_{-\infty}^{t} dt'\mu(t - t')\bar{D}'(r, t') = a\kappa\phi(\bar{z}, t) + a\phi(z, t) + (z - \bar{z})\phi'(z, t) , \tag{3.1.10}$$

giving that, on the boundary,

$$2 \int_{-\infty}^{t} dt'\mu(t - t')u_2'(r, t') = \frac{1}{2i} [(1 - a\kappa)\phi^-(x, t) + (\kappa - a)\phi^+(x, t)] . \tag{3.1.11}$$

This argument applies to the case where the material is in the upper half-plane, since we are taking the surface displacements, given by 3.1.5, to be the limits from above.

If there is no boundary friction, then $a = -1$, giving

$$\bar{\phi}(z, t) = -\phi(z, t) . \tag{3.1.12}$$

Equation (3.1.11) holds with $a = -1$. However, in this case, it is not necessary to assume that $\kappa(t)$ is proportional to a delta function. The imaginary part of (3.1.3c) reads, by virtue of (3.1.12):

$$2 \int_{-\infty}^{t} dt'\mu(t - t')u_2'(x, 0, t')$$

$$= \frac{1}{2i} \int_{-\infty}^{t} dt'[\delta(t - t') + \kappa(t - t')] [\phi^+(x, t') + \phi^-(x, t')] , \tag{3.1.13}$$

which may be rewritten as

$$i \int_{-\infty}^{t} dt'l(t - t')u_2'(x, 0, t') = \phi^+(x, t) + \phi^-(x, t) , \tag{3.1.14}$$

where $l(t)$ is defined by the relation

$$\hat{l}(\omega) = \frac{4\hat{\mu}(\omega)}{1 + \hat{\kappa}(\omega)} = \frac{\hat{\mu}(\omega)}{1 - \hat{v}(\omega)} , \tag{3.1.15}$$

where $\hat{v}(\omega)$ is the Fourier transformed generalized Poisson's ratio, given by (1.9.19). Equation (3.1.14) may be obtained by taking the FT of (3.1.13), removing the complex moduli to the left-hand side and then inverting the FT. Note that we have assumed that $l(t)$ is causal. This follows from the argument of Sect. 1.10 and is anyway necessary on physical grounds.

We have therefore managed to eliminate the embarrassing integral on the right of (3.1.3 c) without resorting to the proportionality assumption. This exceptional case is an example of that referred to in Sect. 2.3, where all the time-dependent viscoelastic quantities can be grouped into a single factor.

3.2 The First and Second Boundary Value Problems

In the case of the first boundary value problem, the stresses are given at every point on the real axis, at all times, and are zero at infinity. We let $p(x, t)$, $s(x, t)$ be the applied pressure and shear on the (upper) half-plane. Then the complex stress is given by

$$\Sigma(x, 0, t) \equiv \Sigma(x, t) = \sigma_{22} - i\sigma_{12} = -p(x, t) + is(x, t) \ . \tag{3.2.1}$$

Equation (3.1.3 b) gives that, on the boundary

$$\phi^+(x, t) - \phi^-(x, t) = \Sigma(x, t) \tag{3.2.2}$$

and $\phi(z, t)$ vanishes at large z. Therefore, from (A2.3.6) we deduce that

$$\phi(z, t) = \frac{1}{2\pi i} \int_{-\infty}^{\infty} dx' \, \frac{\Sigma(x', t)}{x' - z} \ , \tag{3.2.3}$$

which, in fact, constitutes a complete solution of the problem, since every other quantity may now be calculated. In particular, let us determine the boundary displacement derivatives, or $D'(x, 0, t)$ $[\equiv D'(x, t)]$. In this case, even though surface shear exists, it is not necessary to eliminate the integral on the right of the displacement equation (3.1.3 c), since the problem is already solved without reference to this equation. Use of the Plemelj formulae (A2.2.9) gives

$$\int_{-\infty}^{t} dt' \mu(t - t') D'(x, t') = -\Sigma_-(x, t) + \frac{1}{\pi i} \int_{-\infty}^{\infty} dx' \, \frac{\Sigma_+(x', t)}{x' - x} \tag{3.2.4}$$

where

$$\Sigma_+(x, t) = \frac{1}{4} \int_{-\infty}^{t} dt' [\delta(t - t') + \kappa(t - t')] \Sigma(x, t')$$

$$\Sigma_-(x, t) = \frac{1}{4} \int_{-\infty}^{t} dt' [\delta(t - t') - \kappa(t - t')] \Sigma(x, t') \ . \tag{3.2.5}$$

This equation can be solved for $D'(x, t)$ with the aid of (1.2.32), giving

$$D'(x, t) = \int_{-\infty}^{t} dt' \gamma(t - t') \left[-\Sigma_-(x, t') + \frac{1}{\pi i} \int_{-\infty}^{\infty} dx' \, \frac{\Sigma_+(x', t')}{x' - x} \right] \ . \tag{3.2.6}$$

In these equations, and throughout this volume, a principal value integral is understood if the kernel is singular, as it is in these equations unless $\Sigma_+(x, t) = 0$. It is of interest to express the real and imaginary parts of (3.2.4) separately. Let

$$\Sigma_\pm(x,t) = R_\pm(x,t) + iI_\pm(x,t)$$

$$R_\pm(x,t) = -\frac{1}{4}\int_{-\infty}^t dt'[\delta(t-t')\pm\kappa(t-t')]p(x,t') \tag{3.2.7}$$

$$I_\pm(x,t) = \frac{1}{4}\int_{-\infty}^t dt'[\delta(t-t')\pm\kappa(t-t')]s(x,t') \ .$$

Then, (3.2.4) is equivalent to

$$\int_{-\infty}^t dt'\mu(t-t')u_1'(x,t') = -R_-(x,t) + \frac{1}{\pi}\int_{-\infty}^\infty dx'\frac{I_+(x',t)}{x'-x}$$

$$\int_{-\infty}^t dt'\mu(t-t')u_2'(x,t') = -I_-(x,t) - \frac{1}{\pi}\int_{-\infty}^\infty dx'\frac{R_+(x',t)}{x'-x} \ . \tag{3.2.8}$$

If the proportionality assumption is valid, then $R_\pm, I_\pm$ simplify to

$$R_+(x,t) = -(1-v)p(x,t) \ , \qquad R_-(x,t) = (\tfrac{1}{2}-v)p(x,t)$$

$$I_+(x,t) = (1-v)s(x,t) \ , \qquad I_-(x,t) = -(\tfrac{1}{2}-v)s(x,t) \ . \tag{3.2.9}$$

Problem 3.2.1: If $s(x,t) = -fp(x,t)$ where f is a constant and the proportionality assumption holds, show that

$$u_1'(x,t) = -fu_2'(x,t) - (1+f^2)(\tfrac{1}{2}-v)r(x,t)$$

$$u_2'(x,t) - fu_1'(x,t) = \frac{(1-v)(1+f^2)}{\pi}\int_{-\infty}^t dt'\gamma(t-t')\int_{-\infty}^\infty dx'\frac{p(x',t')}{x'-x} \tag{3.2.1p}$$

where

$$r(x,t) = \int_{-\infty}^t dt'\gamma(t-t')p(x,t') \ . \tag{3.2.2p}$$

If there is no surface shear, then equations (3.2.8) decouple, the first relating surface pressure and tangential displacement in a particularly simple way. The second equation is especially important. Remembering (3.1.13 – 15), we write it in the form

$$v(x,t) = \frac{1}{\pi}\int_{-\infty}^\infty dx'\frac{p(x',t)}{x'-x} \tag{3.2.10a}$$

$$v(x,t) = \int_{-\infty}^t dt'l(t-t')u'(x,t') \tag{3.2.10b}$$

where the subscript is dropped from $u_2'(x,t)$ here (and frequently in later sections). Solving for the displacement gives

$$u'(x,t) = \frac{1}{\pi}\int_{-\infty}^\infty dx'\frac{q(x',t)}{x'-x} \tag{3.2.11a}$$

$$q(x, t) = \int_{-\infty}^{t} dt' \, k(t - t') p(x, t') \qquad (3.2.11\text{b})$$

where $k(t)$ is defined by

$$\hat{k}(\omega) = \frac{1 - \hat{v}(\omega)}{\hat{\mu}(\omega)} \, . \qquad (3.2.12)$$

It is causal by virtue of the argument in Sect. 1.10, and is related to $l(t)$ by (2.3.15).

In the second boundary value problem, we have displacements, or rather their tangential derivatives, specified at all times and for every point on the x-axis, so that the left-hand side of the displacement equation (3.1.3c) is known. The integral over $\kappa(t - t')$ is at first sight a hindrance to achieving a reduction to a simple Hilbert problem. However, since the boundary regions are time-independent (in fact $B_u = B$) the Classical Correspondence Principle (Sect. 2.1) applies. Therefore, we can write down the solution of the corresponding elastic problem, and then replace moduli by complex moduli. We put

$$w(x, t) = \int_{-\infty}^{t} dt' \mu(t - t') D'(x, t') \to \mu D'(x, t) \qquad (3.2.13)$$

and $\kappa(t - t') \to \kappa \, \delta(t - t')$, and use (A2.3.25) to write down a solution to (3.1.3c).

$$\phi(z, t) = \frac{1}{\pi \kappa i} \int_{-\infty}^{\infty} dx' \, \frac{w(x', t)}{x' - z} \qquad \text{Im}\{z\} > 0$$

$$\qquad (3.2.14)$$

$$= -\frac{1}{\pi i} \int_{-\infty}^{\infty} dx' \, \frac{w(x', t)}{x' - z} \qquad \text{Im}\{z\} < 0 \, ,$$

which completely solves the elastic problem. The viscoelastic solution is obtained by replacing κ^{-1} by the time integral over the inverse FT of $\hat{\kappa}^{-1}(\omega)$. But let us focus on boundary quantities. In the elastic case (3.2.14), (A2.2.9) and (3.1.3b) give that

$$\Sigma(x, t) = w_{-}(x, t) + \frac{1}{\pi i} \int_{-\infty}^{\infty} dx' \, \frac{w_{+}(x', t)}{x' - x} \, , \qquad \text{where} \qquad (3.2.15)$$

$$w_{\pm}(x, t) = \frac{1 \pm \kappa}{\kappa} w(x, t) \, . \qquad (3.2.16)$$

Going over to the viscoelastic problem, we maintain (3.2.15) but where $w(x, t)$ is given by the middle expression in (3.2.13) and

$$w_{\pm}(x, t) = \int_{-\infty}^{t} dt' \eta_{\pm}(t - t') w(x, t') \, , \qquad \hat{\eta}_{\pm}(\omega) = \frac{1 \pm \hat{\kappa}(\omega)}{\hat{\kappa}(\omega)} \, . \qquad (3.2.17)$$

Let us, as before, separate (3.2.15) into real and imaginary parts. Let

$$w_1(x, t) = \int\limits_{-\infty}^{t} dt' \mu(t - t') u_1'(x, 0, t') \; ,$$

$$w_2(x, t) = \int\limits_{-\infty}^{t} dt' \mu(t - t') u_2'(x, 0, t') \tag{3.2.18}$$

and

$$W_{\pm}^{(i)}(x, t) = \int\limits_{-\infty}^{t} dt' \eta_{\pm}(t - t') w_i(x, t') \; , \quad i = 1, 2 \; . \tag{3.2.19}$$

Then (3.2.15) may be decomposed into

$$\sigma_{22}(x, 0, t) = -p(x, t) = W_{-}^{(1)}(x, t) + \frac{1}{\pi} \int\limits_{-\infty}^{\infty} dx' \, \frac{W_{+}^{(2)}(x', t)}{x' - x}$$

$$\sigma_{12}(x, 0, t) = -s(x, t) = -W_{-}^{(2)}(x, t) + \frac{1}{\pi} \int\limits_{-\infty}^{\infty} dx' \, \frac{W_{+}^{(1)}(x', t)}{x' - x} \; . \tag{3.2.20}$$

Problem 3.2.2: Using the infinite Hilbert transform (Sect. A2.4), deduce (3.2.20) from (3.2.8).

The first boundary value problem could also have been solved by invoking the Classical Correspondence Principle.

If surface shear is zero, we take (3.1.14) as our starting point, which may be solved to obtain [recall the method of deriving (3.2.14)]:

$$\phi(z, t) = \frac{\varepsilon(z)}{2\pi} \int\limits_{-\infty}^{\infty} dx' \, \frac{v(x', t)}{x' - z} \tag{3.2.21}$$

where $v(x, t)$ is given by (3.2.10b) and

$$\varepsilon(z) = \operatorname{sgn}(\operatorname{Im} z) \; . \tag{3.2.22}$$

It follows that the boundary pressure is given by

$$p(x, t) = -\frac{1}{\pi} \int\limits_{-\infty}^{\infty} dx' \, \frac{v(x', t)}{x' - x} \; . \tag{3.2.23}$$

which could have been deduced by inverting the infinite Hilbert transform in (3.2.10a), or alternatively by manipulating (3.2.20).

The relationships between stresses and displacements on the boundary will be useful in later sections, most particularly relation (3.2.10), which is applicable in the frictionless case.

For the next few sections we develop a general theory of mixed boundary value problems and apply it to the case of a rigid indentor in horizontal motion on the half-space. We finish the chapter with a discussion of the problem of a stationary indentor under a varying load.

3.3 The General Mixed Boundary Value Problem

We will consider the problem of a series of rigid indentors pressed into a visco-elastic half-space ($y > 0$) and moving across it. If plane strain conditions are to hold, the indentors must be infinitely long in one direction, taken to be the z direction, and of uniform cross-section. Also the loading distribution must be uniform along each punch. We consider a typical cross-section of this configuration. All subsequent discussion refers to this cross-section of which the material occupies the half-plane $y > 0$.

Let us denote the contact region by $C(t)$, and its complement on the boundary by $C'(t)$. Initially the character of $C(t)$ will not be restricted. If may consist of a series of separate intervals. The general method developed applies in principle to such general contact regions. However, the only detailed solution presented applies to the case where $C(t)$ is a single interval, corresponding to a single load.

The boundary stresses are zero on $C'(t)$. Inside $C(t)$, the vertical displacement conforms to the shapes of the rigid indentors, which are presumed to be known. Therefore, the normal displacement is known in $C(t)$. This is not entirely true, because of the fact noted in Sect. 2.8, that for plane contact problems of this kind, the displacements are divergent at infinity. As noted previously, we eliminate these divergences by using the derivative of the displacement with respect to x, rather than the displacement itself. Therefore, we correct the above remarks, stating that the displacement derivative is known in $C(t)$.

The regions $C(t), C'(t)$ correspond respectively to $B_u(t), B_\sigma(t)$, the boundary regions on which displacements and stresses are specified. This is true in spite of that fact that only the normal displacement derivative, and not the tangential displacement derivative, is given on $C(t)$. We shall see that normal displacement and stress are the fundamental quantities in the problems considered here, from where everything else can be derived. Strictly, contact problems without friction are boundary value problems where the boundary regions depend upon the component, as allowed by (1.8.15). The surface shear is zero over the entire boundary. Limiting frictional contact problems, which we shall also consider, do not fit into the scheme implied by (1.8.15) since the shear component is not specified in the region of contact. However, it is given in terms of the normal pressure and we shall see below that this is adequate for progress to be made.

Let the contact be frictional and let the indentor be moving in the negative x direction. We denote the x derivative of normal displacement on the boundary by $u'(x, t)$ and the complex stress by $\Sigma(x, t)$. From (3.1.6), we have that

$$\Sigma(x, t) = -p(x, t) + is(x, t) = -(1 + if)p(x, t) , \tag{3.3.1}$$

where f is the coefficient of sliding friction. If the contact is lubricated, then f is put to zero. The pressure $p(x, t)$ is a priori unknown in $C(t)$.

From (3.1.3, 11) we deduce the boundary equations

$$\phi^+(x, t) - \phi^-(x, t) = 0 , \qquad x \in C'(t) \tag{3.3.2a}$$

$$\phi^+(x, t) - \eta\phi^-(x, t) = iv(x, t) , \qquad x \in C(t) \quad \text{where} \tag{3.3.2b}$$

$$v(x,t) = \int_{-\infty}^{t} dt' l(t-t') u'(x,t') , \qquad l(t) = \frac{4\mu(t)}{\kappa - a} \tag{3.3.3}$$

and

$$\eta = -\frac{1-a\kappa}{\kappa - a} = \frac{f+ih}{f-ih} , \qquad h = \frac{\kappa+1}{\kappa-1} = \frac{2(1-v)}{1-2v} , \tag{3.3.4}$$

if friction is present. In the frictionless case, the proportionality assumption is not necessary and $l(t)$ is determined by the more general relation (3.1.15). The quantity a is given by (3.1.8). In (3.3.4), v is Poisson's ratio for the material. Note that η has unit modulus. It will be convenient to put

$$\eta = e^{2\pi i \theta} \qquad \text{or} \tag{3.3.5}$$

$$\theta = \frac{\log \eta}{2\pi i} = \frac{1}{\pi} \tan^{-1} \left(\frac{h}{f} \right) . \tag{3.3.6}$$

The quantity θ varies in the interval $[0, \frac{1}{2}]$.

Relation (3.3.2) constitutes a Hilbert problem if $v(x,t)$ is known. This however is not generally the case for problems with varying boundary regions, which is the source of the added difficulty of non-inertial viscoelastic problems over elastic problems. Nevertheless, in this section, we will proceed as if $v(x,t)$ were known.

Before we solve (3.3.2), it is of interest, for later purposes, in Sect. 9, to formulate the same problem but with the hereditary integral moved onto the stresses, as in (2.8.14). We vary the procedure outlined in that section slightly by letting

$$\chi(z,t) = \int_{-\infty}^{t} dt' k(t-t') \phi(z,t') , \tag{3.3.7}$$

where $k(t)$ is defined by (3.2.12). In the case of constant Poisson's ratio, it is proportional to $\gamma(t)$. Instead of (3.3.2), we have

$$\chi^+(x,t) - \chi^-(x,t) = -(1+if) q(x,t) , \qquad x \in C'(t) \tag{3.3.8}$$

$$\chi^+(x,t) - \eta \chi^-(x,t) = i u'(x,t) , \qquad x \in C(t)$$

where $q(x,t)$ is given by (3.2.11 b). There is no guarantee in general that $q(x,t)$ is zero in $C'(t)$. In Sect. 9, we discuss conditions under which it is true.

We now write down a general solution to (3.3.2), based on the solution of the Hilbert problem, given in Sect. A2.3. For the moment, we allow the possibility of singularities at the end points of the contact region.

Let $C(t)$ consist of n intervals $[a_i(t), b_i(t)]$, $i = 1, 2, \ldots, n$. There are $2n$ end points. Let m of these, denoted by $c_1(t), c_2(t), \ldots, c_m(t)$, have no singularities. Then, from (A2.3.22):

$$\phi(z,t) = \frac{X(z,t)}{2\pi} \int_{C(t)} dx' \frac{v(x',t)}{(x'-z) X^+(x',t)} + P(z,t) X(z,t) \tag{3.3.9}$$

where, using θ, defined by (3.3.6),

$$X(z, t) = X_0(z, t) \prod_{i=1}^{m} [z - c_i(t)] \;,$$

$$X_0(z, t) = \prod_{i=1}^{n} [z - a_i(t)]^{-\theta} [z - b_i(t)]^{\theta - 1} \;.$$

(3.3.10)

The branch of $X_0(z, t)$ conventionally chosen is that such that $z^n X_0(z, t)$ approaches unity at large $|z|$. The quantity $P(z, t)$ in (3.3.9) is an arbitrary polynomial, constrained only by the requirement that $\phi(z, t)$ vanish at infinity according to (2.8.11). This solution $\phi(z, t)$ can have singularities at end points, but restricted in their nature by (A2.3.1). The same is therefore true of the stresses. At infinity, $X(z, t)$ behaves as z^{m-n}. If $m \geq n$, the polynomial $P(z, t)$ must be zero. Furthermore, if $m > n$, a solution is possible only if special conditions of the form

$$\int_{C(t)} dx \, \frac{x^r v(x, t)}{X^+(x, t)} = 0, \quad r = 0, 1, 2, \ldots, m - n - 1$$

(3.3.11)

are satisfied. If $n > m$, the polynomial $P(z, t)$ must be determined by extra physical information.

The only case which will be discussed in detail later is that of the smooth indentor, for which $m = 2n$. Let us, without further ado, specialize to this case. The term $P(z, t) X(z, t)$ in (3.3.9) may be dropped. Also:

$$X(z, t) = \prod_{i=1}^{n} [z - a_i(t)]^{1 - \theta} [z - b_i(t)]^{\theta}$$

(3.3.12)

and, from (3.3.11):

$$\int_{C(t)} dx \, \frac{x^r v(x, t)}{X^+(x, t)} = 0 \;, \quad r = 0, 1, 2, \ldots, n - 1 \;.$$

(3.3.13)

It will be useful to have an expression for $v(x, t)$ when $x \in C'(t)$, in other words, outside the contact region. From (3.1.11), it is clear that (3.3.2b) holds at such points. There is no discontinuity so that (3.3.2b) and (3.3.5) give

$$v(x, t) = -2 \sin(\pi\theta) e^{i\pi\theta} \phi(x, t) \;,$$

(3.3.14)

or, more explicitly, letting $X(x, t)$ denote the unique limits of $X(z, t)$,

$$v(x, t) = \frac{-\sin(\pi\theta) e^{i\pi\theta} X(x, t)}{\pi} \int_{C(t)} dx' \, \frac{v(x', t)}{(x' - x) X^+(x', t)} \;.$$

(3.3.15)

Also, the stresses in the contact region $C(t)$ may be written down. With the aid of (3.3.1 − 6) and the Plemelj formulae (A2.2.9), one deduces that

$$p(x, t) = q \left[\frac{X^+(x, t) \sin(\pi\theta)}{\pi} \int_{C(t)} dx' \, \frac{v(x', t)}{(x' - x) X^+(x', t)} + v(x, t) \cos(\pi\theta) \right]$$

$$q = \frac{-i e^{-i\pi\theta}}{1 + if} \;.$$

(3.3.16)

The quantity q is complex. This can be traced back to the fact that $v(x, t)$ is complex. In fact, from (3.3.3)

$$v(x, t) = c w(x, t) \tag{3.3.17}$$

where $w(x, t)$ is a real quantity, given by

$$w(x, t) = \int_{-\infty}^{t} dt' \mu(t - t') u'(x, t) \quad \text{and} \tag{3.3.18}$$

$$c = \frac{4}{\kappa - a} = -\frac{4 \sin(\pi\theta)}{q(\kappa + 1)} \tag{3.3.19}$$

by virtue of (3.3.4–6). Therefore, the complex factor q cancels out.

Observe that (3.3.15, 16a) are essentially the Green's function relations (2.3.9), with the minor difference that displacement derivative, rather than displacement, is now in use.

The contact intervals $[a_i(t), b_i(t)]$ are usually unknown before the problem is solved, while the overall motion of the punches (say the horizontal position of the point of deepest penetration) is generally specified. In order to determine $a_i(t)$, $b_i(t)$, $i = 1, 2, \ldots, n$, we require $2n$ conditions. Equation (3.3.13) supplies n conditions. The remaining relations can be supplied in a number of ways, the simplest being to specify the loads W_k per unit length on each punch. This gives n conditions of the form

$$W_k = \int_{a_k(t)}^{b_k(t)} dx\, p(x, t) = \int_{C(t)} dx\, J_k(x, t) v(x, t) + q \cos(\pi\theta) \int_{a_k(t)}^{b_k(t)} dx\, v(x, t) \tag{3.3.20}$$

from (3.3.16), where

$$J_k(x, t) = q\left(\frac{\sin(\pi\theta)}{\pi X^+(x, t)} \int_{a_k(t)}^{b_k(t)} dx' \frac{X^+(x', t)}{x - x'} \right). \tag{3.3.21}$$

Another alternative, namely to specify the relative penetrations of the punches, is discussed by Muskhelishivili (1963), Chap. 19.

It is clear from these relationships that once the quantity $v(x, t)$, $x \in C(t)$, and the extent of $C(t)$ itself are known, all other quantities of interest can be determined easily.

3.3.1 Single Contact Interval

We will now discuss the case where there is only one indentor and contact region, which will be denoted by $[a(t), b(t)]$, the subscripts being now omitted. The quantity n is equal to unity, and $X(z, t)$ has the form

$$X(z, t) = [z - a(t)]^{1 - \theta} [z - b(t)]^{\theta} \tag{3.3.22}$$

or more specifically, the branch of this function that tends to z at large $|z|$. One deduces that, on the real axis, outside the contact interval:

$$X(x,t) = \begin{cases} n(x,t) & x > b(t) \\ -n(x,t) & x < a(t) \end{cases}$$

$$(3.3.23)$$

$$n(x,t) = |x - a(t)|^{1-\theta} |x - b(t)|^{\theta} .$$

Also, on the contact interval:

$$X^+(x,t) = m(x,t)e^{i\pi\theta}$$

$$(3.3.24)$$

$$m(x,t) = [b(t) - x]^{\theta} [x - a(t)]^{1-\theta} .$$

Equation (3.3.15) gives

$$v(x,t) = \mp \sin(\pi\theta) \frac{n(x,t)}{\pi} \int\limits_{a(t)}^{b(t)} dx' \frac{v(x',t)}{(x'-x)m(x',t)} , \qquad x \in [a(t), b(t)] ,$$

$$(3.3.25)$$

the upper sign referring to $x > b(t)$, the lower sign to $x < a(t)$. Similarly, (3.3.16a) gives

$$p(x,t) = q \left[\frac{\sin(\pi\theta)}{\pi} m(x,t) \int\limits_{a(t)}^{b(t)} dx' \frac{v(x',t)}{(x'-x)m(x',t)} + v(x,t) \cos(\pi\theta) \right]$$

$$(3.3.26)$$

while (3.3.13) reduces to one relationship:

$$\int\limits_{a(t)}^{b(t)} dx \frac{v(x,t)}{m(x,t)} = 0$$

$$(3.3.27)$$

as does (3.3.20), giving with the aid of (3.3.21):

$$W = \int\limits_{a(t)}^{b(t)} dx J(x,t) v(x,t) + q \cos(\pi\theta) \int\limits_{a(t)}^{b(t)} dx v(x,t)$$

$$(3.3.28)$$

where

$$J(x,t) = q \left(\frac{\sin(\pi\theta)}{\pi m(x,t)} \int\limits_{a(t)}^{b(t)} dx' \frac{m(x',t)}{x-x'} \right) = \frac{q}{m(x,t)} [x - a(1-\theta) - b\theta] \quad (3.3.29)$$

by virtue of (A1.2.5). With the help of (3.3.27), condition (3.3.28) becomes

$$W = q \int\limits_{a(t)}^{b(t)} dx \frac{v(x,t)x}{m(x,t)} = -\frac{4\sin(\pi\theta)}{\kappa+1} \int\limits_{a(t)}^{b(t)} dx \frac{w(x,t)x}{m(x,t)} ,$$

$$(3.3.30)$$

the latter equation being written in terms of the real quantity $w(x,t)$, given by (3.3.17).

3.3.2 Frictionless Contact

In the frictionless case, $\theta = \frac{1}{2}$, giving

$$X(z,t) = [z - a(t)]^{1/2} [z - b(t)]^{1/2} .$$

$$(3.3.31)$$

Equation (3.3.25) reduces to

$$v(x,t) = \mp \frac{n(x,t)}{\pi} \int_{a(t)}^{b(t)} dx' \frac{v(x',t)}{(x'-x)m(x',t)}$$

(3.3.32)

$$n(x,t) = |x-a(t)|^{1/2}|x-b(t)|^{1/2} , \qquad m(x,t) = [b(t)-x]^{1/2}[x-a(t)]^{1/2} .$$

The pressure $p(x,t)$, given by (3.3.26), reduces to

$$p(x,t) = -\frac{m(x,t)}{\pi} \int_{a(t)}^{b(t)} dx' \frac{v(x',t)}{(x'-x)m(x',t)} ,$$

(3.3.33)

since, q, given by (3.3.16a), tends to -1 in this limit. Relation (3.3.33) is recognizable as the inverse finite Hilbert transform (for smooth contact; see Sect. A2.4) of (3.2.10a). Finally, (3.3.30) reduces to

$$W = -\int_{a(t)}^{b(t)} dx \frac{v(x,t)x}{m(x,t)} .$$

(3.3.34)

Recall that in the frictionless case, there is no need to assume proportionality. The function $v(x,t)$ is given by (3.2.10b) where $l(t)$ is defined by (3.1.15).

3.4 The General Integral Equation

Let us decompose $v(x,t), x \in C(t)$, as described in Sect. 3.2.4 with the aim of deriving an integral equation for this quantity that is a particular case of the equation given by (2.5.1). We have

$$v(x,t) = \int_{-\infty}^{t} dt' l(t-t')u'(x,t') = \int_{W_u(x,t)} dt' \Pi_u(t,t';x)u'(x,t')$$

$$+ \int_{W_\sigma(x,t)} dt' \Pi_\sigma(t,t';x)v(x,t') , \qquad x \in C(t)$$

(3.4.1)

where Π_u, Π_σ are defined by (2.4.18), with r replaced by x, while $W_u(x,t)$, $W_\sigma(x,t)$ are defined following (2.4.16), where $B_u(t), B_\sigma(t)$ are replaced by $C(t), C'(t)$.

The quantity $v(x,t')$, occurring in the second integral on the right of (3.4.1), is always evaluated outside of $C(t')$, by definition of $W_\sigma(x,t)$. We substitute (3.3.15) for it, to obtain the integral equation

$$v(x,t) = \int_{W_\sigma(x,t)} dt' \int_{C(t')} dx' K(x,x';t,t')v(x',t') + I(x,t) , \qquad x \in C(t) \quad (3.4.2)$$

where

$$K(x,x'; t,t') = \frac{-\sin(\pi\theta)e^{i\pi\theta}\Pi_\sigma(t,t'; x)X(x,t')}{\pi X^+(x',t')(x'-x)}$$

(3.4.3)

$$I(x,t) = \int_{W_u(x,t)} dt' \Pi_u(t,t';x)u'(x,t') .$$

We recall from Sect. 3 that once $v(x, t)$ is known over the contact region $C(t)$, the extent of which has also to be determined, everything else can be calculated by elementary means.

It was pointed out, without detailed discussion, in Sect. 2.5 that the kernel of (3.4.2) possesses no non-integrable singularities. We can see this in a somewhat more detailed manner in the present special case. From the definition of Π_σ, one perceives that the only situations where singularities might arise are at times t' equal to one of the transition points $t_i(x)$. At such times x', at one of the endpoints of $C(t')$, could become equal to x. However, the singularity will be integrable because, at such times, the function $X(x, t')$ is zero. There is another source of difficulty, though. If the factor $X^+(x', t')$ in the denominator of $K(x, x'; t, t')$ is not cancelled, it causes a singularity in the kernel, which may not be square-integrable, thus casting doubt upon the applicability of Fredholm theory as discussed in Sect. 4.1. However, the singularity can be transformed away by a change of variable. An explicit example is discussed in Sect. 3.6.

The general discussion in Sect. 2.5 on methods of solution of (2.5.1) applies here without alteration.

3.5 Moving Load Problems

We discuss the integral equation (3.4.2) in more detail for the problem of indentors in contact with a half-plane under the action of certain loads, and moving across it. In principle, the method outlined in the last section could handle any specified individual motion of the indentors. However, only the simplest will be considered, namely where the indentors are all moving in the same direction, taken to be along the negative x direction.

Consider first the case of a single moving load. Let the lowest point of the indentor have position $x_0(t)$ at time t. This specifies its overall motion. The displacement derivative in the contact region $C(t)$ will then have the form

$$u'(x, t) = f(x - x_0(t)) , \tag{3.5.1}$$

where $f(x)$ is the slope of the punch profile with the origin at the lowest point. The contact region $C(t)$ is in this case an interval $[a(t), b(t)]$. For a typical point $x \in C(t)$, the sets $W_u(x, t)$, $W_\sigma(x, t)$ are given by

$$W_u(x, t) = [t_1(x), t] , \qquad W_\sigma(x, t) = (-\infty, t_1(x)) \tag{3.5.2}$$

where $t_1(x)$ is defined as the (unique) solution of the equation

$$a(t_1(x)) = x, \qquad x \in C(t) . \tag{3.5.3}$$

In other words, it is the time at which the point x entered the contact interval. From (2.4.18) we have that the quantities Π_u, Π_σ, are given by

$$\Pi_u(t, t'; x) = l(t - t') R(t'; t_1(x), t)$$

$$\Pi_\sigma(t, t', x) = T_1(t, t'; x) R(t'; t_2(x), t_1(x)) , \tag{3.5.4}$$

where $T_1(t, t'; x)$ is defined by (2.4.19) with x replacing r. Then from (3.4.2) and (3.4.3) it follows that $v(x, t)$ obeys the equation

$$v(x, t) = \int\limits_{-\infty}^{t_1(x)} dt' \int\limits_{a(t')}^{b(t')} dx' K(x, x'; t, t') v(x', t') + I(x, t) \qquad (3.5.5)$$

where

$$K(x, x'; t, t') = \frac{\sin(\pi\theta) n(x, t') T_1(t, t'; x)}{\pi m(x', t')(x' - x)} , \qquad (3.5.6)$$

by virtue of (3.3.23) and (3.3.24). This form emerges immediately if one uses (3.3.25) in deriving (3.4.2), instead of (3.3.15). Also:

$$I(x, t) = \int\limits_{t_1(x)}^{t} dt' l(t - t') u'(x, t') , \qquad (3.5.7)$$

which is a known quantity, provided the contact interval, and consequently $t_1(x)$, is known. Two subsidiary conditions are required to determine $a(t)$, $b(t)$. These are (3.3.27, 30).

Equation (3.5.5) applies to general, transient conditions. It is not particularly amenable to analytic treatment. However, approximate solutions have been obtained, in the frictionless limit, applicable to times when transient effects have virtually decayed [Golden and Graham (1984)]. The steady-state form of the equation is considerably simpler. It will be discussed later. In this context the term "steady state" implies conditions of uniform motion after transient effects have died away.

Let us consider briefly the case where there is more than one indentor. In each contact interval, we have an equation of the same form as (3.5.1), namely

$$u'(x, t) = f_i(x - x_{0i}(t)) , \qquad i = 1, 2, \ldots, n , \qquad x \in C_i(t) , \qquad (3.5.8)$$

where there are n indentors acting respectively on intervals $C_i(t) = [a_i(t), b_i(t)]$. The number of quantities $t_i(x)$ and their values depend on which contact interval x belongs to. If it lies in the leftmost interval $[a_1(t), b_1(t)]$, then (3.5.2) holds. If it lies in the next one, then there are three $t_i(x)$ and

$$W_u(x, t) = [t_3(x), t_2(x)] \cup [t_1(x), t]$$

$$(3.5.9)$$

$$W_\sigma(x, t) = (-\infty, t_3(x)) \cup (t_2(x), t_1(x)) ,$$

and so on. Equation (3.4.2) now in fact becomes a set of coupled integral equations for the quantities $v^{(i)}(x, t)$, which are the function $v(x, t)$ evaluated in $C_i(t)$.

3.5.1 Steady-State Limit

We consider the single load case in the steady-state limit. Let the indentor have been moving in a negative x direction, at speed V, for a long time. If transient effects have died away then

$$a(t) = a_0 - Vt , \qquad b(t) = b_0 - Vt , \qquad (3.5.10)$$

so that, by virtue of (3.5.3),

$$t_1(x) = \frac{a_0 - x}{V} . \tag{3.5.11}$$

All quantities such as $u'(x, t)$, $v(x, t)$, $p(x, t)$ and so on, will be functions of $x + Vt$ rather than x, t separately. Also, the functions $n(x, t)$, $m(x, t)$, given by (3.3.23, 24), reduce to

$$m(x, t) = m(x + Vt) , \qquad n(x, t) = n(x + Vt) \tag{3.5.12}$$

where

$$m(u) = (u - a_0)^{1-\theta}(b_0 - u)^{\theta} , \qquad n(u) = |a_0 - u|^{1-\theta} |b_0 - u|^{\theta} . \tag{3.5.13}$$

In (3.5.5), we change variables according to

$$x' \to y' = x' + Vt' , \qquad x \to y = x + Vt , \qquad t' \to z = x + Vt' , \tag{3.5.14}$$

giving

$$v(y) = \int_{a_0}^{b_0} dy' K(y, y') v(y') + I(y) \qquad \text{where} \tag{3.5.15}$$

$$K(y, y') = \frac{\sin(\pi\theta)}{\pi} \int_{-\infty}^{a_0} dz \, \frac{T(y, z) n(z)}{V \, (y' - z) m(y')}$$

$$T(y, z) = \int_{z}^{a_0} \frac{du}{V} \, l\left(\frac{y - u}{V}\right) k\left(\frac{u - z}{V}\right) \qquad \text{and} \tag{3.5.16}$$

$$I(y) = \int_{a_0}^{y} \frac{dz}{V} \, l\left(\frac{y - z}{V}\right) u'(z) \tag{3.5.17}$$

where $u'(z)$ is $u'(x, t')$ expressed as a function of $x + Vt'$ only. In the new coordinate system, the contact interval is stationary. Clearly, this is a coordinate system moving with the indentor.

It is convenient to choose length units and origin so that $[a_0, b_0] = [-1, 1]$ and time units such that $V = 1$. This means that any quantity of dimension length, which has value c in our initial units, becomes

$$c_1 = \frac{2c - (a_0 + b_0)}{b_0 - a_0} , \tag{3.5.18}$$

and any time in the theory of value τ in the initial units becomes

$$\tau_1 = \frac{\tau}{\lambda} , \qquad \lambda = \frac{b_0 - a_0}{2V} \tag{3.5.19}$$

in these units. We agree to use the same symbol for all quantities in terms of the new units. We also, for convenience, reintroduce the variables x, x', and y, now denoting space and time coordinates in the new moving, scaled reference system. In the light of (3.5.19), the functions $l(t)$, $k(t)$ in the initial units are replaced by

$$\lambda l(\lambda x) , \qquad \lambda k(\lambda x) \tag{3.5.20}$$

in the new units. These will be denoted by $l(x)$, $k(x)$, respectively. The reason for the factor λ multiplying the overall functions is to ensure that the relation (2.3.15) is preserved. One may check that

$$\int_0^x dx' k(x-x')l(x') = \int_0^x dx' l(x-x')k(x') = \delta(x) , \qquad (3.5.21)$$

using the second relation of (A3.1.1p). Note that, in terms of the physical relaxation and creep functions, under the proportionality hypothesis of Sect. 1.9, one has, by virtue of (1.2.33), (3.1.15) and (3.2.12),

$$l(x) = \frac{1}{1-v}\frac{d}{dx}[H(\lambda x)G(\lambda x)] , \qquad k(x) = (1-v)\frac{d}{dx}[H(\lambda x)J(\lambda x)] . \quad (3.5.22)$$

Equation (3.5.15) becomes

$$v(x) = \int_{-1}^1 dx' K(x,x')v(x') + I(x) , \qquad (3.5.23)$$

where

$$K(x,x') = \frac{\sin(\pi\theta)}{\pi} \int_{-\infty}^{-1} dy \frac{T(x,y)n(y)}{(x'-y)m(x')} \qquad (3.5.24a)$$

$$T(x,y) = \int_y^{-1} dy' l(x-y')k(y'-y) , \qquad |x|<1 , \qquad y<-1 \qquad (3.5.24b)$$

and

$$m(x') = (x'+1)^{1-\theta}(1-x')^\theta , \qquad |x'|<1$$
$$n(y) = (-1-y)^{1-\theta}(1-y)^\theta , \qquad y<-1 . \qquad (3.5.25)$$

Also:

$$I(x) = \int_{-1}^x dz\, l(x-z)u'(z) . \qquad (3.5.26)$$

The subsidiary conditions (3.3.27, 30) become, in the steady-state limit and for this coordinate system,

$$\int_{-1}^1 dx \frac{v(x)}{m(x)} = 0 , \qquad (3.5.27a)$$

$$q \int_{-1}^1 dx \frac{xv(x)}{m(x)} = W_1 , \qquad (3.5.27b)$$

$$W_1 = \frac{2W}{b_0 - a_0} . \qquad (3.5.27c)$$

The pressure, given by (3.3.26) becomes

$$p(x) = q\left[\frac{\sin(\pi\theta)m(x)}{\pi} \int_{-1}^1 dx' \frac{v(x')}{(x'-x)m(x')} + v(x)\cos(\pi\theta)\right] . \qquad (3.5.28)$$

In the frictionless limit, $\theta \to \frac{1}{2}$, so that

$$m(x) = (1 - x^2)^{1/2} , \quad |x| < 1$$
$$n(y) = (y^2 - 1)^{1/2} , \quad y < -1 . \tag{3.5.29}$$

The kernel $K(x, x')$ reduces to

$$K(x, x') = \frac{1}{\pi} \int_{-\infty}^{-1} dy \frac{T(x, y) n(y)}{(x' - y) m(x')} , \tag{3.5.30}$$

while (3.5.27 b) becomes

$$\int_{-1}^{1} dx \frac{v(x) x}{m(x)} = - W_1 . \tag{3.5.31}$$

The pressure function, given by (3.5.28), reduces to

$$p(x) = - \frac{m(x)}{\pi} \int_{-1}^{1} dx' \frac{v(x')}{(x' - x) m(x')} . \tag{3.5.32}$$

It is interesting to observe that if the material is incompressible ($\kappa = 1$) or nearly so, then h, given by (3.3.4), is very large so that $\theta \to \frac{1}{2}$ and the solution has the same form as in the frictionless case. This is a significant simplification. Such an assumption is valid for a wide range of amorphous polymers, at temperatures well above their glass transition temperatures [Walton et al. (1978)].

The expression for the hysteretic friction coefficient also simplifies greatly in this limit, as we shall see in Sect. 3.8.

3.6 Solution for a Single Load

The problem of a single indentor moving across a viscoelastic half-space is of considerable interest, largely because it offers a theoretical framework in which to investigate the phenomenon of hysteretic friction, which is discussed in more detail in Sect. 8. If surface friction is neglected, the problem can be regarded as an indentor sliding across a lubricated half-plane, or a cylinder rolling over the plane. In the latter case, hysteretic friction is manifested as resistance to rolling, often termed rolling friction. This problem was first solved rigorously for a standard linear solid by Hunter (1961). He did so by transforming the equations into elastic form but where displacement and pressure are replaced by these quantities acted upon by differential operators. This amounts to using the differential form of the constitutive relations. Morland (1962) considered the problem by formulating it as a set of dual integral equations. This is a generalization of the method of Sneddon (1951), referred to in Sect. 2.8. It imposes no restriction on the viscoelastic behaviour of the half-plane but there is no question of an analytic solution. Subsequently, Morland (1967, 1968), now using a method closer to that of Hunter – the relationship between the methods is discussed by Golden (1977) – solved the problem of two viscoelastic cylinders rolling on each other. This solution was for the case when viscoelastic behaviour is described by a discrete spectrum model (Sect. 1.6). A limiting case of this problem is the cylinder on a half-plane. In fact, the case of the two finite cylinders is not intrinsically more difficult than this limit, except in its geometric aspects. This is what emerges from Morland's analysis. It has the advantage though, that the infinity, which necessarily arises for the half-plane (Sect. 2.8) does not arise. Harvey (1975) considered the problem of a viscoelastic cylinder rolling, without slipping on a rigid half-plane, using a method related to that of Morland (1967).

Golden (1977) developed a method, essentially that described here, which is not restricted to particular types of materials. Later, Golden (1979a, 1986a), applied the method to the case where limiting friction is present.

For a power law material (Sect. 1.6), an interesting method of solution has been developed by Walton et al. (1978). This has no apparent relationship with the method developed here. They derive, for steady-state conditions, an integral equation, which is a generalized Abel-type equation, essentially generalizing (3.2.11), and obtain explicit solutions. Walton (1984) generalized the method to apply to a material for which the shear relaxation function varies with depth also according to a power law.

This problem has also been considered in detail for an indentor moving over a layer of finite thickness, rather than a half-plane. We mention Alblas and Kuipers (1970), Margetson (1971, 1972), and Nachman and Walton (1978). Kalker (1975, 1977) reviews this topic in a systematic manner.

The discussion will be confined to the steady-state frictionless problem. In fact, however, the method works equally well for problems involving limiting friction [Golden (1979, 1986)]. Also, the transient problem is considered by Golden and Graham (1984).

The punch is assumed to be smooth, with $u'(x)$ given by a polynomial

$$u'(x) = d(x) = \sum_{r=0}^{m} d_r x^r \tag{3.6.1}$$

in the contact region. We are using the dimensionless moving coordinates introduced in Sect. 3.5. The simplest case is that with a parabolic profile, given in dimensional moving coordinates by

$$u'(x) = -\frac{x}{R}, \quad a < x < b, \tag{3.6.2}$$

if the point of deepest indentation is at the origin. The contact interval is denoted by $[a, b]$, which is the interval $[a_0, b_0]$ of the previous section. The subscript is omitted for convenience. The quantity R is the radius of curvature at the origin. Returning to dimensionless coordinates, we have

$$u'(x) = d(x) = d_0 + d_1 x, \quad d_0 = -\frac{b+a}{2R}, \quad d_1 = -\frac{b-a}{2R}, \tag{3.6.3}$$

by virtue of (3.5.18). The quantity d_0 is proportional to the shift away from the origin of the indentor tip in this coordinate system. It is a measure of the asymmetry of the contact patch about the tip of the indentor. This form also applies approximately to the case where the punch is cylindrical and the contact interval $(b-a)$ is small compared with the radius R of the cylinder. As we noted previously, the cylinder can be regarded as sliding or rolling.

The punch profile given by (3.6.3) will be of most interest. However, for the moment, we keep the general form (3.6.1). Let

$$v(x) = \Delta(x) + q(x) \quad \text{where} \tag{3.6.4}$$

$$q(x) = \int_{-\infty}^{x} dx\, l(x-x')d(x') = \int_{0}^{\infty} dy'\, l(y')d(x-y') = \sum_{r=0}^{m} q_r x^r \tag{3.6.5}$$

and

$$\Delta(x) = \int_{-\infty}^{-1} dx'\, l(x-x')\{u'(x') - d(x')\} . \tag{3.6.6}$$

This is a decomposition into a polynominal part $q(x)$ and $\Delta(x)$, which in general has no polynomial part; in an exponential decay model, it decays exponentially with x, as we shall see. The polynomial $q(x)$ is known explicitly apart from the fact that it depends on the contact interval [see for example, (3.6.3)] which is not given a priori, but must be determined as a result of solving the problem.

We rewrite (3.5.23) in the form

$$\Delta(x) = \int_{-1}^{1} dx'\, K(x,x')\Delta(x') + N(x) \quad \text{where} \tag{3.6.7}$$

$$N(x) = \int_{-1}^{1} dx'\, K(x,x')q(x') - q(x) + I(x)$$

$$= \int_{-\infty}^{-1} dy\, T(x,y)Q(y) - \int_{-\infty}^{-1} dy\, l(x-y)d(y) \tag{3.6.8}$$

with the aid of (3.5.30), where

$$Q(y) = \frac{n(y)}{\pi} \int_{-1}^{1} dx'\, \frac{q(x')}{(x'-y)m(x')} . \tag{3.6.9}$$

The quantity $Q(y)$ can be conveniently evaluated in terms of Chebyshev polynomials discussed in Sect. A2.4. Putting [see (A2.4.27)]

$$q(x) = \sum_{l=0}^{m} a_l T_l(x) , \tag{3.6.10}$$

we deduce from (A2.4.28) that for $y < -1$:

$$Q(y) = r(y)n(y) + q(y) , \quad r(y) = \sum_{l=1}^{m} a_l U_{l-1}(y) . \tag{3.6.11}$$

Now

$$\int_{-\infty}^{-1} dy\, T(x,y)q(y) = \int_{-\infty}^{-1} dy\, l(x-y)d(y) , \tag{3.6.12}$$

by virtue of (3.5.21), (3.6.5) and the definition of $T(x,y)$, given by (3.5.24). Therefore

$$N(x) = \int_{-\infty}^{-1} dy\, T(x,y)\, r(y)\, n(y)\ .$$

(3.6.13)

We now specialize to particular viscoelastic models.

3.6.1 Discrete Spectrum Model

The problem will be solved for the case where the viscoelastic half-plane is characterized by a discrete spectrum model (Sect. 1.6). The more general continuous spectrum model is discussed by Golden (1977). The proportionality assumption (Sect. 1.9) will be adopted for the material so that a unique Poisson's ratio exists. Therefore, from (1.6.25, 28, 29), (3.5.20) and (3.5.22), we have

$$l(x) = l_0 \delta(x) + \sum_{i=1}^{N} l_i e^{-a_i x}\ , \qquad k(x) = k_0 \delta(x) + \sum_{i=1}^{N} k_i e^{-\beta_i x}$$

(3.6.14)

where

$$l_0 = \frac{g_0}{1-v} = \frac{1}{1-v} \sum_{i=0}^{N} G_i\ , \qquad k_0 = (1-v) h_0 = (1-v) J_0$$

$$l_i = \frac{1}{1-v} g_i = \frac{-a_i G_i}{1-v}\ , \qquad k_i = (1-v) h_i = (1-v)\beta_i J_i\ .$$

(3.6.15)

In dimensionless coordinates, from (3.5.19),

$$a_i = \frac{b-a}{2V\tau_i}\ , \qquad \beta_i = \frac{b-a}{2V\tau_i'}$$

(3.6.16)

in terms of the decay constants τ_i, τ_i' for relaxation and creep. From (3.5.21), relations analogous to (1.6.1p), (1.6.34) can be deduced:

$$l_0 = \frac{1}{k_0}$$

(3.6.17a)

$$l_0 + \sum_{i=1}^{N} \frac{l_i}{a_i - \beta_j} = 0\ , \qquad j = 1,2,\dots,N$$

(3.6.17b)

$$k_0 + \sum_{j=1}^{N} \frac{k_j}{\beta_j - a_i} = 0\ , \qquad i = 1,2,\dots,N$$

(3.6.17c)

$$k_i = -\left(\sum_{j=1}^{N} \frac{l_j}{(a_j - \beta_i)^2} \right)^{-1}\ .$$

(3.6.17d)

Note that from (3.6.15), the quantities l_i, $i = 1,2,\dots,N$ are negative while the k_i, $i = 1,2,\dots,N$ are positive.

The quantity $T(x,y)$ has the form

$$T(x, y) = \sum_{i,j=1}^{N} A_{ij} \exp\left(-a_i x + \beta_j y\right)$$

$$A_{ij} = \frac{l_i k_j}{a_i - \beta_j} \exp\left(-a_i + \beta_j\right), \quad |x| < 1, \quad y < -1,$$

(3.6.18)

which may be shown with the aid of (3.6.17). The kernel $K(x, x')$, given by (3.5.30), therefore has the form

$$K(x, x') = \frac{-1}{\pi m(x')} \sum_{i,j=1}^{N} A_{ij} e^{-a_i x} J(\beta_j, x')$$

(3.6.19)

where $J(\beta, x)$, defined by (A2.4.31), is evaluated in that section, to the final form (A2.4.37). Clearly, there is no singularity in the kernel at $x = x'$, in agreement with the more general conclusion of Sect. 3.4. However, as noted in that context, the factor $m(x')$ in the denominator of $K(x, x')$ is not cancelled by $J(\beta_j, x')$ so that the kernel has a singularity and is not in fact square-integrable as it stands, which is the requirement that would ensure the applicability of Fredholm theory discussed in Sect. A4.1. Fortunately, the transformation $x = \cos\theta$ gives an integral equation in the variable θ which is non-singular, and Fredholm theory applies. We will not, however, have occasion to call upon the full machinery of this theory. It is clear from (3.6.19) that the kernel is separable (or degenerate or of the Pincherle-Goursat type; see Problem A4.1.1), so that the equation may be reduced without difficulty to a set of algebraic equations, which we shall now proceed to do. From (3.6.6), we have that[1]

$$\Delta(x) = \sum_{i=1}^{N} C_i e^{-a_i x}$$

$$C_i = l_i \int_{-\infty}^{-1} dx' e^{a_i x'} [u'(x') - d(x')] .$$

(3.6.20)

Therefore

$$\int_{-1}^{1} dx' K(x, x') \Delta(x') = \sum_{i,j,k=1}^{N} A_{ij} \Gamma_{jk} C_k e^{-a_i x} \quad \text{with}$$

(3.6.21)

$$\Gamma_{jk} = \frac{1}{\pi} \int_{-\infty}^{-1} dy \int_{-1}^{1} dx' \frac{n(y) \exp\left[\beta_j y - a_k x'\right]}{(x' - y) m(x')} = \Gamma(\beta_j, a_k)$$

(3.6.22)

where $\Gamma(\beta, a)$ is given by (A2.4.30) and more explicitly by (A2.4.42). In detail,

$$\Gamma(\beta_j, a_k) = -\frac{a_k \chi(\beta_j, a_k) + e^{a_k - \beta_j}}{a_k - \beta_j}$$

(3.6.23)

$$\chi(\beta_j, a_k) = -[I_0(a_k) K_1(\beta_j) + I_1(a_k) K_0(\beta_j)] .$$

---

[1] It should be noted that the definition of C_i used here includes l_i, in contrast to the notation adopted by Golden (1977) and to later papers dealing with this problem.

Problem 3.6.1: Show that the matrix $B_{ij} = \exp(-\beta_i + a_j)/(\beta_i - a_j)$ is the inverse of A_{ij}.

Hint: split the denominator of the product into partial fractions and use (3.6.17b), (3.6.17d).

It follows from Problem 3.6.1 that the exponential term in $\Gamma(\beta_j, a_k)$ contributes simply $\Delta(x)$ in (3.6.21), which cancels the left-hand side of (3.6.7). Remembering (3.6.13) and (3.6.18), we see that (3.6.7) reduces to

$$\sum_{i,j=1}^{N} e^{-a_i x} A_{ij} E_j = 0$$

$$E_j = \sum_{k=1}^{N} \frac{C_k a_k \chi(\beta_j, a_k)}{a_k - \beta_j} - \int_{-\infty}^{-1} dy\, e^{\beta_j y} r(y) n(y) \ .$$

(3.6.24)

This must be true for all $|x| < 1$. Therefore, we must have

$$\sum_{j=1}^{N} A_{ij} E_j = 0 \ .$$

(3.6.25)

Since A_{ij} is invertible, it follows that the E_j are zero, giving

$$\sum_{k=1}^{N} \frac{C_k a_k \chi(\beta_j, a_k)}{a_k - \beta_j} = \int_{-\infty}^{-1} dy\, e^{\beta_j y} r(y) n(y) \ .$$

(3.6.26)

So, for a discrete spectrum model, (3.6.7) degenerates to the relatively simple form (3.6.26). This is a set of linear equations for the C_k in terms of the parameters $a_i, \beta_i, i = 1, 2, \ldots, N$. Since from (3.6.17d) it follows that $a_i \neq \beta_j$, $i, j = 1, 2, \ldots, N$, there is no reason to expect difficulties in obtaining unique solutions, except perhaps for very special materials.

The conditions (3.5.27a) for $\theta = \frac{1}{2}$ and (3.5.31) become, in terms of the modified Bessel functions with imaginary argument $I_n(a)$,

$$\sum_{n=1}^{N} C_n I_0(a_n) = -\frac{1}{\pi} \int_{-1}^{1} dx \frac{q(x)}{m(x)}$$

$$\sum_{n=1}^{N} C_n I_1(a_n) - \frac{1}{\pi} \int_{-1}^{1} dx \frac{xq(x)}{m(x)} = \frac{W_1}{\pi}$$

(3.6.27)

by virtue of (3.6.4, 20) and (A1.4.14).

Equation (3.6.5) gives, for a cylindrical punch:

$$q(x) = q_0 + q_1 x \ ,$$

(3.6.28a)

$$q_0 = d_0 \left(l_0 + \sum_{i=1}^{N} \frac{l_i}{a_i} \right) - d_1 \sum_{i=1}^{N} \frac{l_i}{a_i^2} \ ,$$

(3.6.28b)

$$q_1 = d_1 \left(l_0 + \sum_{i=1}^{N} \frac{l_i}{a_i} \right) \ .$$

(3.6.28c)

In this case, from (3.6.10, 11) and (A1.4.11), we have

$$r(x) = q_1 , \tag{3.6.29}$$

so that, with the aid of (A1.4.16), (3.6.26) becomes

$$\sum_{k=1}^{N} \frac{C_k a_k \chi(\beta_j, a_k)}{a_k - \beta_j} = \frac{K_1(\beta_j) q_1}{\beta_j} . \tag{3.6.30}$$

These relations were first given by Morland (1968). The constraints (3.6.27) become

$$q_0 + \sum_{n=1}^{N} C_n I_0(a) = 0 , \qquad -\frac{1}{2} q_1 + \sum_{n=1}^{N} C_n I_1(a) = \frac{W_1}{\pi} . \tag{3.6.31}$$

Equations (3.6.30, 31) together determine the constants C_k and the parameters d_0, d_1 which give the contact region.

An expression for the contact pressure may also be given. From (3.6.4, 20, 28 a), we have

$$v(x) = \sum_{i=1}^{N} C_i e^{-a_i x} + q_0 + q_1 x . \tag{3.6.32}$$

The polynomial contribution to (3.5.32) gives, with the aid of (A1.1.1) and (A1.1.2):

$$p_1(x) = -\frac{m(x)}{\pi} \int_{-1}^{1} dx' \frac{q_0 + q_1 x'}{(x'-x)m(x')} = -q_1 m(x) , \qquad |x| < 1 \tag{3.6.33}$$

while, with the aid of (A1.1.8), the contribution to (3.5.32) of the exponential terms may be evaluated, giving the combined expression

$$p(x) = -q_1 m(x) - \sum_{i=1}^{N} C_i a_i e^{-a_i x} \int_{-1}^{x} e^{a_i y} [y I_0(a_i) - I_1(a_i)] \frac{dy}{m(y)} . \tag{3.6.34}$$

It is easy to check that this expression is bounded, in fact zero, at $x = \pm 1$. It must also be confirmed that the pressure is positive everywhere in the interval $[-1, 1]$. It would seem that in general, this can only be done numerically, for specified configurations.

Let us write down the above equations for the standard linear model. The viscoelastic functions are given by (1.6.1) and (1.6.10). The quantities l_0, l_1, k_0, k_1 are related to the physical coefficients by (3.6.15). Equation (3.6.30) gives that

$$C = \frac{K_1(\beta) q_1 (1 - \beta/a)}{\beta \chi(\beta, a)} \tag{3.6.35}$$

where a, β are the inverse dimensionless decay constants given by

$$a = \frac{b-a}{2V\tau} = -\frac{d_1 R}{V\tau} , \qquad \beta = \frac{b-a}{2V\tau'} = -\frac{d_1 R}{V\tau'} \tag{3.6.36}$$

by virtue of (3.6.3) and (3.6.16). Equations (1.6.10) and (1.6.11) give that

$$1 - \frac{\beta}{a} = \frac{G_1}{G_0 + G_1} \; . \tag{3.6.37}$$

Also, (3.6.28) and (3.6.15) yield

$$q_0 = \frac{1}{1-v}\left(d_0 G_0 + \frac{d_1 G_1}{a}\right) = \frac{1}{1-v}\left(d_0 G_0 - \frac{V\tau G_1}{R}\right) \; , \tag{3.6.38a}$$

$$q_1 = \frac{d_1 G_0}{1-v} \; . \tag{3.6.38b}$$

Equation (3.6.31) gives

$$q_0 = -\frac{K_1(\beta) I_0(a) q_1}{\beta \chi(\beta, a)} \frac{G_1}{G_0 + G_1} \; , \tag{3.6.39a}$$

$$1 + \frac{2 W_1}{\pi q_1} = 2 \frac{K_1(\beta) I_1(a)}{\beta \chi(\beta, a)} \frac{G_1}{G_0 + G_1} \; . \tag{3.6.39b}$$

Relations (3.6.38b) and (3.6.39b) give an implicit equation for d_1 or $(b-a)$. Once this is evaluated, d_0 or $(b+a)$ and C can be calculated without difficulty from (3.6.38), (3.6.39a) and (3.6.35). The quantities $v(x)$ and $p(x)$ are then obtainable from (3.6.32) and (3.6.34). Minor restructuring of (3.6.39a) and (3.6.39b) gives relations identical to those of Hunter (1961), allowing for the different convention for velocity direction.

Some numerical solutions are presented in that reference and, for the case of two decay times, by Morland (1968).

3.7 Small Viscoelasticity Approximation

In Sect. 2.5, we discussed briefly the case where the ratio of viscoelastic (time-dependent terms in the functions $G(t)$, $J(t)$) to purely elastic effects (constant terms in these functions) is small. The possibility of expressing solutions as power series in this ratio was noted. In the present section, we shall consider solutions to first order in this parameter. It turns out that the problem of the moving load greatly simplifies, to the extent that explicit expressions for all quantities of interest may be written down, in contrast to the highly implicit equations which emerged from the exact analysis in the previous section.

We shall confine our attention to the steady state problem, with limiting friction, the general equations for which are given in Sect. 3.5. It should be emphasized however that this approximation, if valid, brings about great simplification in a wide variety of problems.

The equations of interest are therefore (3.5.23 – 28). We let

$$l(x) = l_0(\delta(x) + \varepsilon(x)) \; , \quad k(x) = k_0(\delta(x) - \varepsilon(x)) \; , \quad l_0 k_0 = 1 \tag{3.7.1}$$

where quantities proportional to higher than the first power of $\varepsilon(x)$ are neglected. When $\varepsilon(x)$ is identically zero, the singular viscoelastic functions are proportional to $\delta(x)$. This is the instantaneous elastic limit, which occurs even for non-zero $\varepsilon(x)$ in the short time or large decay time limit. The moduli are the relaxation functions evaluated at zero time. From (3.5.19, 20) we see that this can also be thought of as the large velocity limit. The viscoelastic corrections, which we will consider below, therefore vanish in this limit. In the small decay time or small velocity limit, $\varepsilon(x)$ becomes proportional to a delta function. For spectrum models, this follows from (A3.1.6p). This second (long time) limit is what is conventionally termed the elastic limit, as noted in Sect. 1.4. The results obtained must therefore reduce to elastic form but depending now upon the long time moduli. These remarks are true also for finite viscoelastic corrections.

To first order, it follows from (3.7.1) that $T(x,y)$, defined by (3.5.24b) is given by

$$T(x,y) = \varepsilon(x-y) \ . \tag{3.7.2}$$

Also, for $|x| < 1$

$$v(x) = \int_{-\infty}^{x} dx'\, l(x-x')u'(x') = l_0 \left[u'(x) + \int_{-\infty}^{x} dx'\, \varepsilon(x-x')u'(x') \right] , \tag{3.7.3}$$

where the first term $u'(x)$ is simply the punch profile $d(x)$, and is unchanged in the elastic limit. Substituting this expression into the right-hand side of (3.5.23) gives

$$v(x) = v_0(x) + v_c(x) , \qquad v_0(x) = l_0 d(x)$$

$$\tag{3.7.4}$$

$$v_c(x) = l_0 \int_{-\infty}^{x} dy\, \varepsilon(x-y)u_i'(y)$$

where $u_i'(y)$ is the instantaneous elastic limit of the displacement derivative, given by

$$u_i'(y) = \begin{cases} d(y) , & |y| < 1 \\ \dfrac{n(y)\sin(\pi\theta)}{\pi} \displaystyle\int_{-1}^{1} dx' \dfrac{d(x')}{(x'-y)m(x')} , & y < -1 \end{cases} \tag{3.7.5}$$

and $m(x)$, $n(y)$ are given by (3.5.25). These equations could in fact have been written down without knowledge of (3.5.23); see the elastic limit of (3.3.25). Specializing to a cylindrical indentor, with $d(x)$ given by (3.6.3), we have that

$$u_i'(x) = d(x) + d_1 n(x) , \qquad x < -1 , \tag{3.7.6}$$

by virtue of integrals (A1.1.10) and (A1.3.10). The subsidiary conditions (3.5.27) give that [(A1.3.10−12)]:

$$l_0(d_0 - \eta d_1) + \frac{\sin(\pi\theta)}{\pi} \int_{-1}^{1} dx\, \frac{v_c(x)}{m(x)} = 0$$

$$\tag{3.7.7}$$

$$l_0[\eta(d_1 - d_0) + 2\theta^2 d_1] + \frac{\sin(\pi\theta)}{\pi} \int_{-1}^{1} dx\, \frac{x v_c(x)}{m(x)} = \frac{W_1 \sin(\pi\theta)}{\pi q} ; \qquad \eta = 1 - 2\theta \ .$$

Let us write

$$a = a_i + a_c , \quad b = b_i + b_c . \tag{3.7.8}$$

where $[a_i, b_i]$ is the contact interval in the corresponding instantaneous elastic problem (under load W) and a_c, b_c are the viscoelastic corrections. The quantities a_i, b_i must obey (3.7.7) with $v_c(x)$ put to zero. Denoting

$$d_0 = d_{i0} + d_{c0} , \quad d_{i0} = -\frac{a_i + b_i}{2R} , \quad d_{c0} = -\frac{a_c + b_c}{2R} \quad \text{and} \tag{3.7.9}$$

$$d_1 = d_{i1} + d_{c1} , \quad d_{i1} = -\frac{b_i - a_i}{2R} , \quad d_{c1} = -\frac{b_c - a_c}{2R} \tag{3.7.10}$$

we deduce that [note (3.5.27 c)]

$$d_{i0} = \eta \, d_{i1}$$

$$-d_{i1} = \left(\frac{-W \sin (\pi\theta)}{2\theta(1-\theta)\pi l_0 qR} \right)^{1/2} = \left(\frac{W(1-v)}{2\theta(1-\theta)\pi G(0)R} \right)^{1/2} , \tag{3.7.11}$$

with the aid of (3.3.17, 19) in the instantaneous elastic limit and (1.2.33 a). These are the relations determining the elastic contact interval. The viscoelastic corrections obey the relations

$$l_0(d_{c0} - \eta \, d_{c1}) + L_0 = 0 \tag{3.7.12}$$

and, again recalling (3.5.27 c):

$$l_0[\eta(d_{c1} - d_{c0}) + 2\theta^2 d_{c1}] + L_1 = -\frac{W_1 d_{c1} \sin (\pi\theta)}{\pi d_{i1} q} , \quad \text{where} \tag{3.7.13}$$

$$L_0 = \frac{\sin (\pi\theta)}{\pi} \int_{-1}^{1} dx \frac{v_c(x)}{m(x)} , \quad L_1 = \frac{\sin (\pi\theta)}{\pi} \int_{-1}^{1} dx \frac{v_c(x)x}{m(x)} . \tag{3.7.14}$$

Remembering that

$$2\theta(1-\theta)l_0 d_{i1} = \frac{W_1 \sin (\pi\theta)}{\pi q} , \tag{3.7.15}$$

relation (3.7.13) can be simplified somewhat, giving

$$l_0(d_{c1} - \eta \, d_{c0}) + L_1 = 0 . \tag{3.7.16}$$

Equations (3.7.12, 16) may be solved without difficulty to give

$$a_c = \frac{R}{2(1-\theta)l_0}(L_0 - L_1) , \quad b_c = \frac{R}{2\theta l_0}(L_0 + L_1) . \tag{3.7.17}$$

The pressure is given by (3.5.28). Substituting for $v(x)$ from (3.7.4), we obtain, with the aid of integrals (A1.1.10) and (A1.3.10),

$$p(x) = q \left[l_0 d_1 m(x) + \frac{m(x) \sin (\pi\theta)}{\pi} \int_{-1}^{1} dx' \frac{v_c(x')}{(x'-x)m(x')} + v_c(x) \cos (\pi\theta) \right] . \tag{3.7.18}$$

In the zero friction limit, (3.7.17) becomes

$$a_c = \frac{R}{l_0 \pi} \int_{-1}^{1} dx \, v_c(x) \left(\frac{1-x}{1+x}\right)^{1/2} , \quad b_c = \frac{R}{l_0 \pi} \int_{-1}^{1} dx \, v_c(x) \left(\frac{1+x}{1-x}\right)^{1/2} \quad (3.7.19)$$

and the pressure function (3.7.18) reduces to

$$p(x) = -l_0 d_1 m(x) - \frac{m(x)}{\pi} \int_{-1}^{1} dx' \frac{v_c(x')}{(x'-x)m(x')} \quad (3.7.20)$$

where now $m(x)$ is given by (3.5.29). With the aid of (3.7.4) and (3.7.6), we write $v_c(x)$ a little more explicitly:

$$v_c(x) = l_0 j(x) + d_1 l_0 \int_{-\infty}^{-1} dx' \, \varepsilon(x-x') n(x')$$

$$j(x) = \int_{-\infty}^{x} dx' \, \varepsilon(x-x') d(x') . \quad (3.7.21)$$

The quantity $j(x)$ is a polynomial of degree one. In fact, it is equal to $[q(x) - l_0 d(x)]/l_0$, where $q(x)$ is given by (3.6.5) and in the cylindrical case for a spectrum model, by (3.6.28). We write it as

$$j(x) = j_0 + j_1 x . \quad (3.7.22)$$

Let us now consider the special case of a standard linear model, putting [see (3.6.15, 16)]

$$\varepsilon(x) = -\varepsilon_0 e^{-ax} , \quad \varepsilon_0 = \frac{a G_1}{[G_0 + G_1]} , \quad a = \frac{b_i - a_i}{2V\tau} . \quad (3.7.23)$$

It should be noted that (3.7.23) is essentially as general as an arbitrary discrete or continuous spectrum model, in the present context. This is because all quantities must be linear in the viscoelastic functions, so that the results for a more general model are simply sums of terms of the form that will now be derived. Explicit results can be obtained for the problem with friction, in terms of Whittaker functions. However, these will not be introduced in the present work. We refer to Golden (1979a, 1986a) for further details. In the frictionless case (from (3.7.11) we see that $d_{i0} = 0$):

$$j_0 = \frac{d_{i1} \varepsilon_0}{a^2} , \quad j_1 = -\frac{\varepsilon_0}{a} d_{i1} = -a j_0 \quad (3.7.24)$$

and, also from (3.7.21):

$$v_c(x) = l_0 \left[j(x) - d_{i1} \varepsilon_0 e^{-ax} \int_{-\infty}^{-1} dx' \, e^{ax'} (x'^2 - 1)^{1/2} \right]$$

$$= l_0 [j(x) - d_{i1} \varepsilon_0 e^{-ax} K_1(a)/a]$$

in terms of the modified Bessel Functions with imaginary argument $K_n(a)$. Use has been made of (A1.4.16). From (3.7.19) and (A1.4.12), we deduce that

$$a_c = \frac{R\varepsilon_0 d_{i1}}{a} \left\{ \frac{1}{a} + \frac{1}{2} - K_1(a)[I_0(a) + I_1(a)] \right\}$$

$$b_c = \frac{R\varepsilon_0 d_{i1}}{a} \left\{ \frac{1}{a} - \frac{1}{2} - K_1(a)[I_0(a) - I_1(a)] \right\} . \tag{3.7.25}$$

The pressure, given by (3.7.20), takes the form

$$p(x) = -l_0 \left\{ m(x)[d_1 + j_1] - d_{i1}K_1(a)\varepsilon_0 e^{-ax} \int_{-1}^{x} \frac{dy}{m(y)} e^{ay}[yI_0(a) - I_1(a)] \right\} , \tag{3.7.26}$$

by virtue of (A1.1.1, 2) and (A1.1.8).

An expression for the hysteretic friction coefficient in the small viscoelasticity approximation is derived in the next section.

3.8 Hysteretic Friction

We now give expressions for the coefficient of hysteretic friction, introduced in Sect. 2.11, for the problem of a moving, cylindrical punch discussed in the previous two sections. Let us consider the frictionless case first. We will confine the discussion to steady-state problems. Equation (2.11.12), adapted to plane strain conditions, gives that the coefficient of hysteretic friction has the general form

$$f_H = \frac{1}{W_1} \int_{-1}^{1} dx\, p(x)u'(x)$$

$$W_1 = \int_{-1}^{1} dx\, p(x) = \frac{2W}{b-a} = \frac{-W}{d_1 R} \tag{3.8.1}$$

in terms of dimensionless coordinates, moving with the punch (see Sect. 3.5). The quantity W is the load per unit length. For a cylindrical punch, characterized by (3.6.3), f_H has the form

$$f_H = d_0 + \frac{2\delta}{\pi} \int_{-1}^{1} dx\, p(x)x \tag{3.8.2}$$

where, by virtue of (3.7.11), in the limit $\theta \to \frac{1}{2}$,

$$\delta = \frac{\pi d_1}{2W_1} = -\frac{\pi d_1^2 R}{2W} = -\frac{1}{l_0}\left(\frac{d_1}{d_{i1}}\right)^2 . \tag{3.8.3}$$

Let us now express $p(x)$ in terms of $v(x)$, using (3.5.32). On interchanging the order of integration, we obtain

$$f_H = d_0 + \frac{2\delta}{\pi} \int_{-1}^{1} \frac{dx}{m(x)} v(x)J(x) , \qquad J(x) = \frac{1}{\pi} \int_{-1}^{1} dx' \frac{m(x')x'}{x'-x} \tag{3.8.4}$$

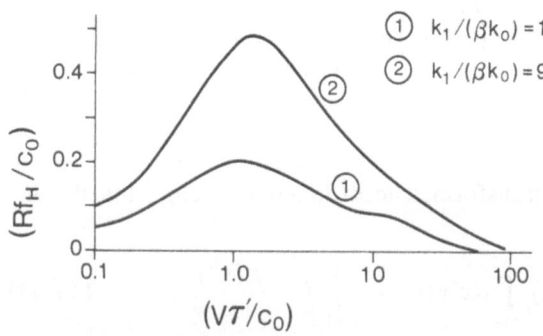

① $k_1/(\beta k_0) = 1$
② $k_1/(\beta k_0) = 9$

Fig. 3.1. Variation of coefficient of friction f_H with velocity: c_0 is the semi-contact width for a stationary cylinder [after Hunter (1961)]

where $m(x)$ is given by (3.5.29). Relations (A1.1.5) or alternatively (A1.1.6) and (A1.4.11) give that

$$J(x) = \tfrac{1}{2} - x^2 = -\tfrac{1}{2} T_2(x) \ . \tag{3.8.5}$$

in terms of the Chebyshev polynomials, discussed in Sect. A2.4. However, by virtue of (3.5.27a), we can write f_H as

$$f_H = d_0 + \frac{2\delta}{\pi} \int_{-1}^{1} \frac{dx}{m(x)} v(x)(e-x^2) \tag{3.8.6}$$

where the value of e is arbitrary. Let us return to the choice $e = \tfrac{1}{2}$. Then, (3.8.5) and the orthogonality relations for the Chebyshev polynomials (A1.4.5) mean that the integral in (3.8.6) will annihilate $q(x)$, the polynominal part of $v(x)$, as given by (3.6.4) and (3.6.28a). It follows that

$$f_H = d_0 + \frac{2\delta}{\pi} \int_{-1}^{1} \frac{dx}{m(x)} \Delta(x) \left(\frac{1}{2} - x^2 \right) \ . \tag{3.8.7}$$

For a discrete spectrum model, $\Delta(x)$ is given by (3.6.20). Substituting this form gives, with the aid of (A1.4.14) and (A1.4.15):

$$f_H = d_0 - \delta \sum_{i=1}^{N} C_i I_2(a_i) \tag{3.8.8}$$

which can be evaluated if (3.6.26, 27) are solved. Numerical results have been presented by Hunter (1961) and Morland (1968). Characteristically, this quantity turns out to be hump shaped, as a function of velocity, in qualitative agreement with experiment. The results of Hunter (1961) are presented in Fig. 3.1.

In the presence of friction, it follows from (2.11.14) and Coulomb's Law (2.11.3) that

$$f_H = \frac{1}{W_1} \int_{-1}^{1} dx \, p(x) \, [u_2'(x) - f u_1'(x)] \tag{3.8.9}$$

in the steady-state limit, and using dimensionless coordinates; recall that the subscripts 1, 2 refer to x, y respectively. The first relation of (3.2.1p) gives that (3.8.9) can be rewritten as

$$f_H = \frac{1+f^2}{W_1} \int_{-1}^{1} dx\, p(x) \left[u_2'(x) + f\left(\frac{1}{2} - v\right) r(x) \right]$$

$$r(x) = \int_{-1}^{x} dx'\, \gamma(x-x') p(x')$$

(3.8.10)

where $\gamma(x)$ is the creep function transformed according to (3.5.20), while the second form of (3.2.1 p) gives that

$$f_H = \frac{(1+f^2)(1-v)}{W_1 \pi} \int_{-1}^{1} dx\, p(x) \int_{-\infty}^{x} dx'\, \gamma(x-x') \int_{-1}^{1} dx'' \frac{p(x'')}{x''-x'} .$$

(3.8.11)

Both of these are of interest. Consider (3.8.10) to begin with. If the material is incompressible, then $v = \frac{1}{2}$, and we have

$$f_H = \frac{1+f^2}{W_1} \int_{-1}^{1} dx\, p(x) u_2'(x) .$$

(3.8.12)

We remarked at the end of Sect. 3.5 that for incompressible materials, the equations of the problem with friction reduced to those of the frictionless case, so that $p(x)$ is given by the form for a lubricated surface. The only effect of the frictional surface is to attach the factor $(1+f^2)$ to f_H. Therefore, if f is small, we can neglect its effects completely since it occurs in second order.

The form given by (3.8.11) is also convenient in that it indicates immediately that f_H must vanish in the instantaneous elastic limit. The point is that $\gamma(x)$ becomes a delta function in this limit so that f_H vanishes, because of the antisymmetry of the Hilbert kernel. We remark that in fact it vanishes in both limits discussed after (3.7.1). Let

$$\gamma(x) = h_0 \delta(x) + \gamma_1(x)$$

(3.8.13)

where the first term is the instantaneous contribution. Then

$$f_H = \frac{(1+f^2)(1-v)}{W_1 \pi} \int_{-1}^{1} dx\, p(x) \int_{-\infty}^{x} dx'\, \gamma_1(x-x') \int_{-1}^{1} dx'' \frac{p(x'')}{x''-x'} ,$$

(3.8.14)

since the instantaneous term drops out immediately. However, in the small decay time limit, $\gamma_1(x)$ is proportional to $\delta(x)$ so that, as stated previously, f_H vanishes in both limits. Put another way, it vanishes at very small and very large velocities. We see therefore, on the basis of this very general argument, that f_H will have, roughly speaking, the hump shape observed experimentally. Of course, any number of maxima and minima could in theory occur in the velocity range. However, for simple choices of the viscoelastic functions, it is reasonable to suppose that only one maximum occurs. As mentioned above, this has been demonstrated numerically in the frictionless case.

3.8.1 Small Viscoelasticity

We consider (3.8.6) in the small viscoelasticity approximation. If e is chosen to be unity, then the portion $v_0(x)$ in (3.7.4) cancels the term d_0, to first order.

This is easy to show, if one recalls that d_0 is a first order quantity so that δ can be replaced by $(-l_0)^{-1}$. It follows that

$$f_H = -\frac{2}{\pi l_0} \int_{-1}^{1} dx\, v_c(x) m(x) \ . \tag{3.8.15}$$

Relation (3.8.15) may be derived more directly, by considering (3.8.14) in the frictionless limit.

Problem 3.8.1: Deduce (3.8.15) from the frictionless limit of (3.8.14), noting that both occurrences of the pressure in (3.8.14) may be replaced by the instantaneous elastic pressure $p_0(x)$, which is proportional to $m(x)$; also, the Hilbert transform of $p_0(x)$ is proportional to the elastic displacement derivative.

Equation (3.8.15) is a convenient general expression for the coefficient of hysteretic friction of a cylindrical indentor on a lubricated, slightly viscoelastic half-plane.

For a standard linear solid, we use (3.7.25) and (A1.4.12) to obtain

$$f_H = \frac{|d_{i1}|\,\varepsilon_0}{a} T(a) \ , \qquad T(a) = \frac{1 - 2I_1(a)K_1(a)}{a} \ . \tag{3.8.16}$$

A plot of the function $T(a)$ is given by Golden (1977). It is zero at small and large a or, put another way, at large and small velocities. This is in accordance with the general observations made in the context of (3.8.11). Recalling (3.7.23), we see that ε_0/a involves only physical coefficients in the relaxation function.

A more general spectrum model would give a sum of terms proportional to $T(a_i)$ for each a_i formed from a decay constant τ_i in accordance with (3.7.23).

A generalization of (3.8.16) for unlubricated contact derived from (3.8.10), is presented by Golden (1986a).

3.9 Increasing and Decreasing Contact Area

In Sect. 2.6, it was pointed out that if the regions of specification of displacement and stress are expanding and contracting respectively, or the other way around, then it is possible to express the solution in a simple manner. In this section, we wish to explore this observation in more detail for plane frictionless contact problems.

Consider the case where the contact area $C(t)$ is contracting, or at least non-expanding, for all t. This is the region on which displacement, or rather its derivative, is specified. It follows that $v(x, t)$ is known in $C(t)$, so that the general solutions of the Hilbert problem (3.3.2) discussed in Sect. 3.3 are in fact final solutions of the problem. These are identical in form to the corresponding elastic solutions but where $v(x, t)$ takes the place of the displacement derivative. This is a special case of the Extended Correspondence Principle discussed in Sect. 2.6.

In $C(t)$, $u'(x, t)$ will conform to the punch shape and will in fact be independent of time. Therefore

$$v(x, t) = v(x) = \int_{-\infty}^{t} dt' l(t - t') u'(x) = l_e u'(x) \tag{3.9.1}$$

where

$$l_e = \int_{0}^{\infty} dt\, l(t) = L(\infty) \tag{3.9.2}$$

if $L(t)$ is related to $l(t)$ by (1.10.1 p)

The quantity $v(x, t)$ outside the contact area may be determined from (3.3.15) for $\theta = \frac{1}{2}$, and $u'(x, t)$, outside $C(t)$, is given by

$$u'(x, t) = \int_{-\infty}^{t} dt' k(t - t') v(x, t') \ . \tag{3.9.3}$$

This integral may involve times when $x \in C(t')$. The pressure is given by (3.3.16) for $\theta = \frac{1}{2}$. Because of (3.9.1), it will be the same as the elastic form.

If the contact region $C(t)$ is expanding (or stationary), it follows that $C'(t)$, the region where stress is specified to be zero, is contracting (or stationary). In this case, it is desirable to use the form (3.3.8). The quantity $q(x, t)$ is zero on $C'(t)$, so the problem is in fact formally identical to (3.3.2) and the elastic problem, except that the contact pressure $p(x, t)$ is replaced by $q(x, t)$. The actual contact pressure is deduced from

$$p(x, t) = \int_{-\infty}^{t} dt' l(t - t') q(x, t') \ , \tag{3.9.4}$$

which is simply (3.2.11 b) inverted. The displacement derivative on $C'(t)$ is given by (3.3.15) for $\theta = \frac{1}{2}$ but with $u'(x, t)$ replacing $v(x,t)$. This is identical to the elastic form, again in accordance with the Extended Correspondence Principle.

Consider the special example where there is one punch, undergoing only vertical movement and with point of deepest penetration at the origin. Let its profile be given or approximated by

$$u'(x, t) = u'(x) = -cx, \quad c = 1/R \tag{3.9.5}$$

where R is the radius of curvature. This covers the case of a cylinder the radius of which is large compared to the contact interval. We assume that no tangential stresses exist. In the case where $C(t)$ is non-expanding, (3.9.1) gives that

$$v(x, t) = v(x) = -l_e cx, \quad x \in C(t) \ . \tag{3.9.6}$$

The contact interval is symmetrical and may be denoted by $[-a(t), a(t)]$. From (3.3.33) and integral (A1.2.3) we have

$$p(x, t) = l_e c [a^2(t) - x^2]^{1/2} \ , \tag{3.9.7}$$

which, as observed above, in a more general context, is the same as in the elastic case. Equation (3.3.32) and integral (A1.2.3) gives that

$$v(x, t) = -l_e c [x - \text{sgn}(x) n(x, t)] , \quad |x| > a(t) . \tag{3.9.8}$$

From (3.9.3), and (3.9.8) we obtain that

$$u'(x, t) = -l_e c \left[k_e x - \text{sgn}(x) \int\limits_{t_1(x)}^{t} dt' k(t - t') n(x, t') \right]$$

$$k_e = \int\limits_{0}^{\infty} dt\, k(t) = \frac{1}{l_e} \tag{3.9.9}$$

where $t_1(x)$ is the time that x leaves $C(t)$. If it was in $C'(t)$ to begin with, then $t_1(x) = -\infty$. The symmetry of the problem gives that (3.3.27) is automatically obeyed. Finally, (3.3.34) and integral (A1.3.12) in the limit $\theta \to \frac{1}{2}$ give a relation between load and contact area:

$$W = \frac{\pi}{2} l_e c a^2(t) , \tag{3.9.10}$$

which is identical to the corresponding elastic relation; see (3.7.11) in the limit $\theta \to \frac{1}{2}$.

In the expanding case, the solution proceeds in an identical manner, except that $l_e q(x, t)$, $l_e u'(x, t)$ play the part of $p(x, t)$, $v(x, t)$, since the basic equations are now (3.3.8). The exception to this is the total load constraint, which still has the form (3.3.34). Therefore we can immediately put

$$q(x, t) = c [a^2(t) - x^2]^{1/2} , \tag{3.9.11}$$

replacing (3.9.7) and

$$u'(x, t) = -c [x - \text{sgn}(x) n(x, t)] \tag{3.9.12}$$

replacing (3.9.8). This latter expression is identical to the elastic result, as remarked previously. Equation (3.9.4) gives that

$$p(x, t) = c \int\limits_{t_1(x)}^{t} dt' l(t - t') [a^2(t') - x^2]^{1/2} , \tag{3.9.13}$$

where $t_1(x)$ is now the time at which x entered the contact region. Let us write this in the form

$$p(x, t) = c \int\limits_{-\infty}^{t} dt' l(t - t') \, \text{Re} \, \{a^2(t') - x^2\}^{1/2} , \tag{3.9.14}$$

which is more convenient for exploring the load condition. We have

$$W = \int\limits_{-a(t)}^{a(t)} dx\, p(x, t) = c \int\limits_{-\infty}^{t} dt' l(t - t') \int\limits_{-a(t')}^{a(t')} dx \, \text{Re} \, \{a^2(t') - x^2\}^{1/2}$$

$$= \frac{c\pi}{2} \int\limits_{-\infty}^{t} dt' l(t - t') a^2(t') , \tag{3.9.15}$$

which depends on the history of the contact region. In general, W will be time-dependent. This relation can be inverted to give

$$\frac{c\pi}{2} a^2(t) = \int_{-\infty}^{t} dt' k(t-t') W(t') \; .$$ (3.9.16)

Problem 3.9.1: Show that if a constant load is applied at $t = 0$ to a viscoelastic material with unique Poisson's ratio, the creep function of which tends to a finite value, then the contact region expands monotonically to some finite interval. Recall (3.2.12).

An early paper by Prokopovici (1956) treated the expanding case for a general punch profile and aging viscoelastic material.

If the history of $C(t)$ involves a finite succession of expanding and contracting phases, then by an iterative procedure, it is possible to apply the methods of this section to derive a solution. This approach is sketched in Sect. 2.6. If $C(t)$, while expanding and contracting, passes continually through the same family of states – such as would occur if an indentor were subjected to a varying normal load – then it will be shown in the next two sections that it is possible to make considerably more detailed statements about the solution.

3.10 The Plane Normal Contact Problem

Consider a rigid smooth punch pressed into a viscoelastic half-space under a varying load. There is no tangential movement of the punch and the contact is lubricated. For simplicity, it will be assumed that the punch is symmetrical, though this is not in fact necessary. We choose the origin so that the contact interval is $[-a(t), a(t)]$. The symmetry assumption then implies that the point of deepest indentation is at the origin. We have the fundamental relationship (3.2.10) between displacement and pressure. In the contact interval, $u'(x, t)$ is the derivative of the punch profile. It is time-independent and will be denoted by $u'(x)$.

As the load varies, the contact interval will pass through a series of states characterized uniquely by $a(t)$, so that one can plot its history by plotting $a(t)$. This problem falls into the category discussed at the end of Sect. 2.6, referred to as involving repetitive expansion and contraction. We now apply the general method developed in that section to this particular case.

Let $a(t)$ pass through a series of maxima and minima before the current time t. First consider the case where it is contracting at time t (Fig. 3.2a). The sets $B_u(t), B_\sigma(t)$ in this case are $C(t), C'(t)$. The times $\theta_l(t)$, given by (2.6.6, 7), are here defined as

$$\theta_l(t) = t_l(a(t))$$ (3.10.1)

in terms of the times $t_l(x)$ introduced in Sect. 2.4, where x replaces r. The period $[\theta_1(t), t]$ belongs to $W_u(t)$ for a contracting phase, remembering that $W_u(t)$ is

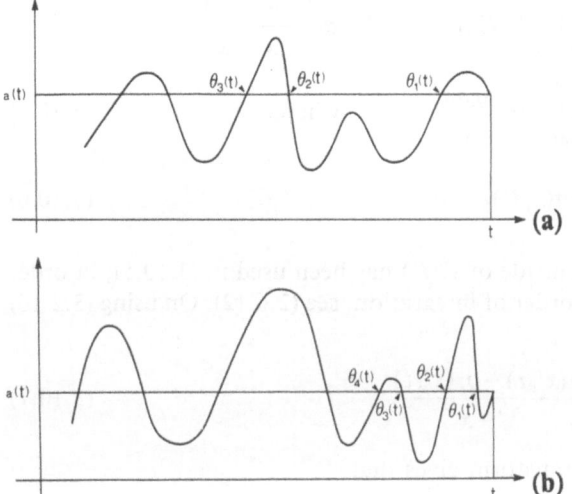

Fig. 3.2. (a) Typical distribution of $\theta_k(t)$, $k = 1,2\ldots$ corresponding to a time t when $a(t)$ is decreasing. **(b)** Typical distribution of $\theta_k(t)$, $k = 1,2\ldots$ corresponding to a time t when $a(t)$ is increasing

the set of times t' such that $C(t) \subseteq C(t')$, or those times at which the contact interval was not less than at time t. The set $W_\sigma(t)$ consists of those times when it was less than at time t. We use the decomposition given by (2.6.8):

$$v(x,t) = \int_{W_u(t)} dt\, \Pi_u(t,t')u'(x,t') + \int_{W_\sigma(t)} dt'\, \Pi_\sigma(t,t')v(x,t') \ , \qquad (3.10.2)$$

where Π_u, Π_σ are given by (2.4.9–11) with the θ_l replacing the t_l. Also the subscript "σ" replaces "v".

If, on the other hand, $C(t)$ is expanding, (Fig. 3.2. b) we have, from (2.6.9), (2.4.12, 13), with the same alterations,

$$u'(x,t) = \int_{W_u(t)} dt'\, \Gamma_u(t,t')u'(x,t') + \int_{W_\sigma(t)} dt'\, \Gamma_\sigma(t,t')v(x,t') \ . \qquad (3.10.3)$$

First consider the case where $C(t)$ is contracting. By definition of $W_u(t)$, if $x \in C(t) = [-a(t), a(t)]$ then $x \in C(t')$ where $t' \in W_u(t)$, so that in the first integral of (3.10.2) we put $u'(x,t') = u'(x)$ and rewrite the integral as

$$\int_{W_u(t)} dt'\, \Pi_u(t,t')u'(x,t') = \Pi_u(t)u'(x)$$

$$\Pi_u(t) = \int_{W_u(t)} dt'\, \Pi_u(t,t') \ . \qquad (3.10.4)$$

Furthermore, for $t' \in W_\sigma(t)$, $C(t') \subseteq C(t)$. Replacing $v(x,t')$ on the right-hand side of (3.10.2) by its expression in terms of the pressure, as given by (3.2.10), we obtain

$$v(x,t) = \Pi_u(t)u'(x) + \frac{1}{\pi} \int\limits_{W_\sigma(t)} dt' \Pi_\sigma(t,t') \int\limits_{-a(t')}^{a(t')} dx' \frac{p(x',t')}{x'-x}$$

$$= \Pi_u(t)u'(x) + \frac{1}{\pi} \int\limits_{-a(t)}^{a(t)} dx' \frac{q_c(x',t)}{x'-x}, \qquad \text{where} \tag{3.10.5}$$

$$q_c(x,t) = \int\limits_{W_\sigma(t)} dt' \Pi_\sigma(t,t')p(x,t') . \tag{3.10.6}$$

The fact that $p(x,t')$ is zero outside of $C(t')$ has been used in (3.10.5), in order to validate the interchange of order of integration; see (2.6.12). On using (3.2.10) again, we obtain

$$\Pi_u(t)u'(x) = \frac{1}{\pi} \int\limits_{-a(t)}^{a(t)} dx' \frac{p(x',t) - q_c(x',t)}{x'-x} . \tag{3.10.7}$$

Inverting the finite Hilbert Transform gives that

$$p(x,t) = q_c(x,t) + \Pi_u(t)P(x,t) \tag{3.10.8}$$

where, on using (A2.4.9), since $p(x,t)$ must be bounded,

$$P(x,t) = -\frac{m(x,t)}{\pi} \int\limits_{-a(t)}^{a(t)} dx' \frac{u'(x')}{(x'-x)m(x',t)} ,$$
$$\tag{3.10.9}$$
$$m(x,t) = [a^2(t) - x^2]^{1/2} .$$

This quantity is known, since $u'(x)$ is simply the slope of the punch profile. In fact, $P(x,t)$ is proportional to the elastic pressure.

Equation (3.10.8) expressed $p(x,t)$ in terms of $q_c(x,t)$, which only involves $p(x,t')$, $t'<t$, and the known quantity $P(x,t)$. Therefore, an explicit solution, after any finite number of maxima and minima, can be constructed.

Problem 3.10.1: Show that the total load $W(t)$ is given by

$$W(t) = \int\limits_{W_\sigma(t)} dt' \Pi_\sigma(t,t') W(t') + \Pi_u(t) W_0(t)$$
$$\tag{3.10.1p}$$
$$W_0(t) = -\int\limits_{-a(t)}^{a(t)} dx \frac{u'(x)x}{m(x,t)}$$

with the help of integral (A1.2.2).

If $a(t)$ is increasing, similar arguments give, instead of (3.10.8), the relations

$$\int\limits_{W_\sigma(t)} dt' \Gamma_\sigma(t,t')p(x,t') = [1 - \Gamma_u(t)]P(x,t)$$
$$\tag{3.10.10}$$
$$\Gamma_u(t) = \int\limits_{W_u(t)} dt' \Gamma_u(t,t') .$$

From (2.4.13, 14), we see that $\Gamma_\sigma(t, t')$ can be rewritten as

$$\Gamma_\sigma(t, t') = k(t - t')R(t', \theta_1(t), t) + \Gamma_\sigma^{(1)}(t, t') , \tag{3.10.11}$$

where $\Gamma_\sigma^{(1)}(t, t')$ is zero for $t' > \theta_2(t)$. Substituting this into (3.10.10) and solving for $p(x, t)$, with the aid of (2.3.15), gives

$$
\begin{aligned}
p(x, t) = & -\int_{\theta_1(t)}^{t} dt' l(t - t') \int_{W_\sigma(t')} dt'' \Gamma_\sigma^{(1)}(t', t'') p(x, t'') \\
& + \int_{\theta_1(t)}^{t} dt' l(t - t')[1 - \Gamma_u(t')] P(x, t')
\end{aligned}
\tag{3.10.12}
$$

which, for $a(t)$ increasing, expresses the pressure at time t in terms of this quantity at earlier times and known functions. A load condition similar to (3.10.1 p) may be deduced from either (3.10.10) or (3.10.12).

In the case of a cylindrical punch, $u'(x)$ is given by (3.9.5), so that, by virtue of (A1.2.3),

$$P(x, t) = c[a^2(t) - x^2]^{1/2} . \tag{3.10.13}$$

Let us consider an example. Let the contact area expand up to a certain point and then contract. Let time t occur after the maximum has been passed. There is only one finite time $\theta_1(t)$. Equations (3.10.6) and (2.4.9) give that

$$q_c(x, t) = \int_{-\infty}^{\theta_1(t)} dt' T_1(t, t') p(x, t') \tag{3.10.14}$$

where $T_1(t, t')$ is given by (2.4.6). The quantity $p(x, t')$ in (3.10.14) has the form (3.9.13). Also, from (2.4.9, 11):

$$\Pi_u(t) = \int_{\theta_1(t)}^{t} dt' l(t - t') . \tag{3.10.15}$$

For a material with definite Poisson's ratio (the proportionality assumption), (3.1.15) and (2.12) give that

$$T_1(t, t') = \dot{G}(t - t') J(0) + \int_{t'}^{\theta_1(t)} dt'' \dot{G}(t - t'') \dot{J}(t'' - t')$$

$$\Pi_u(t) = \frac{G(t - \theta_1(t))}{1 - \nu} \tag{3.10.16}$$

by virtue of (1.2.33).

We can write, with the aid of (A1.3.12), the load constraint (3.10.1 p) in the form

$$W(t) = \int_{-\infty}^{\theta_1(t)} dt' T_1(t, t') W(t') + \tfrac{\pi}{2} \Pi_u(t) c a^2(t) , \tag{3.10.17}$$

where $W(t')$ on the right is given by (3.9.15). The three-dimensional version of this problem is discussed in Sect. 5.2. The more compact formula derived in that section for the pressure applies here also.

3.11 The Steady-State Limit of the Normal Problem

If the loading of the indentor varies periodically with time, we would expect the response of the half-space to reflect this periodicity after a long time, when transient effects have died away. This situation is quite analogous to that of the forced harmonic oscillator subject to frictional resistance. The contact interval will, in particular, vary with the same period but in a manner that is not simply related to the load since from (3.10.17), for example, we see that the relation between the two quantities is not linear. We have therefore

$$p(x,t) = p(x,t+\Delta) \tag{3.11.1a}$$

$$a(t) = a(t+\Delta) \ , \tag{3.11.1b}$$

where Δ is the period of the applied load. These relations are true for all x, t. Let us choose t in the region $[\Delta_1, \Delta_2]$ which includes the origin and where

$$\Delta_2 - \Delta_1 = \Delta \ . \tag{3.11.2}$$

Also, let $a(t)$ pass through a maximum at $t = \Delta_1, \Delta_2$. We will assume that, in the interval $[\Delta_1, t_0]$, $a(t)$ is monotonically decreasing, that is passes through a minimum at t_0 and increases monotonically over $[t_0, \Delta_2]$. The quantities t_0, Δ_1, Δ_2 must remain undetermined for the moment. This type of periodic behaviour is clearly not the most general possible. This is why it must be seen as an assumption. Intuitively, we sense that it is the type of behaviour in $a(t)$ that would be associated with simple sinusoidal behaviour, of period Δ, in the applied load.

First consider $t \in [\Delta_1, t_0]$ during which $a(t)$ is decreasing. We define $t_1(t)$ as being the time in $[t_0, \Delta_2]$ such that $a(t) = a(t_1(t))$. This function is characteristic of the shape of $a(t)$ in the interval $[\Delta_1, \Delta_2]$. In terms of it, the quantities $\theta_i(t)$ of (3.10.1) are given by

$$\theta_1(t) = t_1(t) - \Delta \ , \qquad \theta_2(t) = t - \Delta \ , \qquad \theta_3(t) = t_1(t) - 2\Delta \ , \tag{3.11.3}$$

and so on. Equation (3.11.1a) allows us to write (3.10.8) in the form

$$p(x,t) = \int_t^{t_1(t)} dt' \, \Pi_p(t,t') p(x,t') + \Pi_u(t) P(x,t) \tag{3.11.4}$$

where, from (3.10.4, 6), (2.4.18),

$$\Pi_p(t,t') = \sum_{k=1}^{\infty} T_{2k-1}(t,t'-k\Delta) \ , $$

$$\Pi_u(t) = \int_{t_1(t)-\Delta}^{t} dt' \sum_{k=0}^{\infty} T_{2k}(t,t'-k\Delta) \ . \tag{3.11.5}$$

Now consider the expanding phase, for which $t \in [t_0, \Delta_2]$. In this case, we define $t_1(t)$ by the requirement that it lie in $[\Delta_1, t_0]$ and $a(t_1(t)) = a(t)$. Then

$$\theta_1(t) = t_1(t) \ , \qquad \theta_2(t) = t - \Delta \ , \qquad \theta_3(t) = t_1(t) - \Delta \ , \tag{3.11.6}$$

and so on. The relevant relations in this case are (3.10.10) or the alternative form
(3.10.12). We choose the more compact form (3.10.10), which becomes

$$\int_{t_1(t)}^{t} dt' \, \Gamma_p(t, t') p(x, t') = [1 - \Gamma_u(t)] P(x, t) \tag{3.11.7}$$

where, from (2.4.13),

$$\Gamma_p(t, t') = \sum_{k=0}^{\infty} N_{2k}(t, t' - k\Delta) \, ,$$

$$\Gamma_u(t) = \int_{t-\Delta}^{t_1(t)} dt' \sum_{k=0}^{\infty} N_{2k+1}(t, t' - k\Delta) \, . \tag{3.11.8}$$

3.11.1 The Standard Linear Model

We now explore these relationships for the case of the standard linear solid. A
unique Poisson's ratio v will be assumed. We have, from (3.1.15), (3.2.12) and
(1.6.2p),

$$l(t) = l_0 \delta(t) + l_1 e^{-at} = h\mu(t) \tag{3.11.9}$$

$$k(t) = k_0 \delta(t) + k_1 e^{-\beta t} = \gamma(t)/h \quad \text{where}$$

$$k_0 = \frac{1}{l_0} \, , \quad k_1 = \frac{-l_1}{l_0^2} \, , \quad \beta = a - \frac{k_1}{k_0} \, , \quad h = \frac{1}{1-v} \, . \tag{3.11.10}$$

Note that k_0, l_0, k_1 are positive and l_1 is negative. Also, $\beta < a$.

Our task now is to evaluate the summations occurring in (3.11.5) and (3.11.8)
for the standard linear model. Consider first the contracting phase. From
(2.4.11) and (3.11.9), we have

$$T_0(t, t') = l_0 \delta(t - t') + l_1 \exp[-a(t - t')]$$

$$T_1(t, t') = l_1 k_0 \exp[-a(t - \theta_1) - \beta(\theta_1 - t')] \tag{3.11.11}$$

$$T_2(t, t') = l_1 \exp[-a(t - \theta_1) - \beta(\theta_1 - \theta_2) - a(\theta_2 - t')] \, ,$$

and so on. More generally, for $t \neq t'$ [to exclude the delta function in $T_0(t, t')$]:

$$T_{l+2}(t, t' - \Delta) = T_l(t, t') \exp[-\beta(\theta_l - \theta_{l+1}) - a(\theta_{l+1} - \theta_{l+2})] \, , \quad l \text{ odd}$$

$$\tag{3.11.12}$$

$$= T_l(t, t') \exp[-a(\theta_l - \theta_{l+1}) - \beta(\theta_{l+1} - \theta_{l+2})] \, , \quad l \text{ even} \, .$$

For odd l, it follows from (3.11.3) that

$$\theta_l - \theta_{l+1} = t_1(t) - t$$

$$\theta_{l+1} - \theta_{l+2} = t - t_1(t) + \Delta \, , \tag{3.11.13}$$

while for even l, these are reversed, giving

$$\theta_l - \theta_{l+1} = t - t_1(t) + \Delta \, , \quad \theta_{l+1} - \theta_{l+2} = t_1(t) - t \, . \tag{3.11.14}$$

Therefore, for all l,

$$T_{l+2}(t, t' - \varDelta) = T_l(t, t') E_c(t) ,$$

$$E_c(t) = \exp\{[t - t_1(t)](\beta - a) - a\varDelta\} .$$

(3.11.15)

The first term in the exponent of $E_c(t)$ is positive. However, this is always dominated by the second term, so that $E_c(t) < 1$. Therefore, from (3.11.5)

$$\Pi_p(t, t') = T_1(t, t' - \varDelta)(1 + E_c + E_c^2 + \ldots)$$

$$= \frac{l_1 k_0}{1 - E_c(t)} \exp\{-a[t - t_1(t) + \varDelta] + \beta[t' - t_1(t)]\}$$

(3.11.16)

and

$$\Pi_u(t) = l_0 + \frac{l_1}{a[1 - E_c(t)]} (1 - \exp\{-a[t - t_1(t) + \varDelta]\}) .$$

(3.11.17)

Similarly, for the expansion phase, we have, from (2.4.14):

$$N_0(t, t') = k_0 \delta(t - t') + k_1 \exp[-\beta(t - t')]$$

$$N_1(t, t') = k_1 l_0 \exp[-\beta(t - \theta_1) - a(\theta_1 - t')]$$

(3.11.18)

$$N_2(t, t') = k_1 \exp[-\beta(t - \theta_1) - a(\theta_1 - \theta_2) - \beta(\theta_2 - t')] .$$

and so on. More generally, as before, but now using (3.11.6), we find, for all l, excluding the delta function in $N_0(t, t')$,

$$N_{l+2}(t, t' - \varDelta) = N_l(t, t') E_e(t) , \quad \text{where}$$

(3.11.19)

$$E_e(t) = \exp\{[t - t_1(t)](a - \beta) - a\varDelta\} = E_c(t_1(t)) .$$

(3.11.20)

Let us define for $t > t_0$:

$$E(t) = E_e(t) = E_c(t_1(t)) .$$

(3.11.21)

From (3.11.8), we have

$$\Gamma_p(t, t') = k_0 \delta(t - t') + \frac{k_1 \exp[-\beta(t - t')]}{1 - E(t)}$$

$$\Gamma_u(t) = \frac{k_1 l_0}{a[1 - E(t)]} \exp\{-\beta[t - t_1(t)]\}(1 - \exp\{a[t - t_1(t) - \varDelta]\}) .$$

(3.11.22)

Based on these expressions, we write (3.11.4) and (3.11.7) more explicitly. Take $t > t_0$. Then (3.11.7) may be cast in the form

$$p(x, t) = C_1(t) \int_{t_1(t)}^{t} dt' e^{\beta t'} p(x, t') + H_1(t) P_e(x, t) ,$$

(3.11.23)

where $P_e(x, t)$ is the instantaneous elastic pressure distribution, related to $P(x, t)$ by

$$P_e(x, t) = \frac{1}{k_0} P(x, t) = l_0 P(x, t) , \quad \text{and} \quad C_1(t) = \frac{-rae^{-\beta t}}{1 - E(t)}$$

(3.11.24)

in which

$$r = \frac{k_1 l_0}{a} = \frac{J_1}{J_0 + J_1} < 1 \ . \tag{3.11.25}$$

In this last equation, parameters introduced at the beginning of Sect. 1.6 are used. These must be related to the quantities in (3.11.10) in a manner analogous to (1.6.25, 28 – 30); see (3.6.15). Relations (1.6.10 – 12) or (1.6.2p) are also relevant. Finally,

$$H_1(t) = 1 - \Gamma_u(t) = 1 - \frac{r \exp\{-\beta[t - t_1(t)]\}[1 - \varepsilon(t)]}{1 - E(t)} \quad \text{where} \tag{3.11.26}$$

$$\varepsilon(t) = \exp\{a[t - t_1(t) - \Delta]\} = E(t)\exp\{\beta[t - t_1(t)]\} \ . \tag{3.11.27}$$

We now consider (3.11.4). Let us keep t in the expansion time range $[t_0, \Delta_2]$. Then (3.11.4) may be cast in the form

$$p(x, t_1(t)) = C_2(t) \int_{t_1(t)}^{t} dt' e^{\beta t'} p(x, t') + H_2(t) P_e(x, t) \ , \tag{3.11.28}$$

where we have used the property of elastic solutions that

$$P_e(x, t_1(t)) = P_e(x, t) \ , \tag{3.11.29}$$

and where

$$C_2(t) = \frac{-a r \varepsilon(t) e^{-\beta t}}{1 - E(t)} = \varepsilon(t) C_1(t) \quad \text{and} \tag{3.11.30}$$

$$H_2(t) = 1 - \frac{r[1 - \varepsilon(t)]}{1 - E(t)} \ . \tag{3.11.31}$$

Comparison of (3.11.23) and (3.11.28) gives the following relationship between the pressure function in the expanding and contracting phases. Let $t > t_0$. Then

$$p(x, t_1(t)) = \eta(t) P_e(x, t) + \varepsilon(t) p(x, t) \ , \quad \text{where} \tag{3.11.32}$$

$$\eta(t) = H_2(t) - \varepsilon(t) H_1(t) = (1 - r)[1 - \varepsilon(t)] \ . \tag{3.11.33}$$

Equation (3.11.32) will now be used to transform (3.11.23) into a integral equation for $p(x, t)$ in the expansion phase only. The result is

$$p(x, t) = C_1(t) \int_{t_0}^{t} dt' B(t') p(x, t') + I(x, t) \quad \text{where} \tag{3.11.34}$$

$$B(t') = e^{\beta t'} - \varepsilon(t') e^{\beta t_1(t')} \dot{t}_1(t') = e^{\beta t'}[1 - E(t') \dot{t}_1(t')] \quad \text{and} \tag{3.11.35}$$

$$I(x, t) = H_1(t) P_e(x, t) - C_1(t) \int_{t_0}^{t} dt' \dot{t}_1(t') \eta(t') P_e(x, t') e^{\beta t_1(t')} \ . \tag{3.11.36}$$

The dot notation indicates differentiation with respect to the time argument. The property $t_1(t_0) = t_0$ has been used. Let us differentiate (3.11.34) with respect to time, which gives

$$\frac{d}{dt}\left(\frac{p(x,t)}{C_1(t)}\right) = B(t)p(x,t) + \frac{d}{dt}\left(\frac{I(x,t)}{C_1(t)}\right) \quad \text{or} \tag{3.11.37}$$

$$\dot{p}(x,t) + ap(x,t) = b(x,t) \ , \tag{3.11.38}$$

where the fact that

$$a = -\left(\frac{\dot{C}_1(t)}{C_1(t)} + C_1(t)B(t)\right) \tag{3.11.39}$$

has been incorporated. This result can be verified with the aid of relations such as (see Problem 1.6.2):

$$a - \beta = \frac{k_1}{k_0} = ar \ . \tag{3.11.40}$$

Also, in (3.11.38)

$$b(x,t) = C_1(t)\frac{d}{dt}\left(\frac{I(x,t)}{C_1(t)}\right) \ . \tag{3.11.41}$$

Now

$$\frac{H_1(t)}{C_1(t)} = -\frac{1}{ra}\,e^{\beta t} + \frac{1}{a}\,e^{\beta t_1(t)} + \frac{1-r}{ra}\,E(t)e^{\beta t} \ , \tag{3.11.42}$$

giving that

$$\frac{d}{dt}\left(\frac{H_1(t)}{C_1(t)}\right) - \dot{t}_1(t)\eta(t)e^{\beta t_1(t)} = -\frac{1-r}{r}\,e^{\beta t}[1 - E(t)] \ . \tag{3.11.43}$$

Therefore

$$b(x,t) = \beta P_e(x,t) + H_1(t)\dot{P}_e(x,t) \ . \tag{3.11.44}$$

The solution of (3.11.38) is

$$p(x,t) = e^{-at}\left(\int_{t_0}^{t} dt' e^{at'} b(x,t') + (1-r)e^{at_0}P_e(x,t_0)\right) \tag{3.11.45a}$$

$$= e^{-at}\left(\int_{t_0}^{t} dt'\left\{\beta e^{at'} - \frac{d}{dt'}\,[e^{at'}H_1(t')]\right\}P_e(x,t') + e^{at}H_1(t)P_e(x,t)\right) \ , \tag{3.11.45b}$$

where the initial condition has been included. This is determined from (3.11.34) and (3.11.36)

As it stands, (3.11.45) is an equation for the pressure, in terms of $a(t)$, which is contained in $P_e(x,t)$, and the function $t_1(t)$ which may be determined if $a(t)$ is known. For example, in the case of a cylindrical punch,

$$P_e(x,t) = cl_0[a^2(t) - x^2]^{1/2} \ , \tag{3.11.46}$$

by virtue of (3.10.13) and (3.11.24). However, physically the problem of interest is where the total load $W(t)$ is specified and $a(t)$ is deduced from this. A relation between $W(t)$ and $a(t)$ may be obtained by integrating (3.11.45) over the boundary, giving

$$W(t) = e^{-at} \left[\int_{t_0}^{t} dt' e^{at'} b(t') + (1-r)e^{at_0} W_e(t_0) \right] \tag{3.11.47}$$

where

$$b(t) = \beta W_e(t) + H_1(t) \dot{W}_e(t) . \tag{3.11.48}$$

In the case of a cylindrical punch [see 3.9.10)],

$$W_e(t) = \tfrac{\pi}{2} l_0 c a^2(t) . \tag{3.11.49}$$

Differentiating (3.11.47) (or integrating (3.11.38) over the boundary) gives

$$H_1(t) \dot{W}_e(t) + \beta W_e(t) = \dot{W}(t) + a W(t) \tag{3.11.50}$$

which together with the initial condition, also derived from (3.11.47),

$$W_e(t_0) = \frac{W(t_0)}{1-r} , \tag{3.11.51}$$

determines $W_e(t)$ or $a(t)$ in terms of the specified load $W(t)$ and the function $t_1(t)$. This latter dependence is the complicating feature since $t_1(t)$ depends on $a(t)$. However, we have a further constraint, namely (3.11.32). In integrated form, this reads

$$W(t_1(t)) = \eta(t) W_e(t) + \varepsilon(t) W(t) , \tag{3.11.52}$$

which, combined with (3.11.50, 51), should determine $W_e(t)$, $t_1(t)$ in terms of $W(t)$. Once these quantities are determined, $p(x, t)$ is given by (3.11.45).

Let us solve (3.11.50), defining

$$R(t) = \exp \left(\beta \int_{t_0}^{t} \frac{dt'}{H_1(t')} \right)$$

$$g(t) = \frac{1}{H_1(t)} [\dot{W}(t) + a W(t)] , \tag{3.11.53}$$

in terms of which

$$W_e(t) = \frac{1}{R(t)} \left(\int_{t_0}^{t} dt' R(t') g(t') + W_e(t_0) \right) , \tag{3.11.54}$$

which, when substituted in (3.11.52), gives an equation determining $t_1(t)$, because the form of $W(t)$ is given. Let us first discuss the determination of t_0, Δ_1 and Δ_2. At these points, $\dot{W}_e$ will vanish. Since $t_1(t_0) = t_0$, (3.11.50), together with (3.11.51) and (3.11.40) give that $\dot{W}(t_0)$ also vanishes. In other words, the minimum point of the applied load coincides with the minimum point of the contact interval. The points Δ_1, Δ_2 are characterized by the fact that

$\Delta_1 = t_1(\Delta_2)$ together with (3.11.2). It follows that $\varepsilon(\Delta_2) = 1$ and $\eta(\Delta_2)$ vanishes so that (3.11.52) gives simply the periodicity condition. We must seek elsewhere for information on the form of $W_e(\Delta_2)$. Consider (3.11.23) at $t = \Delta_2$. This together with (3.11.26) gives that

$$P_e(x, \Delta_2) = p(x, \Delta_2) - C_1(\Delta_2) \int_{\Delta_1}^{\Delta_2} dt' e^{\beta t'} p(x, t') = \frac{1}{k_0} \int_{-\infty}^{\Delta_2} dt' k(\Delta_2 - t') p(x, t') \ .$$

$$(3.11.55)$$

This first form is the one of immediate practical interest here. However, the second form is worth noting also. It may be derived by expanding the term $1 - E(\Delta_2)$ in the denominator of $C_1(\Delta_2)$, where we note that $E(\Delta_2) = \exp(-\beta \Delta)$. Integrating the first form over x gives

$$W_e(\Delta_2) = W(\Delta_2) - C_1(\Delta_2) \int_{\Delta_1}^{\Delta_2} dt' e^{\beta t'} W(t') \ . \tag{3.11.56}$$

Also, $\dot{W}_e(\Delta_2)$ vanishes so that we deduce from (3.11.50) the condition determining Δ_2

$$(\beta - a) W(\Delta_2) - \beta C_1(\Delta_2) \int_{\Delta_1}^{\Delta_2} dt' e^{\beta t'} W(t') = \dot{W}(\Delta_2) \ . \tag{3.11.57}$$

Equations (3.11.51, 55) are special cases of results shown to be valid for a periodic loading of general viscoelastic bodies by Graham and Golden (1988).

Let us now explore these equations for the case where $W(t)$ has the simple sinusoidal form

$$W(t) = K[d - \cos(\omega t)] \ , \quad d \geq 1 \ , \quad \Delta = 2\pi/\omega \ , \tag{3.11.58}$$

where K is a positive constant. Its minimum occurs at the origin so that

$$t_0 = 0 \ . \tag{3.11.59}$$

Condition (3.11.57) becomes

$$\tan(\omega \Delta_2) = \frac{\omega \beta r}{\beta^2 + \omega^2(1 - r)} = \tan \phi(\omega) \ , \tag{3.11.60}$$

where $\phi(\omega)$ is the loss angle defined by (1.6.7) for a standard linear solid. It is the solution of this equation in the region $[\pi, 3\pi/2]$ that gives $\omega \Delta_2$, so that $\omega \Delta_2 = \pi + \phi(\omega)$. This result has been obtained in another way by Graham and Golden (1988).

Equation (3.11.52) may be solved iteratively, with $W_e(t)$ given by (3.11.54), to obtain $t_1(t)$. For small values of r, the iteration converges rapidly from the initial guess $t_1(t) = (1 - 2\pi/\omega \Delta_2)t$, while for large r, it is slower. Some numerical examples are presented on Figs. 3.3, 4. In Fig. 3.3, the behaviour of the semi-contact interval $a(t)$, or more correctly, the dimensionless combination $a(t)c$ ($c = 1/R$, where R is the radius of curvature), is plotted for several values of r. In Fig. 3.4, the pressure distribution, in units of l_0, is plotted against the

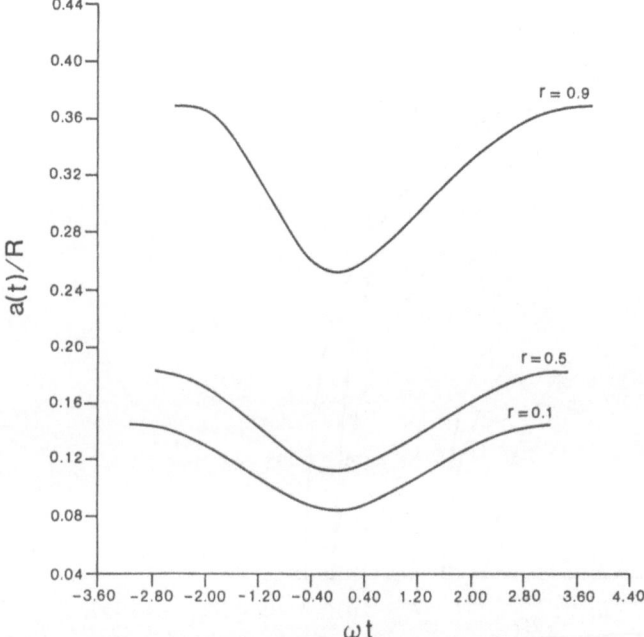

Fig. 3.3. The behaviour of the dimensionless semi-contact interval over a complete period, for three values of r and $a/\omega = 1$, $d = 2$, $K/(l_0 R) = 0.01$.

dimensionless position variable xc, at various times, for $r = 0.5$. For these curves, a/ω is unity, $d = 2$ and cK/l_0 is assigned a value 0.01.

From Fig. 3.3, we see that contact area increases with r and, as might be anticipated, the asymmetry between positive and negative times also increases. Fig. 3.4 indicates that for positive times, when the contact area is increasing, the pressure distribution tends away from the familiar elliptical shape towards a more uniform shape with a hump near $|x| = a(t)$, and a sharp decline beyond that. This effect is more marked for larger values of r; see Golden and Graham (1987b) for more detailed data and further discussion.

3.12 Summary

Contact problems on a viscoelastic half-space under plane strain conditions are considered in this chapter.

I. Method of Solution. The method of solution is based on the viscoelastic Kolosov-Muskhelishvili equations, adapted to a half-space. Explicit solutions to the first and second boundary value problems are presented in detail. In these cases no restrictions on material behaviour are necessary. In the case of mixed boundary value problems where surface friction is present, it is necessary to make the proportionality assumption. Limiting frictional contact problems are

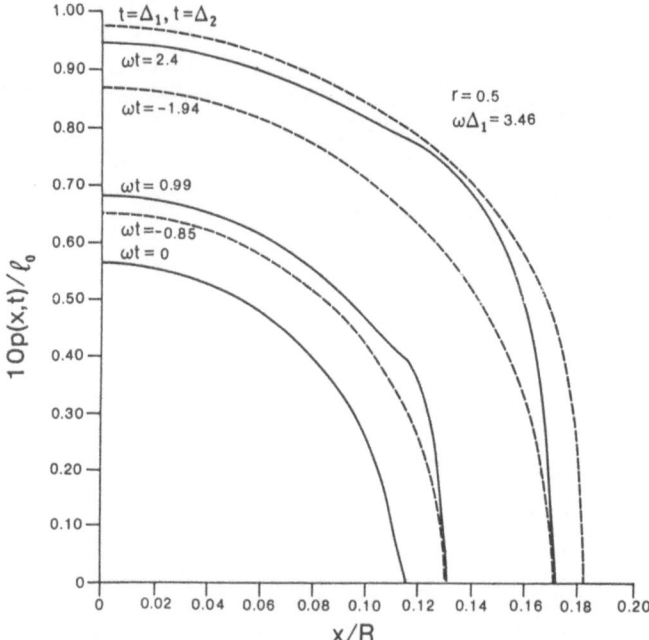

Fig. 3.4. The dimensionless pressure distribution at various times during the cycle, for $r = 0.5$. The broken lines are for negative times, when the contact area is decreasing, and for times of maximum contact area. The pairs of lines which meet on the horizontal axis, one broken, one continuous, are for times related through the function $t_1(t)$. All these curves refer to the case when $a/\omega = 1$, $d = 2$ and $K/l_0 R = 0.01$

considered and also the interesting special case of frictionless contact, for which the proportionality assumption is not necessary.

II. Mixed Boundary Value Problems. The viscoelastic Kolosov-Muskhelishvili equations reduce to a Hilbert problem, the solution of which gives all quantities of interest in terms of $v(x, t)$, defined by (3.3.3) for frictional contact or with $l(t)$ given by the more general relation (3.1.15) for lubricated contact. Use of the decomposition of hereditary integrals, developed in Chap. 2, enables one to write down the integral equation [(3.5.5) for $v(x, t)$], which is applicable, in principle, to a variety of problems and specifically to tangentially moving contact problems involving one or more indentors.

III. Single Indentor. Considerable progress can be made in solving the problem of a single indentor in motion across the surface. When the velocity is uniform and steady-state conditions have been established, the integral equation and subsidiary constraints – including one determining total applied load – give a system of implicit equations for $v(x, t)$ and the contact interval. The pressure may then be determined without difficulty. For a discrete spectrum model and a cylinder in frictionless contact, these are a set of algebraic equations given by (3.6.30 – 34); and for a standard linear model they reduce to (3.6.35 – 39).

If the viscoelastic effects are assumed to be small, then these equations are explicitly solvable to the same order and all quantities can be determined explicitly [Sect. 3.7, for example (3.7.27)].

IV. Hysteretic Friction. The hysteretic friction coefficient due to a moving indentor in lubricated contact with the half-plane is given by (3.8.1); and in the frictional case, by (3.8.9) or alternatively by (3.8.10) or (3.8.11). On an incompressible half-plane, it reduces to (3.8.12) which has the same form as for lubricated contact, apart from a factor $(1 + f^2)$. These expressions cannot be evaluated until the implicit equations governing the problem are solved, since the pressure is required.

In the small viscoelasticity approximation, however, the entirely explicit formula (3.8.15), or for standard linear solid, (3.8.16), can be given for lubricated contact. The expression for a more general spectrum model is a sum of terms of the form given by (3.8.16).

V. Plane Normal Contact. If no tangential motion is present but the load is varying over time, causing an increasing and decreasing contact region, the problem may be solved without recourse to the general integral equation (3.5.5), but instead by the methods outlined in Sect. 2.6. The special cases of non-expanding and non-contacting contact regions are discussed in general terms [see (3.9.1 – 4) and earlier results referenced]. For the case of one punch, explicit formulae are given, in the non-expanding case, by (3.9.6 – 10) and in the non-contracting case by (3.9.11 – 16).

VI. Alternative Expansion and Contraction. The general problem is discussed in Sect. 3.10 and the steady-state limit in Sect. 3.11. The detailed solution for a standard linear solid and a sinusoidal load is contained in (3.11.45 – 60). Numerical solutions may be obtained without difficulty.

4. Plane Non-inertial Crack Problems

Some simple fracture mechanics problems for viscoelastic media are considered in this chapter. Much of the emphasis will be on qualitative differences in behaviour from corresponding elastic configurations, in other words specifically viscoelastic phenomena. We refer to standard texts on fracture mechanics for a treatment of the elastic problems; for example, Liebowitz (1968), Sneddon and Lowengrub (1969), Lardner (1974) and Cherepanov (1979).

There are two alternative approaches to crack problems in elastic or viscoelastic media, which might be termed the microscopic and the macroscopic approach, respectively. The microscopic approach is based on the dislocation concept [Lardner (1974), for example] while the macroscopic approach views a crack simply as a cavity in the medium. Both are finally equivalent, of course.

The microscopic approach has the advantage of being physically intuitive and also, in some respects, mathematically convenient in that integral equations can often be written down with a minimum of theoretical development. However, the dislocation concept at times appears unnecessary and burdensome. Also, with this approach, it is difficult to bring into play the full power and elegance of the mathematical methods, based on complex variable techniques which have been developed for plane elastic (and easily adaptable to viscoelastic) problems.

We therefore adopt the second approach, omitting the introduction and use of the dislocation concept.

Since this chapter deals only with the non-inertial approximation, problems of unstable crack propagation are not considered here. Such topics are dealt with, briefly, in Chap. 7. We do however deal with growth initiation criteria in Sect. 4.6, 7. Growing cracks are dealt with, but only in the non-inertial approximation. This would correspond to stable growth, due to some kind of fatigue mechanism perhaps, or to large internal frictional damping effects. Fatigue is not discussed specifically. We note however work on fatigue crack growth [Golden and Graham (1984)], where further references may be found.

One feature of interest in crack problems is that the exact nature of the boundary conditions is not always a priori obvious, but has to be determined in the course of solving the problem. This is particularly true for viscoelastic media.

4.1 Problem Formulation

We confine our attention to the case where the cracks lie along a single line in an infinite viscoelastic medium. This line is taken to be the x-axis. The problem of main interest is where stresses are applied at large distances in the y direction from the origin. These include, in general, both tensile (or compressive) and shear stresses. They may depend upon time.

Consider the stress distribution in such a problem. Let us subtract from this the applied stresses at infinity, thus giving the distribution for the problem where the stresses tend to zero at infinity and have uniform applied stresses, of the same magnitude and opposite sign, on the open crack faces. It is clear that this distribution also obeys the dynamical equation (1.8.17), since the applied stresses, which are independent of position, contribute nothing. This latter problem is more convenient in the context of the methodology introduced in the present chapter, and will be adopted as the statement of the problem. The stress distribution of final interest may be deduced trivially, once the problem is solved.

Problems involving a crack in a field of bending will also be considered, where the neutral axis is the y-axis. For such problems, the stress is zero on the open crack face and linear in x along the x-axis, at large distances from the origin. By subtracting this linear term, we obtain a linear stress distribution on the open crack face and no divergent stresses at infinity. Also, the equilibrium equations still hold.

The method which we shall use here in fact applies to quite general distributions of stress over the open crack face, so there is no difficulty about the space dependence of the stresses in the case of a crack in a field of bending. In the development of the methodology, no assumptions will be made concerning the stress distribution on the crack face. This generality is not entirely spurious in that it will be useful for theoretical reasons in Sect. 4.7.

Our approach is based on the viscoelastic Kolosov-Muskhelishvili equations, as in the case of contact problems. Consider the general form, given by (2.8.9). These relations must hold over the entire complex plane, where $\phi(z, t)$, $\psi(z, t)$ are analytic everywhere, except on some or all of the real axis, and go to zero at infinity as $1/z^2$, according to (2.8.12).

We recall that, in Sect. 3.1, $\psi(z, t)$ was eliminated by requiring essentially that the discontinuity in $\phi(z, t)$ over the real axis be zero where the stresses are zero. A somewhat different tactic is employed in the present case. First, let us state our boundary conditions explicitly. All stresses are zero at infinity. Off the crack face, the displacements are continuous everywhere, in particular along the x-axis. Let the region of the x-axis, on the crack face, be $F(t)$, made up of $0(t)$ and $C(t)$, two disjoint sets, $0(t)$ being the region on which the crack face is open and $C(t)$ being the region on which it is closed. On $0(t)$, we have

$$u_2(x, 0^+, t) - u_2(x, 0^-, t) > 0$$

$$\sigma_{22}(x, 0^+, t) = \sigma_{22}(x, 0^-, t) = -p(x, t) \tag{4.1.1}$$

$$\sigma_{12}(x, 0^+, t) = \sigma_{12}(x, 0^-, t) = -s(x, t) ,$$

where $0^\pm$ denotes y approaching zero from above ($+$) and below ($-$). The quantities $p(x, t)$, $s(x, t)$ are the specified pressure and shear on the crack face. On $C(t)$, frictional forces between the faces will be neglected. We have therefore

$$u_2(x, 0^+, t) - u_2(x, 0^-, t) = 0$$

$$\sigma_{22}(x, 0^+, t) = \sigma_{22}(x, 0^-, t) < -p(x, t) \qquad (4.1.2)$$

$$\sigma_{12}(x, 0^+, t) = \sigma_{12}(x, 0^-, t) = -s(x, t) \ .$$

In the elastic problem, if $p(x, t)$ has the same sign everywhere, it is usually positive. Tensile stresses on the crack face would simply cause instant closure. In the viscoelastic case, closure would not always be instant, and applied stresses which change sign over time are of some interest.

Note that the stresses are continuous at every point on the x-axis, even across the open crack. From (2.8.9), this has the consequence that

$$\phi^+(x, t) + \bar{\phi}^-(x, t) + x\bar{\phi}^{-\prime}(x, t) + \bar{\psi}^-(x, t)$$
$$= \phi^-(x, t) + \bar{\phi}^+(x, t) + x\bar{\phi}^{+\prime}(x, t) + \bar{\psi}^+(x, t) \qquad (4.1.3)$$

at every point on the x-axis, where $\phi^\pm(x, t)$ are the limits of $\phi(z, t)$ from above and below the real axis. We write (4.1.3) as

$$\phi^+(x, t) - \bar{\phi}^+(x, t) - x\bar{\phi}^{+\prime}(x, t) - \bar{\psi}^+(x, t)$$
$$= \phi^-(x, t) - \bar{\phi}^-(x, t) - x\bar{\phi}^{-\prime}(x, t) - \bar{\psi}^-(x, t) \ , \qquad (4.1.4)$$

so that the function $\{\phi(z, t) - \bar{\phi}(z, t) - z\bar{\phi}'(z, t) - \bar{\psi}(z, t)\}$ is analytic everywhere. It is also zero at infinity, by virtue of (2.8.12). It follows from Liouville's Theorem (Sect. A2.1) that it is zero everywhere so that [c.f. (3.1.2)]:

$$\bar{\psi}(\bar{z}, t) = \phi(\bar{z}, t) - \bar{\phi}(\bar{z}, t) - \bar{z}\bar{\phi}'(\bar{z}, t) \ , \qquad (4.1.5)$$

and we write (2.8.9) as

$$\sigma_{11} + \sigma_{22} = 2[\bar{\phi}(\bar{z}, t) + \phi(z, t)] = 4 \operatorname{Re}\{\phi(z, t)\} \qquad (4.1.6a)$$

$$\sigma_{22} - i\sigma_{12} = \phi(z, t) + \phi(\bar{z}, t) + (z - \bar{z})\bar{\phi}'(\bar{z}, t) \qquad (4.1.6b)$$

$$2d(r, t) = [\kappa * \phi](z, t) - \phi(\bar{z}, t) - (z - \bar{z})\bar{\phi}'(\bar{z}, t) \quad \text{where} \qquad (4.1.6c)$$

$$d(r, t) = \int_{-\infty}^{t} dt' \mu(t - t') D'(r, t') \qquad (4.1.7)$$

and the convolution notation (1.2.32) has been used. Approaching the x-axis from above and below in (4.1.6c), and subtracting gives

$$2[d(x, 0^+, t) - d(x, 0^-, t)] = (1 + \kappa) * [\phi^+(x, t) - \phi^-(x, t)] \quad \text{or} \qquad (4.1.8)$$

$$\frac{1}{2} \int_{-\infty}^{t} dt' l(t - t') \Delta'(x, t') = \phi^+(x, t) - \phi^-(x, t) \ , \qquad (4.1.9)$$

where $l(t)$ is defined by (3.1.15) and $\Delta'(x, t)$ is the x derivative of the complex displacement across the gap, given by

$$\Delta'(x, t) = D'(x, 0^+, t) - D'(x, 0^-, t) \ . \qquad (4.1.10)$$

This difference is of more immediate physical interest than the $D'(x, 0^\pm, t)$ individually. It is interesting to recall, in this context, the discussion in Sect. 3.1, leading to (3.1.4). From (4.1.9) we see that in the present case, it is possible to remove the dependence on $\kappa(t)$ from the right-hand side without any special assumptions.

If the left-hand side of (4.1.9) is zero at a given value of x, then $\phi(z, t)$ is continuous over the x-axis at that point. This will be true if $\Delta'(x, t')$ is zero for all $t' \leq t$, in particular off the crack face.

The region of the x-axis off the crack face will sometimes be denoted by $F'(t)$, the complement of $F(t)$ on the x-axis.

Some standard terminology will occasionally be used. If the stresses on the crack face are purely normal, the crack is said to be subject to opening mode or Mode I displacement, or it is simply referred to as a Mode I crack. If the stresses are purely shear, the crack is subject to sliding mode or Mode II displacement, while if the stresses are perpendicular to the plane, we have tearing mode or Mode III displacement [Irwin (1960), Sih and Liebowitz (1968), Sneddon and Lowengrub (1969) for example]. In this Chapter, we consider mainly Mode I displacement and, to a certain extent, Mode II. Tearing mode cracks, which are typically the simplest to analyze, are considered briefly in Chap. 7, in the context of inertial problems.

4.2 Fully Open Cracks that are Stationary or Growing

In this case, $C(t)$ is empty and $F(t)$ is expanding or stationary. This is an example of the type of problem covered by the Extended Correspondence Principle discussed in Sect. 2.6 so that we expect to obtain solutions closely related to the corresponding elastic solutions. From (4.1.6b) we have that

$$\Sigma(x, t) = -p(x, t) + \mathrm{i}s(x, t) = \phi^+(x, t) + \phi^-(x, t) \ , \qquad x \in F(t) \ . \tag{4.2.1}$$

Also, for $x \in F'(t)$, off the crack face, the left-hand side of (4.1.9) is known to be zero for all t, since if $x \in F'(t)$, it follows that $x \in F'(t')$, $t' \leq t$ and the complex displacement $\Delta(x, t')$ is zero for all $t' \leq t$. Therefore, the function $\phi(z, t)$ is continuous on $F'(t)$. This condition, together with (4.2.1), constitutes a Hilbert problem as specified by (A2.3.2) for $\eta = -1$, the solution of which is, from (A2.3.22):

$$\phi(z, t) = \frac{X(z, t)}{2\pi \mathrm{i}} \int_{F(t)} dx' \ \frac{\Sigma(x', t)}{X^+(x', t)(x' - z)} + P(z, t)X(z, t) \ , \tag{4.2.2}$$

where $P(z, t)$ is a polynomial, as yet undetermined, and $X(z, t)$ is given by

$$X(z, t) = \left\{ \prod_{i=1}^{n} [z - a_i(t)][z - b_i(t)] \right\}^{-1/2} , \tag{4.2.3}$$

the union of the intervals $[a_i(t), b_i(t)]$ being $F(t)$. We choose the branch of $X(z, t)$ such that $z^n X(z, t) \to 1$ as $|z| \to \infty$. Note that we have allowed singularities at the end points. This is because it is not clear a priori whether they can be excluded.

We will now focus on the case of a single crack, since this is the only configuration which will be discussed in detail in later sections. In this case, $n = 1$

and the correct behaviour at infinity is obtained by choosing $P(z, t) = 0$. Note that for $n > 1$, this is not true. Let us put

$$X(z, t) = \{[z - a(t)] [z - b(t)]\}^{-1/2} . \tag{4.2.4}$$

On the real axis, for $x \in [a(t), b(t)]$:

$$X^+(x, t) = -X^-(x, t) = \frac{1}{im(x, t)} \tag{4.2.5}$$

$$m(x, t) = \{[x - a(t)] [b(t) - x]\}^{1/2} ,$$

and for $x \notin [a(t), b(t)]$

$$X(x, t) = \pm \frac{1}{n(x, t)} , \quad n(x, t) = |[x - a(t)] [x - b(t)]|^{1/2} . \tag{4.2.6}$$

It is positive for $x > b(t)$ and negative for $x < a(t)$, as a consequence of choosing the branch of $X(z, t)$ as specified above. We write $\phi(z, t)$ as

$$\phi(z, t) = \frac{X(z, t)}{2\pi} \int_{F(t)} dx' \frac{\Sigma(x', t) m(x', t)}{x' - z} . \tag{4.2.7}$$

Note that it is singular at the end points of the cracks. This implies that the stresses will also be singular there. Consider now the expression for the derivative of the gap. From (4.2.5, 7) and the Plemelj formula (A2.2.10):

$$\phi^+(x, t) - \phi^-(x, t) = \frac{1}{\pi im(x, t)} \int_{a(t)}^{b(t)} dx' \frac{\Sigma(x', t) m(x', t)}{x' - x} , \tag{4.2.8}$$

which is a principal value integral. Using (4.1.9), we obtain

$$\Delta'(x, t) = \frac{2}{\pi i} \int_{t_1(x)}^{t} dt' k(t - t') \frac{1}{m(x, t')} \int_{a(t')}^{b(t')} dx' \frac{\Sigma(x', t') m(x', t')}{x' - x} \tag{4.2.9}$$

where $k(t)$ is defined by (3.2.12) and $t_1(x)$ is the time at which x crossed the crack tip. Integrating (4.2.9) from either crack tip and making a change in integration orders we find, using (A1.2.1), that

$$\Delta(x, t) = \frac{2i}{\pi} \int_{t_1(x)}^{t} dt' k(t - t') \int_{a(t')}^{b(t')} dy' \Sigma(y', t') M(x, y'; t') , \tag{4.2.10}$$

where

$$M(x, y'; t') = m(y', t') \int_{x}^{b(t')} \frac{dy}{(y' - y) m(y, t')} . \tag{4.2.11}$$

A direct calculation, with the aid of (A1.2.8), confirms that

$$M(x, y'; t') = \log_e \left| \frac{\sqrt{y' - a(t')} \sqrt{b(t') - x} - \sqrt{b(t') - y'} \sqrt{x - a(t')}}{\sqrt{y' - a(t')} \sqrt{b(t') - x} + \sqrt{b(t') - y'} \sqrt{x - a(t')}} \right|$$

$$= M(y', x; t') . \tag{4.2.12}$$

Equations (4.2.11, 12) imply that

$$M(a(t'),x;t') = M(b(t'),x;t') = M(x,a(t');t') = M(x,b(t');t') = 0 , \quad (4.2.13)$$

while, by differentiating (4.2.12) with respect to time, we find that

$$\dot{M}(x,y';t) = \frac{1}{b-a} \left(\dot{a} \sqrt{\frac{b-x}{x-a}} \sqrt{\frac{b-y'}{y'-a}} - \dot{b} \sqrt{\frac{x-a}{b-x}} \sqrt{\frac{y'-a}{b-y'}} \right) . \quad (4.2.14)$$

Equations (4.2.13, 14) are used in Sect. 4.6. The complex stress $\Sigma(r,t)$, off the crack surface, is given by (4.1.6b). On the real axis, this has the form $\phi^+(x,t) + \phi^-(x,t)$. From (4.2.6, 7) and the Plemelj formula (A2.2.10), we obtain

$$\Sigma(x,t) = \pm \frac{1}{\pi n(x,t)} \int_{a(t)}^{b(t)} dx' \frac{\Sigma(x',t)m(x',t)}{x'-x} , \quad x \notin [a(t),b(t)] , \quad (4.2.15)$$

the upper sign referring to $x > b(t)$. This is the same as in elastic theory. Note that it is independent of the material parameters. It is easily checked that this is true everywhere, which, it will be perceived, is a consequence of Michell's Theorem (1899), mentioned in Sect. 2.10. The complex stress intensities are given by

$$K_1(b) - iK_2(b) = \lim_{x \to b} \{[2(x-b)]^{1/2} \Sigma(x,t)\}$$
$$= -\frac{1}{\pi \sqrt{c(t)}} \int_{a(t)}^{b(t)} dx' \Sigma(x',t) \left(\frac{x'-a(t)}{b(t)-x'} \right)^{1/2} \quad (4.2.16)$$

and

$$K_1(a) - iK_2(a) = -\frac{1}{\pi \sqrt{c(t)}} \int_{a(t)}^{b(t)} dx' \Sigma(x',t) \left(\frac{b(t)-x'}{x'-a(t)} \right)^{1/2} \quad (4.2.17)$$

where

$$c(t) = \frac{b(t)-a(t)}{2} . \quad (4.2.18)$$

Problem 4.2.1: If the middle point of the crack face is chosen as the origin and $\Sigma(x,t)$ is even in x over the crack face, show that (4.2.16) and (4.2.17) give the same result, namely, choosing $b(t) = -a(t) = c(t)$,

$$K_1 - iK_2 = -\frac{\sqrt{c(t)}}{\pi} \int_{-c(t)}^{c(t)} dx' \frac{\Sigma(x',t)}{(c^2(t)-x'^2)^{1/2}} . \quad (4.2.1p)$$

The most interesting special case is where $\Sigma(x,t)$ is independent of x on the crack face. We write it as $\Sigma(t)$. Then, (4.2.9) gives, with the aid of integral (A1.2.2),

$$\Delta'(x,t) = -2i \int_{t_1(x)}^{t} dt' \frac{k(t-t')}{m(x,t')} \left(\frac{b(t')+a(t')}{2} - x \right) \Sigma(t') , \quad (4.2.19)$$

which can be integrated to give

$$\Delta(x,t) = -2\mathrm{i} \int\limits_{t_1(x)}^{t} dt' k(t-t') \Sigma(t') m(x,t') \ , \tag{4.2.20}$$

where $t_1(x)$ is the time at which the crack reaches the point x. The complex stress $\Sigma(x,t)$, given by (4.2.15), becomes [integral (A1.2.2)]:

$$\Sigma(x,t) = \Sigma(t) \left(1 \pm \frac{\dfrac{b(t)+a(t)}{2} - x}{n(x,t)} \right) \ , \qquad x \in [a(t), b(t)] \ . \tag{4.2.21}$$

One deduces that the complex stress intensities are given by

$$K_1(b) - \mathrm{i} K_2(b) = K_1(a) - \mathrm{i} K_2(a) = -c^{1/2}(t)\Sigma(t) \ . \tag{4.2.22}$$

In particular, for purely normal stresses acting on the crack face, $\Sigma(t) = -p(t)$ and

$$K_1(a) = K_1(b) = c^{1/2}(t)p(t) \ . \tag{4.2.23}$$

We will show in Sect. 4.4 that a stationary crack may be open even when $p(t) < 0$, which does not happen in elastic theory. It follows that in viscoelastic theory, the stress intensity factor K_1 may be zero or negative. In elastic theory, it may be zero but not negative.

The special case where $a(t) = -b(t)$ may be extracted trivially from these formulae.

The results presented in this section were first given in a special case by Kachanov (1961) and more generally by Graham (1970), using an alternative form of the elastic results, namely that described by Sneddon and Lowengrub (1969).

4.3 Monotonically Closing Cracks

In the previous section, we explored the case of expanding or fixed length cracks which, in terms of the discussion of Sect. 2.6, corresponds to $B_u(t)$ contracting or stationary. It is of interest to ask whether $B_\sigma(t)$ can contract monotonically, because if so, it should be possible also to write down closed from solutions in this case. The phenomenon of crack rehealing will not be considered, so the only way that this can happen is by crack closure, without adhesion.

This leads us to consider partially closed cracks and the limiting case where the crack is completely closed. A partially closed crack will be defined as one such that the stress intensity factor vanishes at at least one end of the open region.

The discussion will be confined to the case where shear is absent (Mode I) so that displacements will be continuous on the closed portions of the crack face. Let the open region be $[a_1(t), b_1(t)]$, which is contracting. We will look for solutions with the property that no singularity occurs in the stress at points $a_1(t)$, $b_1(t)$. This is the property, that distinguishes the partially closed crack from the fully open crack, which, as we found in Sect. 4.2, has singularities at the end points. It is more convenient in this case to use the alternative form of the generalized Kolosov-Muskhelishvili equations given at the end of Sect. 2.8, with the hereditary integrals thrown onto the stresses. Let us define $\Lambda(z, t)$ by

$$\Lambda(z, t) = \frac{1}{4} \int_{-\infty}^{t} dt' [\delta(t-t') + \kappa(t-t')] \chi(z, t') = \int_{-\infty}^{t} dt' k(t-t') \phi(z, t') ,$$

(4.3.1)

where $k(t)$ is defined by (3.2.12). Equations (4.1.6b, 9), adapted to the present problem, give, on the x-axis,

$$-q(x, t) = \Lambda^{+}(x, t) + \Lambda^{-}(x, t) , \quad \tfrac{1}{2} i \Delta_2'(x, t) = \Lambda^{+}(x, t) - \Lambda^{-}(x, t) \quad (4.3.2)$$

where $q(x, t)$ is related to $p(x, t)$ by (3.2.11 b) and $\Delta_2(x, t)$ is the normal gap, defined by

$$\Delta_2'(x, t) = u_2'(x^{+}, t) - u_2'(x^{-}, t) , \quad u_2'(x^{\pm}, t) = u_2'(x, 0^{\pm}, t) \quad (4.3.3)$$

where $u_2'(x, 0^{\pm}, t)$, is the normal displacement on the x-axis. Since $q(x, t)$ in (4.3.2) is known for all $x \in [a_1(t), b_1(t)]$, this is a well-defined Hilbert problem, with solution similar to (4.2.7) and given by

$$\Lambda(z, t) = -\frac{X(z, t)}{2\pi} \int_{a_1(t)}^{b_1(t)} dx' \frac{q(x', t) m(x', t)}{x' - z} .$$

(4.3.4)

Instead of (4.2.9) we have

$$\Delta_2'(x, t) = \frac{2}{\pi m(x, t)} \int_{a_1(t)}^{b_1(t)} dx' \frac{q(x', t) m(x', t)}{x' - x} ,$$

(4.3.5)

and (4.2.15) is replaced by

$$\sigma_{22}(x, 0, t) = \mp \int_{t_1(x)}^{t} dt' l(t-t') \frac{1}{\pi n(x, t')} \int_{a_1(t')}^{b_1(t')} dx' \frac{q(x', t') m(x', t')}{x' - x}$$

$$+ \int_{-\infty}^{t_1(x)} dt' l(t-t') q(x, t') , \quad x \notin [a_1(t), b_1(t)] ,$$

(4.3.6)

where $t_1(x)$ is the time at which x leaves the open portion of the crack and $m(x', t')$, $n(x, t')$ are given by (4.2.5, 6) with $a_1(t)$, $b_1(t)$ replacing $a(t)$, $b(t)$, respectively; we write this as

$$\sigma_{22}(x, 0, t) = \mp \left(\frac{l_0}{\pi n(x, t)} \int_{a_1(t)}^{b_1(t)} dx' \frac{q(x', t) m(x', t)}{x' - x} \right.$$

$$+ \int_{t_1(x)}^{t} dt' \frac{l_1(t-t')}{\pi n(x, t')} \int_{a_1(t')}^{b_1(t')} dx' \frac{q(x', t') m(x', t')}{x' - x} \right)$$

$$+ \int_{-\infty}^{t_1(x)} dt' l_1(t-t') q(x, t')$$

(4.3.7)

where $l_1(t)$ is $l(t)$ minus its singular portion $l_0\delta(t)$. The first term is singular unless

$$\int_{a_1(t)}^{b_1(t)} dx\, q(x,t) \left(\frac{x-a_1(t)}{b_1(t)-x}\right)^{1/2} = \int_{a_1(t)}^{b_1(t)} dx\, q(x,t) \left(\frac{b_1(t)-x}{x-a_1(t)}\right)^{1/2} = 0 \ . \qquad (4.3.8)$$

These conditions cannot hold unless $q(x,t)$ changes sign on the open crack face. In the important case where $p(x,t)$ is independent of x, they certainly cannot hold. However, in the case of a crack in a field of bending where $p(x,t)$ is linear in x, it is a different matter, as we shall see in Sect. 4.8.

We now ask whether, if (4.3.8) cannot be enforced, it is inevitable that the stress have a singularity at the end points $a_1(t)$, $b_1(t)$. Specifically, is it possible that the second term could cancel the singularity, by developing a singularity itself at these end points? Let us generalize the question and ask under what circumstances this term can have a singularity. For definiteness, let $x > b_1(t)$. We first observe that if $b_1(t)$ is monotonically decreasing, and is never constant, then the second term will have no singularity. We can change the time variable in (4.3.7) to $b_1(t')$ under this monotonicity assumption, and the square root singularity is integrated away. However, if $b_1(t') = b_0$ for $\theta_1 < t' < \theta_2$, then the second term in (4.3.7) has a portion singular at b_0, of the form

$$\frac{1}{\pi(x-b_0)^{1/2}} \int_{\theta_1}^{\theta_2} dt'\, \frac{l_1(t-t')}{[x-a_1(t')]^{1/2}} \int_{a_1(t')}^{b_0} dx'\, \frac{q(x',t')m(x',t')}{x'-x} \ . \qquad (4.3.9)$$

If $a_1(t')$ is also a constant, equal to a_0, then we get a term similar to the first part of (4.3.7) but with $[a_0, b_0]$ replacing $[a_1(t), b_1(t)]$ and

$$\int_{\theta_1}^{\theta_2} dt'\, l_1(t-t')q(x',t') \qquad (4.3.10)$$

replacing $l_0 q(x',t)$. If the crack was stationary at $[a_1(t), b_1(t)]$ for a period prior to t, in other words, if $\theta_2 = t$, then the second term in (4.3.7) will have singularities at these end points. However, cancellation will occur only if equations (4.3.8) are satisfied when $q(x,t)$ is replaced by

$$\int_{\theta_1}^{t} dt'\, l(t-t')q(x,t') \ . \qquad (4.3.11)$$

This cannot happen if the quantity defined by (4.3.11) does not change sign over the region of integration. We conclude that in particular for $p(x,t) = p(t)$, the crack may only be totally open or totally closed.

These observations apply to a history of monotonic closure. The question whether partial closure could occur after an arbitrary history will be considered in Sect. 4.5.

4.4 Stationary Cracks

The problem of a stationary crack which has always been open is relatively trivial. Since it is covered by the Classical Correspondence Principle, we expect that its solution is closely related to the elastic solution. This may be perceived easily in (4.2.9, 15). The latter has the same form for fixed and expanding cracks. In (4.2.9), we can bring the time integration inside the space integration and apply it to the quantity $\Sigma(x, t')$ which will be the only time-dependent factor. However, the problem is not so trivial if crack closure can occur. Let us take

$$p(x, t) = p(t) , \quad b(t) = -a(t) = c , \tag{4.4.1}$$

where c is a constant. Also shear stresses on the crack face will be neglected. Equations (4.2.20) and (4.2.21) give

$$\Delta_2(x, t) = 2m(x) \int_{-\infty}^{t} dt' k(t-t')p(t') , \quad |x| < c$$

$$\sigma_{22}(x, 0, t) = -p(t) \left(1 - \frac{|x|}{n(x)}\right), \quad |x| > c \tag{4.4.2}$$

where

$$m(x) = (c^2 - x^2)^{1/2} , \quad n(x) = (x^2 - c^2)^{1/2} , \tag{4.4.3}$$

provided the crack has been open up to time t. The function $\Delta_2(x, t)$ is the normal gap, given by (4.3.3), which cannot be negative, by virtue of the boundary conditions (4.1.1, 2). The stress intensity factor, (4.2.23), has the form

$$K_1 = c^{1/2} p(t) . \tag{4.4.4}$$

We now consider the problem where $p(t)$ oscillates from positive to negative signs [Graham (1976), Graham and Sabin (1981)]. From (4.4.2), it is clear that the crack will remain completely open, while

$$q(t) = \int_{-\infty}^{t} dt' k(t-t')p(t') \tag{4.4.5}$$

remains positive. When $q(t)$ vanishes, the crack is closed completely and $p(t)$ is no longer given. This is consistent with the observations of Sect. 4.3. We put

$$p(t) = \begin{cases} p_0(t), & \text{crack open} \\ p_c(t) > p_0(t), & \text{crack closed,} \end{cases} \tag{4.4.6}$$

where $p_0(t)$ is known and $p_c(t)$ must be determined. Consider now a time t after an arbitrary number of closures have taken place. We argue that (4.4.2) still holds, where for the periods of closure, $p(t)$ is replaced by $p_c(t)$, as yet undetermined. This may be seen by supposing $p_c(t)$ to be given and solving the problem exactly as in Sect. 4.2. The question of whether the crack is open or closed is irrelevant.

Let the crack be open for a portion of $(-\infty, t)$, denoted by $W_\sigma(t)$, and closed during the complementary set of times $W_u(t)$. For $t' \in W_\sigma(t)$, $p(t')$ is given — to be equal to $p_0(t')$, while for $t' \in W_u(t)$, $q(t')$ is given — to be zero. This is precisely the situation discussed at the beginning of Sect. 2.4. We can take the results of that section over directly.

Consider first the case where the crack is closed at time t. Our object is to find $p_c(t)$. We write

$$p_c(t) = \int_{-\infty}^{t} dt' \, l(t-t') q(t') = \int_{W_\sigma(t)} dt' \, \Pi_\sigma(t,t') p_0(t') \,, \tag{4.4.7}$$

using the inverse form of (4.4.5) [see (3.2.10b, 11b)] and also (2.4.3), combined with the fact that $q(t')$ is zero for $t' \in W_u(t)$. We have replaced the subscript "v" in (2.4.3) by "σ". Equation (2.4.9) gives us that

$$p_c(t) = \sum_{r=1,3,5,\ldots} \int_{\theta_{r+1}}^{\theta_r} T_r(t,t') p_0(t') dt' \,, \tag{4.4.8}$$

where $[\theta_2, \theta_1]$, $[\theta_4, \theta_3]$ etc. are the time intervals for which the crack is open, and the functions $T_l(t,t')$ are given by (2.4.11) (with time θ_r replacing t_r, $r = 1,2,\ldots$). This is an explicit expression for $p_c(t)$, once the times $\theta_1, \theta_2, \ldots$ are known. The time when the crack next opens is the solution θ_s to the relation

$$p_c(\theta_s) = p_0(\theta_s) \,. \tag{4.4.9}$$

Next, consider a time t at which the crack is open. In order to calculate the gap $\Delta_2(x,t)$, we need to know $q(t)$. Let us apply the expansion (2.4.12) to (4.4.5), giving

$$q(t) = \int_{W_\sigma(t)} dt' \, \Gamma_\sigma(t,t') p_0(t') = \sum_{r=0,2,4,\ldots} \int_{\theta_{r+1}}^{\theta_r} dt' \, N_r(t,t') p_0(t') \,, \quad \theta_0 = t \,, \tag{4.4.10}$$

which is an explicit expression for $q(t)$, once the θ_r are known. Note that in (4.4.8) we have numbered the θ_r so that odd values of r correspond to times of closing and even values to times of opening. In (4.4.10), that situation is reversed. The time θ_c at which the crack closes again is characterised by the condition that $q(\theta_c)$ vanishes. It remains zero until the next time of opening. Thus

$$q(\theta_c) = q(\theta_s) = 0 \,. \tag{4.4.11}$$

Problem 4.4.1: Show, using (4.4.8) and (4.4.10), that the condition that $q(\theta_c^-)$ vanishes implies that the pressure is continuous at a time of crack closure, in other words, $p_c(\theta_c^+) = p_0(\theta_c)$. Similarly show that continuity of the pressure at θ_s implies continuity of displacement, in other words, $q(\theta_s^+) = 0$.
Hint: by splitting $l(t)$, $k(t)$ into singular and non-singular parts, analogously to (3.8.13), show that $T_1(\theta_c^+, t') = -l_0 k_1(\theta_c^+ - t')$ where $k_1(t)$ is the non-singular portion of $k(t)$ and θ_c^+ is the limit as $t \to \theta_c$ from above; use the fact that $T_1(\theta_c^-, t') = \delta(\theta_c - t')$, where θ_c^- is the limit from below, which follows from (2.3.15). Similarly show that $N_1(\theta_s^+, t') = -k_0 l_1(\theta_s^+ - t')$ where $l_1(t)$ is the non-singular portion of $l(t)$.

It is clear that one can establish the times of opening and closing, and from these, the complete solution, by starting at the beginning of the deformation history and working through the various phases with the aid of the above formulae.

4.4.1 The Standard Linear Model

The functions $T_l(t, t')$, $N_l(t, t')$ have been given explicitly for such materials by (3.11.12, 18). With the aid of these, expressions (4.4.8) and (4.4.10) can be made entirely explicit in terms of the θ_r. Let us assume a simple sinusoidal form for $p_0(t)$:

$$
\begin{aligned}
p_0(t) &= p_m \sin(\omega t) , & t > 0 \\
&= 0 , & t < 0 .
\end{aligned}
\tag{4.4.12}
$$

We will write down expressions which give the time of first closing and reopening. Let the time of first closing be $\theta_{1c} = (\pi + \varepsilon)/\omega$, determined by (4.4.10) and (4.4.11), where θ_1 in (4.4.10) is zero. Then

$$
q(t) = k_0 p_m C(\phi, \gamma, \eta, T)
$$

$$
C(\phi, \gamma, \eta, T) = \frac{1}{1 + \eta^2} \{[(1 + \eta^2 + \gamma)^2 + \gamma^2 \eta^2]^{1/2} \sin(T - \phi) + \gamma \eta \, e^{-T/\eta}\} \tag{4.4.13}
$$

in terms of the dimensionless variables

$$
\gamma = \frac{k_1}{\beta k_0} , \qquad \eta = \frac{\omega}{\beta} , \qquad T = \omega t , \tag{4.4.14}
$$

and the loss angle ϕ defined by the property that

$$
\tan \phi = \frac{\gamma \eta}{1 + \eta^2 + \gamma} , \tag{4.4.15}
$$

which may be shown to be identical to that given by (1.6.7). Noting that

$$
\sin \phi = \frac{\gamma \eta}{[(1 + \eta^2 + \gamma)^2 + \gamma^2 \eta^2]^{1/2}} , \tag{4.4.16}
$$

we can rewrite $C(\phi, \gamma, \eta, T)$ in the form

$$
C(\phi, \gamma, \eta, T) = \frac{\gamma \eta}{1 + \eta^2} \left(\frac{\sin(T - \phi)}{\sin \phi} + e^{-T/\eta} \right) . \tag{4.4.17}
$$

Condition (4.4.11) gives the equation for ε:

$$
\sin(\varepsilon - \phi) = \sin(\phi) e^{-(\pi + \varepsilon)/\eta} , \tag{4.4.18}
$$

which shows that $\varepsilon > \phi$ and is therefore positive. Also from (4.4.8) and (3.11.11), we see that

$$
\begin{aligned}
p_c(t) &= p_m l_1 k_0 \exp[-a(t - \theta_{1c})] \int_0^{\theta_{1c}} dt' \exp[-\beta(\theta_{1c} - t')] \sin(\omega t') \\
&= p_m \exp[-a(t - \theta_{1c})] \sin(\omega \theta_{1c}) ,
\end{aligned}
\tag{4.4.19}
$$

by virtue of the vanishing of $q(\theta_{1c})$ and (3.11.10), so that the condition for reopening at θ_{1s} is

$$\exp\left[-a(\theta_{1s}-\theta_{1c})\right]\sin(\omega\theta_{1c}) = \sin(\omega\theta_{1s}) \ . \tag{4.4.20}$$

This will be satisfied at $\theta_{1s} = (2\pi - \varepsilon_1)/\omega$ where ε_1 is positive and less than ε. Thus the crack closes after $p_o(t)$ becomes negative and opens before it becomes positive again. It follows from (4.4.4) that the stress intensity factor takes positive, negative and zero values. Note that, according to (4.4.2), the crack, when closed, will still have a singularity at the end points. This is a specifically viscoelastic effect also. The pressure $p_c(t)$ vanishes in the elastic limit.

Problem 4.4.2: Using (4.4.8), (3.11.11), show that, during a closed phase beginning at θ_c and after an arbitrary previous history [Graham and Sabin (1981)]:

$$p_c(t) = p_c(\theta_c)e^{-a(t-\theta_c)} \ . \tag{4.4.1 p}$$

4.4.2 Steady-State Solution

It will be assumed that a sufficiently long time has passed since the stress was first applied to allow steady-state conditions to be established. We can use techniques similar to those described in Sect. 3.11, except that the present case is easier in some respects. Consider first the case where the crack is closed. Let this be the first cycle of closure after $t = 0$, partly occupying the time interval $[\pi/\omega, 2\pi/\omega]$. We expect θ_1 to be somewhat larger than π/ω. Let us denote it by θ_c. Also, θ_2 will be in the vicinity of $t = 0$. We denote it by θ_s. For steady-state conditions,

$$\theta_{r+2} = \theta_r - \Delta \ , \quad r = 1, 2, \dots \ , \quad \Delta = \frac{2\pi}{\omega} \ . \tag{4.4.21}$$

Equation (4.4.8) can be written as

$$p_c(t) = p_m \int_{\theta_s}^{\theta_c} dt' \, \Pi(t, t') \sin(\omega t')$$

$$\Pi(t, t') = \sum_{k=0}^{\infty} T_{2k+1}(t, t' - k\Delta) \ . \tag{4.4.22}$$

Consider now the time period just after $t = 0$ for which the crack is open, roughly $[0, \pi/\omega]$. In this case, $\theta_1 = \theta_s$ and $\theta_2 = \theta_c - \Delta$. Equation (4.4.10) can be written as [see (2.4.13)]:

$$q(t) = p_m \int_{\theta_s}^{t} dt' \, k(t - t') \sin(\omega t') + p_m \int_{\theta_s}^{\theta_c} dt' \, \Gamma(t, t') \sin(\omega t')$$

$$\Gamma(t, t') = \sum_{k=1}^{\infty} N_{2k}(t, t' - k\Delta) \ . \tag{4.4.23}$$

We now give more explicit expressions for these quantities in the case of a standard linear solid. Equation (3.11.12) gives that, for odd l,

$$T_{l+2}(t, t'-\Delta) = T_l(t, t')E$$
$$E = \exp\left[-(\beta-a)(\theta_c-\theta_s)-a\Delta\right] ,$$

(4.4.24)

so that, from (4.4.22) and (3.11.11):

$$p_c(t) = \frac{p_m}{1-E} \int_{\theta_s}^{\theta_c} dt'\, T_1(t, t')\sin(\omega t')$$

$$= \frac{p_m A}{1-E} \int_{\theta_s}^{\theta_c} dt'\, e^{-at+\beta t'}\sin(\omega t') , \qquad \text{where}$$

(4.4.25)

$$A = l_1 k_0 \exp\left[-(\beta-a)\theta_c\right] .$$

(4.4.26)

Also (3.11.19) gives that, for even l

$$N_{l+2}(t, t'-\Delta) = N_l(t, t')E , \qquad t \neq t'$$

(4.4.27)

so that, from (4.4.23):

$$\Gamma(t, t') = \frac{E}{1-E} k_1(t-t') ,$$

(4.4.28)

where $k_1(t-t')$ denotes $k(t-t')$ without the instantaneous singular delta function. Therefore,

$$q(t) = p_m k_0 \sin(\omega t) + p_m k_1 \int_{\theta_s}^{t} dt'\, e^{-\beta(t-t')}\sin(\omega t')$$

$$+ \frac{p_m k_1 E}{1-E} \int_{\theta_s}^{\theta_c} dt'\, e^{-\beta(t-t')}\sin(\omega t') .$$

(4.4.29)

The conditions that $q(\theta_c)$ and $q(\theta_s)$ vanish take the form

$$k_0 e^{\beta\theta_c}\sin(\omega\theta_c) = -\frac{k_1 I}{1-E}$$

(4.4.30a)

$$k_0 e^{\beta\theta_s}\sin(\omega\theta_s) = -\frac{k_1 E I}{1-E}$$

(4.4.30b)

$$I = \int_{\theta_s}^{\theta_c} dt'\, e^{\beta t'}\sin(\omega t') .$$

(4.4.30c)

Combining (4.4.30a) and (4.4.25) gives that (see Problem 1)

$$p_c(\theta_c) = p_m \sin(\omega\theta_c) ,$$

(4.4.31)

which confirms the continuity of the pressure function and, with the aid of (4.4.1p) tells us that $p_c(t)$ has the simple form

$$p_c(t) = p_m e^{-a(t-\theta_c)}\sin(\omega\theta_c) .$$

(4.4.32)

The condition determining the *next* time of opening, namely $\theta_s+\Delta$, has the form

$$\exp\left[-a\left(\theta_s-\theta_c+\varDelta\right)\right]\sin\left(\omega\,\theta_c\right)=\sin\left(\omega\,\theta_s\right),\tag{4.4.33}$$

which in fact is identical to the particular case given by (4.4.20). A second constraint on θ_c and θ_s is obtained by combining (4.4.30a) and (4.4.30b) to result in

$$\beta\gamma I=-\left[e^{\beta\theta_c}\sin\left(\omega\,\theta_c\right)-e^{\beta\theta_s}\sin\left(\omega\,\theta_s\right)\right]\quad\text{or}\tag{4.4.34}$$

$$e^{\beta\theta_c}\sin\left(\omega\,\theta_c-\phi\right)=e^{\beta\theta_s}\sin\left(\omega\,\theta_s-\phi\right),\tag{4.4.35}$$

where γ and ϕ are defined by (4.4.14) and (4.4.15). The quantity $q(t)$, during the period when the crack is open is given in terms of dimensionless variables by

$$q(t)=\frac{\gamma p_m k_0}{1+\eta^2}\left\{\frac{\eta\sin\left(T-\phi\right)}{\sin\phi}+\frac{(1+\eta^2)^{1/2}e^{-T/\eta}}{1-E}\right.$$
$$\left.\times\left[E e^{T_c/\eta}\sin\left(T_c-\psi\right)-e^{T_s/\eta}\sin\left(T_s-\psi\right)\right]\right\},\tag{4.4.36}$$

where ψ is defined by

$$\tan\psi=\frac{\omega}{\beta}=\eta\quad\text{and}\tag{4.4.37}$$

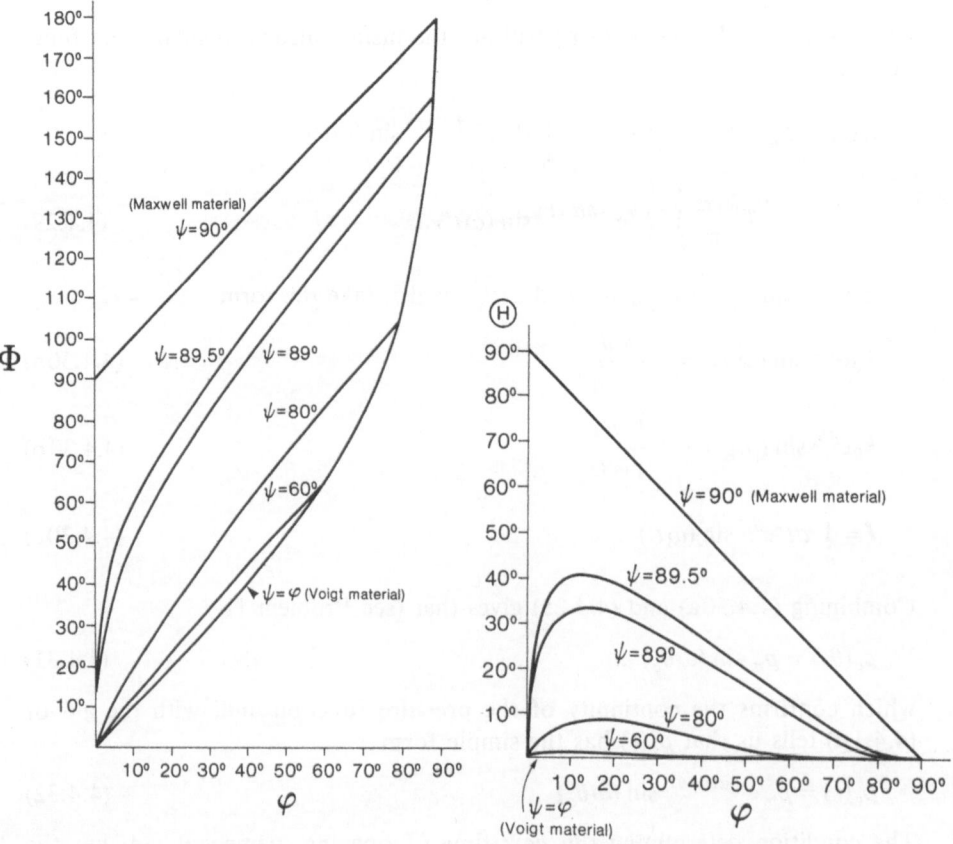

Fig. 4.1. Variation of Φ with ϕ **Fig. 4.2.** Variation of Θ with ϕ

$$T_c = \omega \theta_c \ , \qquad T_s = \omega \theta_s \ . \tag{4.4.38}$$

With the aid of (4.4.14, 15, 37) and (3.11.10), we find that

$$\tan (\psi - \phi) = \frac{\omega}{a} \ , \tag{4.4.39}$$

so that $\phi \le \psi \le \frac{\pi}{2}$. The limit $\psi = \phi$ corresponds to a Voigt material while for a Maxwell material $\psi = \frac{\pi}{2}$.

If we set

$$\Phi = T_c - \pi \ , \qquad \Theta = - T_s \ , \tag{4.4.40}$$

then Φ is a measure of the "angular delay" in the crack closing, once the applied stress has changed to compression, while Θ is the "angular advance" in the crack opening, before the applied stress changes from compression to tension. In terms of these quantities, equations (4.4.33, 35) become

$$e^{-\Theta \cot (\psi - \phi)} \sin (\Theta) = e^{-(\pi - \Phi) \cot (\psi - \phi)} \sin (\Phi) \ , \tag{4.4.41}$$

$$e^{\Phi \cot (\psi)} \sin (\Phi - \phi) = e^{-(\pi + \Theta) \cot (\psi)} \sin (\Theta + \phi) \ . \tag{4.4.42}$$

The solutions of (4.4.41, 42) in terms of ϕ and ψ have been derived by Graham and Sabin (1981) and are presented in Figs. 4.1, 2.

4.5 Growing Cracks under a General Loading History

Consider a crack occupying $[a(t), b(t)]$ at time t. Let its previous history of opening and closing be arbitrary, except that whenever it was open it was growing. First, we ask whether it can partially close under a position-independent normal stress $p(t)$ applied to its open part. The answer is in the negative, and this may be seen from the discussion in Sect. 4.3. Let the first partial closing begin at t_1. Up to this time, the crack can be regarded as open under a stress $p(x, t)$ which can be determined once a complete solution to the problem is known. It is shown below to be non-negative in the absence of rehealing effects. The same holds for $q(x, t)$, defined by (3.2.11 b), since the singular and non-singular parts of $k(t)$ are non-negative, at least if the proportionality assumption holds. One is therefore effectively dealing with a situation similar to that discussed in Sect. 4.3. It follows that the crack must be completely closed or completely open.

Consider a crack, open and growing at time t, after an arbitrary number of closures. Once again, we can regard it as having been open always under a stress history $\Sigma(x, t)$, the normal component of which is not known until the problem is solved. Then (4.2.9, 15) hold so that the stress will have the same form as in the elastic case and is independent of material properties. This is a particular manifestation of the Generalized Partial Correspondence Principle, proved in Sect. 2.6.

Expressions (4.2.16, 17) will hold for the stress intensity factor. Also, assume that while the crack is open, it is strictly growing at both ends, never stationary.

Then, using an argument similar to that in Sect. 4.3, we see that the dominant behaviour of the normal gap derivative $\Delta'_2(x, t)$ given by (4.3.3), which may be deduced from (4.2.9), is

$$\Delta'_2(x, t) \underset{x \to a(t)}{\approx} \frac{2k_0}{\pi m(x, t)} \int_{a(t)}^{b(t)} dx' \frac{p(x', t) m(x', t)}{x' - x} \approx \frac{2k_0 \sqrt{c(t)}\, K_1(a(t))}{m(x, t)} , \quad (4.5.1)$$

by using the notation of (4.2.17, 18). It follows that

$$\Delta_2(x, t) \underset{x \to a(t)}{\approx} \frac{2k_0 K_1(a(t))}{\sqrt{c(t)}}\, m(x, t) . \quad (4.5.2)$$

The corresponding result at $x = b(t)$ is

$$\Delta_2(x, t) \underset{x \to b(t)}{\approx} \frac{2k_0 K_1(b(t))}{\sqrt{c(t)}}\, m(x, t) . \quad (4.5.3)$$

These quantities can never be negative and since the crack is open, the stress intensity factors $K_1(a(t))$ and $K_1(b(t))$, which in this case are given by (4.2.18, 23), must be positive. It follows that $p(t)$ must be positive.

Note that if the crack is stationary for a period before and including t, the right-hand sides of (4.5.2, 3) are replaced by hereditary integrals, with the consequence that $K_1(a(t))$, $K_1(b(t))$ and $p(t)$ may not be confined to positive values. If the open crack stops at a certain point and begins to expand again after a certain time lapse, a singularity is created at this point, the strength of which decays with time.

Consider now the periods when the crack is closed. If, at time t, $a(t)$ and $b(t)$ are decreasing and increasing respectively, (which on physical grounds is unlikely, but we do not a priori exclude the possibility) then, following the above calculation, we find that

$$K_1(a(t)) = K_1(b(t)) = 0 , \quad (4.5.4)$$

where $K_1(a(t))$ and $K_1(b(t))$ are generated through (4.2.17, 18) in terms of the pressure function when the crack is closed. If, when closed, the crack is not extending, for a period before t, then the dominant term in $\Delta'_2(x, t)$ is

$$\Delta'_2(x, t) \sim \frac{2}{m(x)\pi} \int_a^b dx' \frac{q(x', t) m(x')}{x' - x} = 0 , \quad (4.5.5\,\mathrm{a})$$

$$m(x) = [(x - a)(b - x)]^{1/2} , \quad \text{and} \quad (4.5.5\,\mathrm{b})$$

$$q(x, t) = \int_{t_0}^{t} dt'\, k(t - t') p(x, t') . \quad (4.5.6)$$

The quantity t_0 is the time at which the crack becomes stationary. It follows from (4.5.6) that

$$p(x, t) = \int_{t_0}^{t} dt'\, l(t - t') q(x, t') \quad (4.5.7)$$

by virtue of an interchange of order of integration and the fact that $l(t)$, $k(t)$ are inverses of each other in the sense of (2.3.15). Note that (4.5.5a) holds for all $t' \in [t_0, t]$, so operating on (4.5.5a) with $l(t)$ gives again that the stress intensity factors must be zero.

From the above analysis we may conclude that, provided the crack is growing whenever it is open, $p(x, t)$ is non-negative. When the crack is open, $p(x, t) = p(t)$ where $p(t)$ is positive. As soon as $p(t)$ becomes negative, the crack closes entirely. Then $p(x, t)$ represents the non-negative pressure acting across the crack face.

These results are due to Graham and Sabin (1978), who derive them by means of a more explicit method.

It is interesting to note the fundamental qualitative differences between extending and stationary cracks. These are traceable to the fact that the dominant singular term for an extending crack comes from the delta function part of the hereditary integral, while this is not so for a stationary crack. Note that this "instantaneous" property of singular terms, in the case of extending cracks, leads to properties similar to those found in the elastic case, while stationary viscoelastic cracks behave quite differently to the elastic case.

In the next section, dealing with propagation criteria, the interesting similarity between elastic and extending viscoelastic cracks is further manifested.

4.6 The Griffith Criteria for Crack Extension in Viscoelasticity

Consider a crack in the interval $[a, b]$, subject to a complex stress $\Sigma = -p + is$, that is independent of position but may vary with time. Griffith (1921, 1925) (who did not include the shear term) argued that if $c = (b - a)/2$ is larger than a certain critical value c_0, rapid unstable crack extension will take place. The value of c_0 is determined by the following argument. A small increase in crack length δc would require energy $4T\delta c$ where T is the energy per unit length required to produce free surface. Let us consider an elastic material for a moment. Such an extension of the crack length can be shown to result in a release of energy. Goodier [in Liebowitz (1968)] shows that for the case of a fixed applied tension, the strain energy of a linear elastic body increases if crack extension occurs, but only half the work done by the boundary forces goes to this increase. The other half is available for creating new surface. Beyond a certain crack length, c_0, the release of energy will match or more than match the energy required to produce new surface. Therefore, one expects crack extension which will be unstable since each extension makes further extension even more energetically possible.

In order to discuss the linear viscoelastic case, we adopt a more formal approach, which incidentally also applies to the elastic case in the limit, and illustrates Goodier's result. Following Golden and Graham (1987c), we write down an isothermal energy balance relation of the form

$$\int_B ds\, \sigma_{ij} n_j \dot{u}_i = \dot{S} + \dot{H} + 2T(\dot{b} - \dot{a}) , \tag{4.6.1}$$

where ds is a surface element (becoming a line element in the plane strain case), $\dot{S}$ is the rate of increase of stored mechanical enery and $\dot{H}$ is the rate of dissipation of mechanical energy. These are respectively given by (1.9.11) and (1.9.12), integrated over the entire body. The crack, which is allowed to grow with time, occupies the interval $[a(t), b(t)]$ of the x-axis and the quantity T, which is assumed to be a constant, is the surface energy per unit length. Since the quantity on the left is the rate of work of the boundary forces, this relation may be taken as self-evident [see Erdogan (in Liebowitz (1968)), for example]. Recalling the identity given by (1.3.4) and taking account of (1.3.5, 6), it is clear that $\dot{S}(t) + \dot{H}(t)$ can be replaced by the space integral of $\sigma_{ij}\dot{\varepsilon}_{ij}$, and we have:

$$\int_B ds\,\sigma_{ij}(r,t)n_j(r)\dot{u}_i(r,t) = \int_V dv\,\sigma_{ij}(r,t)\dot{\varepsilon}_{ij}(r,t) + 2T[\dot{b}(t) - \dot{a}(t)] \ . \tag{4.6.2}$$

A centrally important observation is that (4.6.2) is an identity, in the absence of crack extension, namely the energy identity discussed in Sects. 1.3 and 1.8. In the presence of an extending crack however, it becomes non-trivial. The former observation implies that all terms will cancel except those that involve in some essential manner a derivative of the crack length.

Let us now adapt our notation more explicitly to the crack problem. The boundary B consists of the two crack faces on which the position and time-dependent stresses are assumed to be equal with opposite sign and $\Sigma(x,t) = (-p + is)(x,t)$. Thus

$$\int_B ds\,\sigma_{ij}(r,t)n_j(r)\dot{u}_i(r,t) = (-)\int_{a(t)}^{b(t)} dx\,[\sigma_{12}(x,0,t)\dot{\Delta}_1(x,t) + \sigma_{22}(x,0,t)\dot{\Delta}_2(x,t)]$$

$$= -\,\mathrm{Im}\left\{\int_{a(t)}^{b(t)} dx\,\Sigma(x,t)\dot{\Delta}(x,t)\right\} , \tag{4.6.3}$$

where Δ_1 and Δ_2 are the components of displacement across the gap, given by

$$\Delta_i(x,t) = u_i(x,0^+,t) - u_i(x,0^-,t) , \quad i = 1,2 \tag{4.6.4}$$

and $\Delta(x,t) = (\Delta_1 + i\Delta_2)(x,t)$ is given by (4.2.10).

We decompose $\Delta(x,t)$ into an instantaneous portion and a remainder, to obtain

$$\Delta(x,t) = \Delta_0(x,t) + \Delta_r(x,t) , \quad \text{where} \tag{4.6.5}$$

$$\Delta_0(x,t) = \frac{2ik_0}{\pi}\int_{a(t)}^{b(t)} dy'\,\Sigma(y',t)M(x,y';t) , \tag{4.6.6}$$

$$\Delta_r(x,t) = \frac{2i}{\pi}\int_{t_1(x)}^{t} dt'\,k_1(t-t')\int_{a(t')}^{b(t')} dy'\,\Sigma(y',t')M(x,y';t') ;$$

$k_1(t)$ is the non-singular part of $k(t)$ [analogous to $\gamma_1(x)$ in (3.8.13); see also Problem 4.4.1], and $M(x,y';t)$ is given by (4.2.11). Then

$$\dot{\Delta}(x,t) = \dot{\Delta}_0(x,t) + \dot{\Delta}_r(x,t) , \tag{4.6.7}$$

where, by virtue of (4.2.13),

$$\dot{\Delta}_0(x,t) = \frac{2ik_0}{\pi} \int\limits_{a(t)}^{b(t)} dy' \left[\dot{\Sigma}(y',t) M(x,y';t) + \Sigma(y',t) \dot{M}(x,y';t) \right] \qquad (4.6.8\,\mathrm{a})$$

$$\dot{\Delta}_r(x,t) = \frac{2ik_1(0)}{\pi} \int\limits_{a(t)}^{b(t)} dy' \Sigma(y',t) M(x,y';t)$$

$$+ \frac{2i}{\pi} \int\limits_{t_1(x)}^{t} dt' \dot{k}_1(t-t') \int\limits_{a(t')}^{b(t')} dy' \Sigma(y',t') M(x,y';t') \,. \qquad (4.6.8\,\mathrm{b})$$

We now consider the first term on the right-hand side of (4.6.2). We use (1.1.4) to rewrite it as

$$\int\limits_{V} dv \, \sigma_{ij}(r,t) \frac{\partial}{\partial x_j} \dot{u}_i(r,t) \,. \qquad (4.6.9)$$

In normal circumstances, and in particular for a crack that is not expanding, Green's Theorem, and the fact that $\sigma_{ij,j}$ vanishes, gives that this is equal to the boundary term on the left of (4.6.2). However the integral contains divergent terms, as a result of the crack extension, which makes it meaningless without reinterpretation. The displacements off the crack can be written down from (4.1.6c) and (4.1.7) by transferring the hereditary integral to the right-hand side of (4.1.6c). The explicit expression for $\phi(z,t)$ is given by (4.2.7) where $X(z,t)$ is the complex function $\{[z-a(t)][z-b(t)]\}^{-1/2}$. All that is essential for our purposes is that it possesses square root singularities at the crack ends. Similarly, the stresses at points off the crack face may be evaluated with the aid of (4.1.6a, b) and $\phi(z)$ in its explicit form. These also have square root singularities at the crack tips. Let us write at a general point

$$u_i(r,t) = u_{i0}(r,t) + u_{ir}(r,t) \,, \qquad (4.6.10)$$

where $u_{i0}(r,t)$ is the instantaneous portion of the hereditary integral that gives the displacement, and $u_{ir}(r,t)$ is the remainder, consisting of an integral with a smooth kernel over the history of the real and imaginary portions of the right-hand side of (4.1.6c). Now the quantity $\partial/\partial x_j \dot{u}_{ir}(r,t)$ will possess at most a square root singularity. This is so because the time derivative will act only on the kernel as in (4.6.8b). This singularity, combined with that in the stresses, gives rise to a linear singularity which is integrable since the integral in (4.6.9) is two-dimensional. Thus, Green's theorem can be applied and what is obtained is the non-instantaneous portion of (4.6.3). The non-instantaneous terms in fact cancel out of the equation. This observation is the crucial one for our purposes. It means that specifically viscoelastic effects do not contribute.

Consider now the instantaneous portion of (4.6.9). It may be checked that, due to the expansion of the crack $\partial/\partial x_j \dot{u}_{i0}(r,t)$ contains a singularity of order 3/2 which, combined with that in the stresses, gives rise to a non-integrable singularity. Thus, this portion of the integral is meaningless in its present form. We now proceed to rewrite it in a form that avoids divergent integrals.

The instantaneous portion of the strain tensor ε_{ij} has the same form as that of the elastic problem with moduli equal to the instantaneous moduli of the viscoelastic material. Furthermore, as noted in Sect. 4.2, the stresses in this problem are the same as the elastic stresses and independent of material parameters. They may in particular be written as

$$\sigma_{ij} = G_{ijkl}(0)\varepsilon_{kl}^{(0)} , \tag{4.6.11}$$

where we have, for a moment, adopted the general anisotropic notation for the sake of compactness. In (4.6.11), $G_{ijkl}(0)$ are the instantaneous moduli and $\varepsilon_{kl}^{(0)}$ is the instantaneous portion of the strain tensor in the present problem. Now we can write the instantaneous portion of (4.6.9) in the form

$$\int_V dv\, G_{ijkl}(0)\,\varepsilon_{ij}^{(0)}(r,t)\dot{\varepsilon}_{kl}^{(0)}(r,t) = \frac{1}{2}\frac{d}{dt}\int_V dv\, G_{ijkl}(0)\,\varepsilon_{ij}^{(0)}(r,t)\varepsilon_{kl}^{(0)}(r,t) . \tag{4.6.12}$$

The integral on the right is now convergent, since the time differentiation is removed outside the integral sign. Application of Green's Theorem gives

$$\frac{1}{2}\frac{d}{dt}\int_V dv\, G_{ijkl}(0)\,\varepsilon_{ij}^{(0)}(r,t)\varepsilon_{kl}^{(0)}(r,t) = \frac{1}{2}\frac{d}{dt}\int_{a(t)}^{b(t)} dx\,\sigma_{ij}(r,t)n_j(r)u_{i0}(r,t)$$

$$= -\frac{1}{2}\frac{d}{dt}\,\mathrm{Im}\left\{\int_{a(t)}^{b(t)} dx\,\bar{\Sigma}(x,t)\Delta_0(x,t)\right\} , \tag{4.6.13}$$

on recalling the manipulations that led up to (4.6.3). Remembering that the non-instantaneous portion of (4.6.3) has already been cancelled, we finally write (4.6.2) in the form

$$-\mathrm{Im}\left\{\frac{2ik_0}{\pi}\int_{a(t)}^{b(t)} dx\,\bar{\Sigma}(x,t)\int_{a(t)}^{b(t)} dy'\,[\dot{\Sigma}(y',t)M(x,y';t)+\Sigma(y',t)\dot{M}(x,y';t)]\right\}$$

$$= (-)\frac{d}{dt}\,\mathrm{Im}\left\{\frac{ik_0}{\pi}\int_{a(t)}^{b(t)}\bar{\Sigma}(x,t)\int_{a(t)}^{b(t)} dy'\,\Sigma(y',t)M(x,y';t)\right\}$$

$$+ 2T[\dot{b}(t)-\dot{a}(t)] . \tag{4.6.14}$$

Using (4.2.13, 14), it is found that the terms not involving $\dot{a}(t), \dot{b}(t)$ cancel and we are left with

$$\frac{\pi k_0}{2}\{[K_1^2(b)+K_2^2(b)]\dot{b}(t)-[K_1^2(a)+K_2^2(a)]\dot{a}(t)\} = 2T[\dot{b}(t)-\dot{a}(t)] , \tag{4.6.15}$$

in terms of the stress intensity factors K_i, $i=1,2$ defined by (4.2.16, 17). We conclude that the conditions for crack growth at the respective crack tips are:

$$\pi k_0[K_1^2(b)+K_2^2(b)] = 4T , \quad \pi k_0[K_1^2(a)+K_2^2(a)] = 4T . \tag{4.6.16}$$

If the middle point of the crack face is chosen as the origin, and $\Sigma(x,t)$ is even in x, these conditions reduce to the single condition

$$\pi k_0[K_1^2(c)+K_2^2(c)] = 4T , \tag{4.6.17}$$

where $c(t) = b(t) = -a(t)$ and K_1, K_2 are given by (4.2.1p). In particular if $\Sigma = \Sigma(t)$ is independent of x, it follows from (4.2.22) that (4.6.17) becomes

$$\pi k_0 |\Sigma(t)|^2 c(t) = 4T .$$ (4.6.18)

Problem 4.6.1: If the middle point of the crack face is chosen as the origin and $\Sigma(x,t) = \Sigma(t)$, by proceeding from (4.2.20), show that condition (4.6.2) is equivalent to condition (4.6.18).

Strictly, (4.6.1) should be an inequality stating that the left-hand side is greater than or equal to the right, in which case conditions (4.6.16 – 18) become inequalities. These conditions have the same form as the Griffith criterion for crack extension for an elastic body with k_0, which is an instantaneous inverse modulus, replacing the elastic inverse modulus. If a unique Poisson's ratio exists, then

$$k_0 = J_1(0)(1-v)$$ (4.6.19)

in terms of the instantaneous limit of the creep function $J_1(t)$. One can see intuitively why the instantaneous limit is the crucial one in this context by observing that it is the regions in the vicinity of the crack tips which are important and that as the crack progresses into previously uncut parts of the material, the displacement near the crack tips, which will determine energy input, is proportional to the instantaneous response of the medium.

Note that in (4.6.14), the term on the right, proportional to $[\dot{b}(t) - \dot{a}(t)]$, is half the corresponding term on the left. This is a manifestation of the theorem of Goodier [in Liebowitz (1968)] for elastic bodies, mentioned at the beginning of the section.

What has been shown here is that energy considerations for a viscoelastic medium in the non-inertial approximation give no more than a Griffith instability criterion similar to that for an elastic medium. One cannot therefore hope to obtain a condition determining crack velocity from a non-inertial energy equation. The status of conditions which emerge by using approximate solutions [Christensen (1979), Christensen and Wu (1981)] has been discussed by Christensen and McCartney (1983).

Problem 4.6.2: Use (2.8.1p) to show that the plane strain state

$$u_x = x\varepsilon_{xx} + 2y\varepsilon_{xy} ; \quad u_y = y\varepsilon_{yy} ; \quad u_z = 0 ,$$

where the strain components ε_{ij} are related to the stresses

$$\sigma_{22} = p(t) , \quad \sigma_{11} = q(t) , \quad \sigma_{12} = -s(t) , \quad \sigma_{31} = \sigma_{32} = 0$$ (4.6.2p)

through (1.9.17b) may be superimposed on the plane strain solutions used in this section to produce a solution for a stress-free crack in an infinite body acted upon by the stresses (4.6.2p) at infinity. Then verify that condition (4.6.2) leads to (4.6.18) if the body is loaded in the latter fashion.

Conditions having the same form as (4.6.18) where given by Graham (1970) and Kostrov and Nikitin (1970). In a later paper [Graham (1975)], it was shown that the same criterion applied to a crack extending, after an arbitrary previous history. Earlier, M. Williams (1965), (1967) had presented an equation for the history of growth of a viscoelastic crack, which, in contrast to (4.6.18), depended explicitly on time. Crack growth was viewed as an ablation of the viscoelastic body. Nuismer (1974) pointed out that no account had been taken of the energy of the ablated material and if this is done, the resulting condition has the same form as (4.6.18). Kostrov and Nikitin (1970) and Blackburn (1971) considered the possibility that the surface energy might depend upon velocity. [See also Knauss (1970).] Under this assumption, it is possible to give a growth rate for the crack, since, in contrast to the constant surface energy case treated above, the velocity does not cancel from the equation. Gurtin (1979) has provided a theoretical foundation for this assumption.

Earlier authors proposed theories which predict the speed of viscoelastic crack propagation, by assigning a detailed structure to the crack tip, both under unstable and subcritical or fatigue failure. We mention Knauss (1970, 1974), Knauss and Dietmann (1970), Mueller and Knauss (1971) and a review by Knauss (1973); also the work of Wnuk (1971–73b). Later, there was the work of Schapery (1974–79), McCartney (1977–79) and Golden and Graham (1984). See also Kanninen and Popelar (1985). Majidzadeh et al. (1976) discuss various models in the context of application to pavement design. McCartney (1987) is concerned with crack extension criteria for fibre-reinforced composites.

As presented here, the argument applies to cracks in brittle materials. In fact, up to now, in this chapter, we have considered only such cracks. However, the Griffith formalism, with suitable reinterpretation, may be applied to materials which are not completely brittle. Orowan (1950) and Irwin (1948) argued that the plastic flow, even in the fracture of fairly brittle materials, absorbs far more energy than the surface creation process. To take account of this, one simply replaces $2T$ by a term characterizing the plastic flow, taking it to be a material constant as a first approximation (but see Lardner (1974), p. 205). This constant is Irwin's crack extension force.

4.7 Barenblatt's Theory of Brittle Cracks

An alternative model of brittle fracture is that of Barenblatt (see Barenblatt (1962), which includes earlier references). The motivation for this theory was to eliminate the stress singularities at the crack tips which are a feature of Griffith's theory. Such singularities are clearly unphysical. No real material could withstand very high stresses without yielding or failing in some manner. Barenblatt put forward the idea that, in the vicinity of the crack tip, there are large, cohesive stresses acting, which are counteracting the tendency of the crack to extend. These forces nullify the stress singularity.

For simplicity, we will restrict ourselves to the case of purely normal forces. Let the crack occupy $[-c(t), c(t)]$. We denote the cohesive forces as $-f(x, t)$. This function will be zero except in a small region near each tip, which in fact may be regarded as beyond the tip itself. Let the crack plus cohesive region occupy $[-a(t), a(t)]$ where $a(t) > c(t)$. For convenience, let us regard $p(x, t)$ as given over the small regions $c < |x| < a$, even though they are off the crack face.

In practice, $f(x, t)$ will be very large over these regions so the value of $p(x, t)$ is irrelevant. We now impose the condition that no stress singularity exists, due to the presence of $f(x, t)$. We can express this condition by simply saying that the combination $p(x, t) - f(x, t)$ has stress intensity factor, as given by (4.2.1p), equal to zero, or

$$\int_{-a(t)}^{a(t)} dx \, \frac{p(x, t)}{\sqrt{a^2(t) - x^2}} - \int_{-a(t)}^{a(t)} dx \, \frac{f(x, t)}{\sqrt{a^2(t) - x^2}} = 0 \; . \tag{4.7.1}$$

In the first integral, we can replace $a(t)$ by $c(t)$ with negligible error since $p(x, t)$ is a smooth function. Using (4.2.1p) again allows us to write (4.7.1) in the form

$$\frac{\pi}{\sqrt{c}} K_1(t) = \int_{-a(t)}^{a(t)} dx \, \frac{f(x, t)}{\sqrt{a^2 - x^2}} \tag{4.7.2}$$

or, changing variables to $u = a - x$ and putting $f(x, t) = g(u, t)$,

$$K_1(t) \approx \frac{\sqrt{2}}{\pi} \int_0^d du \, \frac{g(u, t)}{\sqrt{u}} \; , \qquad d = a - c \; , \tag{4.7.3}$$

since $g(u, t)$ is non-zero only over $0 < u < d$. We have assumed that $f(x, t)$ is an even function of x.

It is worth noting that exactly this type of reasoning has been used to incorporate regions of plastic flow at crack tips, for ductile materials [Dugdale (1960), Bilby et al. (1963), Lardner (1974), Golden and Graham (1984)].

In Barenblatt's theory, the crack extension criterion may be stated as follows. As stress increases, $g(u, t)$ will change until a critical state is reached where fracture is about to begin. In this state, $g(u, t) = g_m(u)$ and $d = d_m$ where it is *assumed* that $g_m(u)$, d_m are material parameters. This is actually qualitatively the same as (4.6.18), which may be re-expressed easily in terms of stress intensity factors. What is not obvious though, and what was shown by Willis (1967a), [see also Goodier in Liebowitz (1968)], is that the two conditions are identical in the elastic case. Lardner (1974) gives a concise coverage of these topics, summarizing the work of Willis (1967a). Here we will briefly discuss the equivalent result for the viscoelastic case.

Let us first write down an approximate expression for the displacement derivative near $x = \pm a(t)$. By virtue of condition (4.7.1), the relation (4.2.9), in this context, can be put in the form

$$\Delta_2'(x, t) = u_2'(x, 0^+, t) - u_2'(x, 0^-, t)$$

$$= \frac{2}{\pi} \int_{t_1(x)}^t dt' \, k(t - t') m(x, t') \int_{-a(t')}^{a(t')} dx' \, \frac{[p(x', t') - f(x', t')]}{(x' - x) m(x', t')} \; . \tag{4.7.4}$$

This may be shown by subtraction of terms proportional to the right-hand side of (4.2.16) and (4.2.1p) with range replaced by $[-a(t'), a(t')]$ from the space integral in (4.2.9). These terms are zero in the present context. We see that the fact that the stress intensity factor corresponding to $p(x, t) - f(x, t)$ is zero implies that $\Delta_2'(x, t)$ will be non-singular at the end points. This is not surprising, given that there will be no stress singularities. Now, for x near $\pm a(t)$, $f(x',t)$ will dominate in the integral in (4.7.4), by virtue of an argument of Willis (1967a), which we omit. If the crack propagates in the interval $[t, t+\delta t]$, then we can say that $f(x,t) = g(u,t) = g_m(u)$ and also that the singular term in $k(t)$ will dominate because $t_1(x)$ approaches t as $x \to +a(t)$. Taking d to be small, we finally obtain

$$\Delta_2'(x,t) = \frac{2k_0}{\pi} \int_0^{d_m} du' \left(\frac{u}{u'}\right)^{1/2} \frac{g_m(u')}{u'-u} , \quad u = a(t) - x \ll 1 . \tag{4.7.5}$$

The work done in bringing about complete separation of the surfaces (at one tip) is

$$2T = \int_0^{d_m} du \Delta_2'(u) g_m(u) , \tag{4.7.6}$$

where $\Delta_2'(u) = -\Delta_2'(x,t)$. The former will not depend on t, near the tip. Equations (4.7.5, 6) give that

$$2T = -\frac{2k_0}{\pi} \int_0^{d_m} du\, g_m(u) \int_0^{d_m} du' \left(\frac{u}{u'}\right)^{1/2} \frac{g_m(u')}{u'-u} . \tag{4.7.7}$$

Interchanging u, u' merely inverts the square root function and changes the sign, leaving the overall value unchanged. Adding gives that

$$4T = \pi K_{1c}^2 k_0 \tag{4.7.8}$$

by virtue of (4.7.3) where K_{1c} is the critical stress intensity factor. Equation (4.2.23) gives that this is equivalent to (4.6.18).

4.8 Crack in a Field of Bending

Consider a fixed length crack occupying $[-c, c]$ along the x-axis where the material is subject to a bending moment [1] $m(t)$, the neutral axis being the y-axis. The sign convention is that when $m(t)$ is positive, there will be compression across the positive x-axis, at least at points far from the origin.

This type of problem was considered in the elastic case by Bowie and Freese (1976) and Comninou and Dundurs (1979). A central feature of interest was that the crack closed over part of the interval, including one end. The end which closed depended on the sign of $m(t)$ and switched instantly, once the sign of this quantity changed. For the case under consideration here, the closed interval was $[c/3, c]$ for $m(t)$ positive, and the reflection of this in the y-axis for $m(t)$ negative. We emphasize that when closure occurs, it is instantaneous.

Let us now consider the viscoelastic case. The boundary conditions are (4.1.1, 2) with $s(x,t) = 0$ and $p(x,t) = -m(t)x$, under the convention that the stresses at infinity are transferred to the crack face.

Let $m(t)$ be applied from $t = 0$ onwards, with a positive value initially. Guided by the results of the elastic problem, we anticipate that closure will occur over a portion of the positive x-axis, or, more correctly, the crack will open over the

---

[1] Strictly bending moment divided by moment of inertia, see Sokolnikoff (1956), Chap. 4.

region $[-c, c]$ excluding an interval at the positive end, since in fact it is initially closed. As in Sect. 4.3, a necessary condition for closure is that the stress intensity factor, at the point of the opening of the crack adjacent to the closed interval, be zero. Let this point be denoted as b. Then (4.2.16) gives us that [see (A 1.3.4)]:

$$\int_{-c}^{b} dx\, x \left(\frac{x+c}{b-x}\right)^{1/2} = \frac{b+c}{8}(3b-c) = 0 \quad \text{or} \tag{4.8.1}$$

$$b = c/3 \tag{4.8.2}$$

as in elastic theory. The open region of the crack is $[-c, c/3]$. We can calculate the gap from (4.2.9), with $t_1(x) = 0$, since the crack is fixed length. Using (4.8.2) and (A 1.2.4) gives that the normal gap defined by (4.3.3) has the form

$$\Delta'_2(x, t) = -M(t)f_0(x) , \quad -c \le x \le c/3 , \tag{4.8.3}$$

where

$$f_0(x) = \frac{2}{3}\left(\frac{(c/3)-x}{x+c}\right)^{1/2}(2c+3x)$$

$$\tag{4.8.4}$$

$$M(t) = \int_{0}^{t} dt'\, k(t-t')m(t') .$$

It follows that

$$\Delta_2(x, t) = M(t)(c/3-x)^{3/2}(x+c)^{1/2} , \quad -c < x < c/3 . \tag{4.8.5}$$

From (4.2.10) and (A 1.2.4), we deduce that $p(x, t)$, on the closed portion of the crack is given by

$$p(x, t) = m(t)[-x+h(x)] \quad \text{with}$$

$$h(x) = \left(\frac{2c}{3}+x\right)\left(\frac{x-c/3}{x+c}\right)^{1/2} . \tag{4.8.6}$$

This clearly includes the contribution from the bending moment. Boundary conditions (4.1.2) are satisfied since $h(x)$ is positive. This solution holds up to time t_1 when

$$m(t_1) = 0 , \tag{4.8.7}$$

since when $m(t)$ becomes negative, the stress given by (4.8.6) fails to obey the boundary condition (4.1.2) and the closed portion must open. However, the quantity $M(t_1)$ will still be positive for a viscoelastic material so that the crack is still open at the negative end. Therefore, from this time, the crack is completely open. We deduce again from (4.2.9) and (A 1.2.4) that

$$\Delta'_2(x, t) = \int_{0}^{t} dt'\, k(t-t')v(x, t')$$

$$v(x, t') = \frac{-m(t')(c^2-2x^2)}{(c^2-x^2)^{1/2}} , \quad t' > t_1 \tag{4.8.8}$$

$$= -m(t')f_0(x) , \quad 0 < t' < t_1 ,$$

giving that

$$\Delta'_2(x, t) = -M_1(t)f_0(x)d_1(x) - N_1(t)f_1(x) , \qquad t > t_1$$

$$M_1(t) = \int_0^{t_1} dt' \, k(t - t')m(t') \tag{4.8.9}$$

$$N_1(t) = \int_{t_1}^{t} dt' \, k(t - t')m(t') = M(t) - M_1(t) \qquad \text{and}$$

$$d_1(x) = \begin{cases} 1 , & x \in [-c, c/3] \\ 0 , & x \in [c/3, c] \end{cases}$$

$$f_1(x) = \frac{c^2 - 2x^2}{(c^2 - x^2)^{1/2}} . \tag{4.8.10}$$

The separation, for this fully open phase, has the form

$$\Delta_2(x, t) = M_1(t)(c/3 - x)^{3/2}(x + c)^{1/2}d_1(x) - N_1(t)x(c^2 - x^2)^{1/2} , \qquad |x| < c . \tag{4.8.11}$$

This solution will be valid until the crack begins to close again. There are two possibilities, which we refer to as cases (i), (ii). Case (i) is where the crack closes on the negative x-axis, while case (ii) is where, before this can happen, closure occurs again on the positive x-axis. We now consider these options in more detail.

Case (i). We look for a time t_2 such that at some point $x, g(x, t)$ given by (4.8.11) vanishes, or

$$\frac{((c/3) - x)^{3/2}}{x(c - x)^{1/2}} = \frac{N_1(t_2)}{M_1(t_2)} . \tag{4.8.12}$$

From (4.8.9), we see that $M_1(t)$ is positive, since $k(t)$ is a positive function (this is clear at least when Poisson's ratio is unique, since then $k(t)$ is related to the derivative of the creep function). Similarly, remembering (4.8.7), we have that $N_1(t)$ is negative so the right-hand side of (4.8.12) is negative and decreasing. Therefore, no solution can occur for positive x. It will occur first at $x = -c$, since this is the point on the negative axis at which the function on the left is maximum. Therefore, the condition is

$$\frac{4}{3}\left(\frac{2}{3}\right)^{1/2} = \frac{-N_1(t_2)}{M_1(t_2)} = 1 - \frac{M(t_2)}{M_1(t_2)} . \tag{4.8.13}$$

Of course, $g(x, t)$ vanishes anyway at $x = -c$. Condition (4.8.13) is really the requirement that its derivative vanish also, since the factor $(x + c)^{1/2}$ cancels in (4.8.12) on taking the ratio.

A priori, it is not clear whether the crack closes suddenly, as in the elastic case, or gradually. This is an interesting question, the answer of which emerges from the solution. Either way, the region on which $p(x, t)$ is known shrinks and so the considerations of Sect. 4.3, rather than Sect. 4.2, apply. Let the crack have closed to $a(t)$ at time $t > t_2$. Then, on $[a(t), c]$,

$$q(x,t) = \int_0^t dt' \, k(t-t') p(x,t') = -M(t)x + M_1(t) d_2(x) h(x)$$

$$d_2(x) = \begin{cases} 1 , & x \in [c/3, c] \\ 0 , & x \in [a(t), c/3] , \end{cases}$$

(4.8.14)

where $h(x)$ is defined by (4.8.6). We must impose the condition that no singularity occurs at $a(t)$. It follows from (4.3.6) [see also (4.3.8)] that this condition has the form

$$\int_{a(t)}^c dx \, q(x,t) \left(\frac{c-x}{x-a} \right)^{1/2} = 0 \quad \text{or}$$

(4.8.15)

$$\frac{M(t)}{M_1(t)} = \frac{I_2(a(t))}{I_1(a(t))} ,$$

(4.8.16)

where, with the aid of (A1.3.2):

$$I_1(a(t)) = \frac{1}{\pi} \int_a^c dx \left(\frac{c-x}{x-a} \right)^{1/2} x = \frac{c-a}{8} (c+3a)$$

$$I_2(a(t)) = \frac{1}{\pi} \int_{c/3}^c dx \left(\frac{c-x}{x-a} \right)^{1/2} \frac{((2c/3)+x)(x-c/3)^{1/2}}{(x+c)^{1/2}} .$$

(4.8.17)

The quantity $I_2(a(t))$ is an unwieldly function of $a(t)$ and best left in integral form. Equation (4.8.16) is an implicit equation determining $a(t)$, which may be solved numerically, as will be shown below. It is easy to show that if $a(t) = -c$ in (4.8.16), we obtain (4.8.13).

The form of (4.8.16) strongly suggests that the crack will close gradually rather than suddenly. This is confirmed in an example, later. Such behaviour is in sharp contrast to the behaviour of an elastic material.

Case (ii). If (4.8.13) cannot be satisfied then, when $m(t)$ becomes positive again, one can have a situation where $N_1(t)$ in (4.8.11) becomes zero and the crack closes again, suddenly, at the positive end, without closing at all at the negative end. A (frequency-dependent) class of standard linear solids with this property is distinguished in the next section. The fact that this can happen a second time suggests that a steady-state solution may exist, consisting of repeated closures of this kind.

4.8.1 A Standard Linear Solid

We now consider these results for $l(t), k(t)$ given by (3.11.9, 10), and for $m(t)$ varying sinusoidally as

$$m(t) = m_0 \sin(\omega t) , \quad t > 0 .$$

(4.8.18)

From (4.8.4), we have

$$M(t) = k_0 m_0 C(\phi, \gamma, \eta, T) ,$$

(4.8.19)

where the function $C(\phi, \gamma, \eta, T)$ is defined by (4.4.13) and the dimensionless quantities ϕ, γ, η, T by (4.4.14) and (4.4.15). At time

$$t_1 = \pi/\omega \; , \tag{4.8.20}$$

the quantity $m(t)$ vanishes while $M(t_1)$ is still positive. From (4.8.9), we have

$$M_1(t) = \frac{k_0 m_0 \gamma \eta e^{-T/\eta}}{1 + \eta^2} (e^{\pi/\eta} + 1) \; . \tag{4.8.21}$$

Let us define $R(T)$ as

$$R(T) = \frac{M(t)}{M_1(t)} = \frac{e^{T/\eta} \dfrac{\sin(T - \phi)}{\sin \phi} + 1}{e^{\pi/\eta} + 1} \; , \tag{4.8.22}$$

where we note that ϕ cannot be treated as independent of η (or γ). It varies in the range $[0, \frac{\pi}{2}]$. The limit $\gamma \to \frac{\pi}{2}$ can be obtained by allowing η, γ to become large, holding $\eta = \sqrt{\gamma}$.

Condition (4.8.13) may be written as

$$R(T) = 1 - \frac{4}{3} \left(\frac{2}{3} \right)^{1/2} = -0.088662 \; . \tag{4.8.23}$$

Assuming that an acceptable solution exists, say for $T = T_2$, condition (4.8.16), determining the interval of crack closure, has the form

$$R(T) = \frac{I_2(a(t))}{I_1(a(t))} = I(s(t)) \; , \quad s(t) = -\frac{a(t)}{c} \; , \tag{4.8.24}$$

where $I_1(a), I_2(a)$ are defined by (4.8.17). The quantity $I(s(t))$ is independent of the viscoelastic properties of the medium. It is plotted on Fig. 4.3. Note that it is negative for $s \geq 1/3$. As s approaches $1/3$, $I(s)$ tends to minus infinity. The quantity $R(T)$ cannot become infinite for finite T. Therefore, the elastic limit $s = 1/3$ will not be reached.

Examples of the function $R(T)$ are shown on Fig. 4.4. Further plots are given by Golden and Graham (1984). These are for various values of η, γ and for T in the region $[\pi, 2\pi]$. As a function of T, it descends from unity at $T = \pi$ to negative values, passing through a minimum and rising again, though in some cases this minimum occurs near $T = 2\pi$. The descent is particularly rapid for small values of η, γ.

In all cases shown and in many others examined, $R(T)$ drops below the constant on the right of (4.8.23), which we will refer to as K, so that the crack closes at the negative end. This closure is smooth rather than sudden, the point $a(t)$ being determined by comparing Figs. 4.3, 4.4. It will be perceived, however, that the closure is not particularly gradual, except perhaps for large values of η, γ, because of the very small slope of $I(s)$ for values of s above about 0.5. Very rapid closure will occur until this value is reached, and perhaps for longer. The final closing action will be relatively slow and to a value of s greater than $1/3$, as mentioned. The smaller the values of η, γ, the closer this point will be to the elastic value.

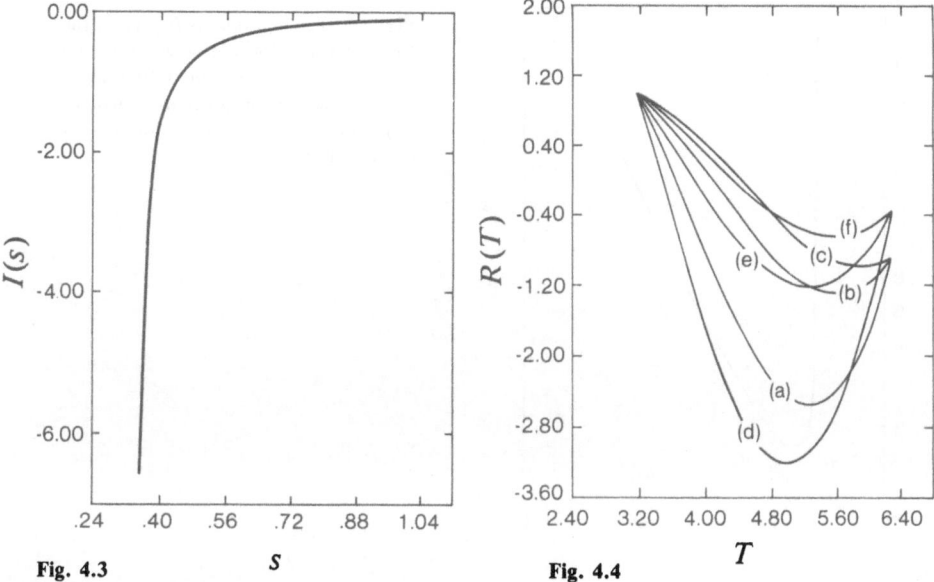

Fig. 4.3

S

Fig. 4.4

T

Fig. 4.3. The function $I(s)$ for relevant values of s

Fig. 4.4. The function $R(T)$ over the interval $[\pi, 2\pi]$ for **(a)** $\eta = 5.0$, $\gamma = 2.0$; **(b)** $\eta = 5.0$, $\gamma = 5.0$; **(c)** $\eta = 5.0$, $\gamma = 10.0$; **(d)** $\eta = 10.0$, $\gamma = 2.0$; **(e)** $\eta = 10.0$, $\gamma = 5.0$; **(f)** $\eta = 10.0$, $\gamma = 10.0$

As the value of $R(T)$ begins to ascend, the crack will begin to reopen smoothly, but this will not be complete before $\theta = 2\pi$ is reached. Beyond this point, the picture becomes increasingly complex and will not be explored.

Finally, it is of interest to distinguish the class of solids for which (4.8.23) cannot be satisfied. This is case (ii), described above. The minimum value of $R(T)$ occurs at

$$T_m = 2\pi + \phi - \delta , \quad \delta = \tan^{-1}\eta , \qquad (4.8.25)$$

which is always less than or equal to 2π. At this point,

$$R(T_m) = \frac{\sin \phi - \sin \delta e^{T_m/\eta}}{\sin \phi (e^{\pi/\eta} + 1)} , \quad \sin \delta = \frac{\eta}{(1 + \eta^2)^{1/2}} , \qquad (4.8.26)$$

which we observe is frequency-dependent. The equation $R(T_m) = K$ can be solved numerically to give the relationship between η, γ shown in Fig. 4.5. Points inside this curve correspond to materials and frequencies for which closure does not take place at the negative end before the positive end clamps down a second time. More detailed discussion of the curve may be found in Golden and Graham (1984). The application of the Extended Correspondence Principle to this problem was considered by Graham (1982).

Problem 4.8.1: Observe that the problem considered in this section provides an example of a receding contact problem as described in Sect. 2.9, and use it to illustrate the results given there.

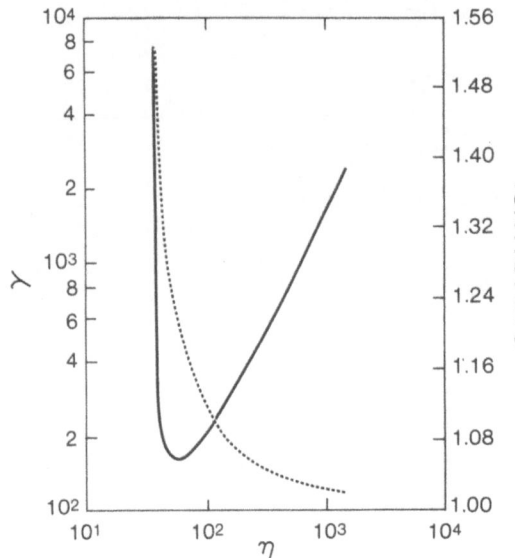

Fig. 4.5. The *continuous line* gives the functional relationship between η, γ implied by $R(T_m) = K$. The dotted line plots the loss angle ϕ against η where $\phi(\gamma, \eta)$ is calculated with values of η, γ on the curve

4.9 Summary

In this chapter, stress distributions and displacements are calculated for viscoelastic bodies loaded in plane strain and containing cracks. Conditions are derived for the extension of those cracks.

To a degree greater than their elastic counterparts, stress analysis problems for cracked viscoelastic bodies are subject to difficulties arising from the facts that cracks may extend with time and that boundary conditions may not be definitely known until the problem is solved. Valid solutions should involve no material overlap over the crack face. Furthermore, at points where the crack is closed, the normal traction should be a pressure while at points where it is open, this quantity is usually prescribed.

As in the previous chapter, we rely upon an adaptation of the Kolosov-Muskhelishvili equations. Problems for which the normal and shear forces are prescribed on the surfaces of open, and possibly growing, cracks may be solved without any restriction on viscoelastic material behaviour. The case of a single crack is studied in detail and explicit formulae are derived for stress intensity factors and displacements across the crack surface.

The mode of closure of a shear-free crack is considered and conditions [for example, (4.3.8)] derived for its closing. It is useful to distinguish stationary cracks from those that grow.

The problem of a stationary crack (i.e. a crack that remains constant in length) subject to an alternating time-dependent normal applied load has been studied for a standard linear solid in Sect. 4.4. It is found that the crack is always either entirely open or entirely closed. Furthermore, in contrast with elastic theory, the crack remains open for some time after the load becomes compressive, and opens up again before the load becomes tensile. Also, the stress singularities at the crack end become compressive. Detailed results are presented in Figs. 4.1, 2.

Under this loading regime, the state of being either entirely open or entirely closed carries over to the case of a crack that is growing whenever it is open. Furthermore, in this case, the crack will be open if an only if the load is tensile and stress singularities at the crack ends never become compressive. This is in line with the results of elastic theory.

The similarity between elastic and extending viscoelastic cracks carries over to the criteria for crack extension. In Sect. 4.6, the implications of an energy balance criterion for the growth of a crack in a viscoelastic body are explored. The condition obtained depends only on the initial values of the relaxation functions and takes the same form as that derived by Griffith for an elastic body. There follows, in Sect. 4.7, a demonstration valid in viscoelasticity, that the above condition may be expressed in terms of critical values of the stress intensity factors, which in turn are expressed in terms of distributions of cohesive forces at the crack ends. In this way, Barenblatt's elastic fracture criterion has been extended to viscoelasticity.

In Sect. 4.8, the problem of a stationary crack in a field of pure bending, whose neutral axis is perpendicular to the crack at its mid point, is studied. In elasticity, this problem involves instantaneous closure over a specific part of the crack face, depending on the sign of the applied bending moment. However, under similar loading conditions, a crack in a viscoelastic body may not close down and, if it does, the region of closure may vary with time. Furthermore, as in the case of a stationary crack under normal loading, stress singularities at the crack ends may become compressive. Numerical calculations are presented in Figs. 4.3 – 5.

5. Three-dimensional Contact Problems

It is to be expected that three-dimensional boundary value problems will present greater difficulties than plane problems. In particular, with the far wider choice of boundary regions on which to specify displacement and stress, one rapidly meets problems that are unsolvable – at least analytically. This is true even for elastic materials. In fact, the contact problem with an elliptical contact area is the most general problem that allows an explicit analytic solution – for elastic materials [Galin (1961), Lur'e (1964)], in the case of half-space problems. This corresponds to an ellipsoidal indentor, according to classical Hertz theory. The theory can be extended to cover contact between two gently curved bodies. The solution is valid only for quasi-static conditions.

In the absence of tangential motion, the equivalent viscoelastic problem is solvable, just as in the plane case – discussed in Sect. 3.10 – and by essentially the same methods. However, the problem is worth considering in some detail because, in contrast with the plane case, the indentation is determinate. This enables one to discuss impact problems, which are of considerable interest. These topics are covered in Sects. 5.2 and 5.3.

Inertial normal contact problems in three dimensions have been considered by Sabin (1975, 1987). By means of an integral transform technique, he reduces the problem to a set of dual integral equations, which are in turn reduced to a single Volterra integral equation. This is solved numerically. He obtains the interesting result that the contact pressure is not significantly different from that in the non-inertial problem. A similar observation had been made earlier, in connection with the elastic problem, by Tsai (1971).

A further and fundamental difficulty emerges, however, in connection with viscoelastic contact problems, if tangential motion is present. Distortion of the contact area occurs so that nothing is known a priori about its shape, which greatly increases the difficulty of finding solutions. In the elastic or normal viscoelastic problem, the shape of the contact area, though not known quantitatively before the problem is solved, can usually be determined, or at least intuitively guessed at, in a qualitative sense, from symmetries possessed by the indentor. A trivial example would be the case of rotational symmetry which gives circular contact patches. This greatly helps the solution process. The distortion that arises in the presence of tangential motion is of course associated with the phenomenon of hysteretic friction, discussed in Sect. 2.11.

The problem of a moving indentor has been considered by Golden (1982) in an approximate manner, by simply assuming that the contact region is elliptical for an ellipsoidal or spherical punch. The difficulty is that the physical significance of this assumption is not totally apparent. It is valid at both low and high velocities, as argued in the original paper. To what extent it is a good approximation for intermediate velocities is not clear. Therefore, discussion of this work is not included here. However, expressions for the hysteretic friction coefficient in the small viscoelasticity approximation are given in Sect. 5.4.

The problem of a circular specified load of fixed radius moving at uniform velocity over two- and three-layer viscoelastic media, in the context of pavement design, is considered by Battiato et al. (1977) and Battiato et al. (1982), using the Correspondence Principle.

In the non-inertial case, penny-shaped cracks have been considered by Graham (1970). No new result of physical or methodological significance emerges that has not been observed for plane cracks. Sabin (1975) treats inertial penny-shaped cracks.

5.1 Generalized Boussinesq Formula

We will write down the displacement-traction relationship on the boundary that will form the basis of the considerations of this chapter. This is essentially the solution of the stress boundary value problem, discussed in Sect. 3.2 in the plane case. We shall neglect surface shear,. however, so that the required relationship is a generalization to Viscoelasticity of the classical Boussinesq relationship. Its form follows directly from the elastic result by invoking the Classical Correspondence Principle. A more explicit derivation may be found in Hunter (1961) and also Golden (1978), who includes a shear traction term. Letting

$$\sigma_{zz}(x, y, 0, t) = -p(x, y, t) , \tag{5.1.1}$$

we have, for a half-space occupying $z > 0$:

$$v(r, t) = \int_{-\infty}^{t} dt' \, l(t-t')u(r, t') = \frac{1}{2\pi} \int_{C(t)} ds \frac{p(r', t)}{|r'-r|}$$

$$\tag{5.1.2}$$

$$r = (x, y, 0); |r'-r| = [(x-x')^2 + (y-y')^2]^{1/2} ,$$

where $l(t)$ is defined by (3.1.15) and ds is a surface element on the contact region $C(t)$. With certain modifications, essentially this formula applies to contact between two viscoelastic bodies with gently curving boundaries [Graham (1965), Golden (1978)].

A direct, formal approach towards the solution of the mixed boundary value problem, even for an elastic half-space, is a complex undertaking [Sneddon (1951), Galin (1961), Lur'e (1964)]. For a viscoelastic medium, such a direct approach would not, in general, be practical. Therefore, we adopt an indirect strategy, basing our reasoning on (5.1.2) and relying heavily on analogy with known elastic solutions.

As in the plane case, there will be an apparently total emphasis on boundary quantities, to the exclusion of discussion of the behaviour of the various quantities in the interior of the body. There is no fundamental reason for this other than that from the phenomenological point of view, surface quantities are often more interesting. Anyway, once these are known, stresses and strains all over the body are easily calculated – with the aid of the Classical Correspondence Principle.

5.2 The Normal Contact Problem under Varying Load

Consider the problem of a rigid indentor, pressed into a viscoelastic half-space, under a varying load. Friction between the indentor and the half-space is neglected here; Graham (1980) considers this problem, in the case of complete and partial adhesion, and Graham and Sneddon (1981) examine the problem with rotational friction. As noted above, our approach will be based on (5.1.2). The strategy will be to reduce it to a form analogous to the elastic equation, and then use the familiar elastic solutions to determine the viscoelastic quantities.

As the load varies, it will be assumed that the contact patch will pass through a one-parameter family of states, as shown schematically in Fig. 3.2. This assumption will be justified later on the basis that it enables the problem to be solved. Furthermore, it will be shown that the one-parameter family of states is in fact the family of possible elastic states. The fact that $C(t)$ is a one-parameter family means that the explicit formalism developed for repetitive expansion and contraction in Sects. 2.6 and 3.10 may be used, as opposed to the more general method summarized in Sect. 2.6 in the context of the Extended Correspondence Principle, which is applicable to any situation where the boundary regions are expanding and contracting in time.

This problem was first considered by Lee and Radok (1960) for the special case where $C(t)$ is non-decreasing. Hunter (1960) and Graham (1965 a) gave the solution for the case where $C(t)$ possesses a single maximum, while Efimov (1966a) considers the problem beyond the second maximum. Solutions in the general case, where $C(t)$ has any number of maxima, were given by Ting (1966, 1968) and Graham (1967). The corresponding problem for an aging material was solved by Sabin and Graham (1980). We consider the non-aging problem here, using the same method as in Sect. 3.10, which is equivalent to that of Ting (1966, 1968) and Graham (1967), but quite different in superficial details (see (2.11.6)).

In the contact region, the normal displacement, which we denote by $u(r, t)$, dropping the subscript z, will have the form

$$u(r, t) = D(t) - S(r) , \qquad S(0) = 0 , \tag{5.2.1}$$

where $S(r)$ characterizes the shape of the indentor, and $D(t)$ is its penetration into the half-space at the origin – which in practice is generally chosen to be the point of maximum penetration.

Before considering a general history of loading, two simple cases will be discussed. First consider the situation where $C(t')$ is non-decreasing at all times t' up to the current time t. This is covered by the Extended Correspondence Principle discussed in Sect. 2.6; see also Sect. 3.9. Rather than invoke the Principle directly, it is instructive to give an explicit solution. Let us replace $C(t)$

in (5.1.2) by the entire boundary and then transfer the hereditary integral onto the pressure, giving

$$q(r, t) = \int_{-\infty}^{t} dt' \, k(t - t') p(r, t') \, , \qquad (5.2.2)$$

the quantity $k(t)$ being defined by (3.2.12). Any point that is outside the contact area at time t has always been outside it, so that $q(r, t) = 0$, $r \in C'(t)$. Therefore, we can take the region of integration of the space integral to be $C(t)$, giving

$$u(r, t) = D(t) - S(r) = \frac{1}{2\pi} \int_{C(t)} ds' \, \frac{q(r', t)}{|r - r'|} \, . \qquad (5.2.3)$$

Equation (5.2.3) is formally identical to the elastic equation. Let us envisage the same indentor on an elastic half-space with shear modulus and Poisson's ratio denoted by μ_e and ν_e, respectively. Then

$$q(r, t) = k_e \, p_e(r, t) \qquad (5.2.4)$$

$$k_e = \frac{1 - \nu_e}{\mu_e} \, ,$$

where $p_e(r, t)$ is the pressure distribution on an elastic medium characterized by k_e, when the contact region is $C(t)$. We deduce that the pressure $p(r, t)$ is given by

$$p(r, t) = k_e \int_{-\infty}^{t} dt' \, l(t - t') p_e(r, t') \, . \qquad (5.2.5)$$

Integrating this equation over the boundary gives a relation between the loads on the viscoelastic and elastic half-spaces, respectively and k_e. We are free to choose k_e, but then the load $W_e(t)$ on the elastic half-space is fixed. In certain later contexts, k_e will be chosen to be equal to k_0. Another choice of k_e is

$$k_e = \int_{0}^{\infty} dt' \, k(t') = \frac{1 - \hat{\nu}(0)}{\hat{\mu}(0)} \qquad (5.2.6)$$

which is the long-time elastic limit of $K(t)$ where $k(t) = d/dt(H(t) K(t))$ [cf. (1.2.33)]. If the proportionality assumption holds, then k_e is $(1 - \nu)/\mu_e$ where μ_e is given by the long time limit of the shear relaxation function $G(\infty)$. The indentation $D(t)$ has the form

$$D(t) = D_e(t) \, , \qquad (5.2.7)$$

where $D_e(t)$ is the elastic indentation corresponding to $p_e(r, t)$ and $C(t)$.

The second case is where $C(t')$ is non-increasing for all $t' < t$. This is also covered by the Extended Correspondence Principle. All points in the contact region at time t have always been there. Therefore, since $S(r)$ is time-independent, (5.1.2) can be written as

$$l_e \, (D_l(t) - S(r)) = \frac{1}{2\pi} \int\limits_{C(t)} ds' \frac{p(r', t)}{|r' - r|}$$

$$l_e = \int\limits_0^\infty dt' \, l(t') \tag{5.2.8}$$

$$D_l(t) = \frac{1}{l_e} \int\limits_{-\infty}^t dt' \, l(t - t') D(t') \; .$$

We deduce that the pressure is given by

$$p(r, t) = p_e \, (r, t) \tag{5.2.9}$$

where $p_e \, (r, t)$ is the pressure distribution on an elastic half-space, characterized by l_e, when the contact region is $C(t)$; and the penetration is determined by $D_l(t) = D_e \, (t)$ so that

$$D(t) = l_e \int\limits_{-\infty}^t dt' \, k(t - t') D_e \, (t') \; . \tag{5.2.10}$$

The quantity l_e is the inverse of k_e, defined by (5.2.6). Note, however, that we are not restricted to this choice. Another coefficient could be inserted in (5.2.8).

Consider now a general loading history. We follow closely the procedure developed in Sect. 3.10. In the case where $C(t)$ is contracting at time t, we have, instead of (3.10.5) and (3.10.6):

$$v(r, t) = v_c(r, t) + \frac{1}{2\pi} \int\limits_{C(t)} ds' \frac{q_c(r', t)}{|r' - r|} \; , \tag{5.2.11}$$

where

$$q_c(r, t) = \int\limits_{W_\sigma(t)} dt' \, \Pi_\sigma(t, t') p(r, t') \tag{5.2.12a}$$

$$v_c(r, t) = \int\limits_{W_u(t)} dt' \, \Pi_u(t, t')[D(t') - S(r)] \tag{5.2.12b}$$

$$= D_c(t) - S(r) \, \Pi_u(t) \; , \tag{5.2.12c}$$

using the notation of (3.10.4) and

$$D_c(t) = \int\limits_{W_u(t)} dt' \, \Pi_u(t, t') D(t') \; . \tag{5.2.13}$$

Instead of (3.10.7), we have

$$v_c(r, t) = \frac{1}{2\pi} \int\limits_{C(t)} ds' \frac{p(r', t) - q_c(r', t)}{|r' - r|} \; , \tag{5.2.14}$$

where $v_c \, (r, t)$ is known in $C(t)$ and has a form similar to (5.2.1), namely that given by (5.2.12c). Equation (5.2.14) therefore has the standard elastic form and we deduce that

$$p(r, t) = q_c \, (r, t) + k_e \, \Pi_u(t) p_e \, (r, t) \; . \tag{5.2.15}$$

The occurrence of the factor $\Pi_u(t)$ is a consequence of the fact that $v_c(r, t)$, given by (5.2.12c), has such a factor multiplying the shape function $S(r)$. Equation (5.2.15) defines the pressure at time t inductively. The point is that $q_c(r, t)$, given by (5.2.12), depends on $p(r, t')$, $t' \le \theta_1(t)$. This is clear from the definition of $\Pi_\sigma(t, t')$ given by (2.4.18). Therefore, if $p(r, t')$ is known up to the last maximum of the contact area, then (5.2.15) determines its form during the contracting phase. In the present context, the $\theta_l(t)$, $l = 1, 2, \ldots$, are defined by [see (2.6.7), (3.10.1)]:

$$\theta_l(t) = t_l(a(t)) \tag{5.2.16}$$

in terms of the times $t_l(r)$ introduced in Sect. 2.4 [see (2.4.19)], where $a(t)$ is the distance from the origin to the point on the boundary of $C(t)$ in a pre-defined, fixed direction. The parameter $a(t)$ uniquely characterizes $C(t)$.

In (5.2.15), we can as before choose k_e freely, for example according to (5.2.6), but then the elastic load $W_e(t)$ is given in terms of the viscoelastic load $W(t)$ by the integrated form of (5.2.15).

The depth of penetration is determined by the condition

$$D_c(t) = \int_{W_u(t)} dt'\, \Pi_u(t, t')D(t') = \Pi_u(t)\, D_e(t) , \tag{5.2.17}$$

where, as before, $D_e(t)$ is the elastic penetration corresponding to the pressure distribution $p_e(r, t)$.

If $C(t)$ is increasing at time t, then instead of (3.10.3), we have

$$D(t) - D_u(t) - S(r)[1 - \Gamma_u(t)] = \frac{1}{2\pi} \int_{C(t)} ds'\, \frac{q_\sigma(r', t)}{|r' - r|}$$

$$q_\sigma(r, t) = \int_{W_\sigma(t)} dt'\, \Gamma_\sigma(t, t')p(r, t') \tag{5.2.18}$$

$$D_u(t) = \int_{W_u(t)} dt'\, \Gamma_u(t, t')D(t')$$

where $\Gamma_u(t)$ is given by (3.10.10b). Thus, in analogy with (3.10.10a):

$$\int_{W_\sigma(t)} dt'\, \Gamma_\sigma(t, t')p(r, t') = k_e[1 - \Gamma_u(t)]p_e(r, t) . \tag{5.2.19}$$

This can be recast in a form corresponding to (3.10.12) which shows it explicitly to be an equation for $p(r, t)$, once again assuming that this quantity is known up to the last minimum. The equation determining the penetration is

$$D(t) = \int_{W_u(t)} dt'\, \Gamma_u(t, t')D(t') + D_e(t)[1 - \Gamma_u(t)] . \tag{5.2.20}$$

These results allow one to build up knowledge of the pressure and penetration after any finite number of maxima and minima. The inductive procedure can be initiated using (5.2.5, 7, 9) and (5.2.10).

Thus a solution to the problem can be found on the basis of our initial assumption regarding the shape of the contact area. This in effect justifies that assumption. Since the solution was found by reducing the problem to elastic form, we may deduce that the one-parameter family of contact patches is the family of elastic contact patches.

The simplest case of (5.2.15) and (5.2.17) will now be discussed in more detail, partly to provide a more explicit illustration of these results, and partly because it is required in the next section in the context of impact problems. We assume that $C(t)$ is decreasing, having previously passed through a single maximum. The quantity $\Pi_u(t)$ has the form [see (2.4.18) and (3.10.4)]:

$$\Pi_u(t) = \int_{\theta_1(t)}^{t} dt'\, l(t-t') = \int_{0}^{t-\theta_1(t)} dt'\, l(t') = l_e\, F[t-\theta_1(t)] \qquad (5.2.21\,\text{a})$$

$$F(t) = \frac{1}{l_e} \int_{0}^{t} dt'\, l(t') \ . \qquad (5.2.21\,\text{b})$$

Also, $p_e(r, t) = p_e(r, \theta_1(t))$. Using (2.4.18) and (2.4.6) to give Π_σ (noting that $\theta_2(t) = -\infty$) and (5.2.5) to give the form of $p(r, t')$ before the maximum point, we obtain

$$p(r, t) = \int_{-\infty}^{\theta_1(t)} dt'\, j(t, t')\, p_e(r, t')$$

$$j(t, t') = F(t-\theta_1(t))\, \delta(t' - \theta_1(t)) + k_e\, l(t-t'), \quad t > t' \ . \qquad (5.2.22)$$

Equation (2.3.15) has been used, after a change in the order of integration.

Problem 5.2.1: Show that (5.2.22) can be expressed as

$$p(r, t) = k_e \int_{-\infty}^{\theta_1(t)} dt'\, L(t-t')\, \dot{p}_e(r, t') \qquad (5.2.1\text{p})$$

where [see (1.10.1p)]:

$$l(t) = L(0)\, \delta(t) + \dot{L}(t)\, H(t) \ . \qquad (5.2.2\text{p})$$

This is essentially the result of Hunter (1960) and Graham (1965a).

From (5.2.17) and (2.4.18), we have that the degree of penetration $D(t)$ obeys the equation

$$F(t-\theta_1(t))\, D_e(t) = k_e \int_{\theta_1(t)}^{t} dt'\, l(t-t')\, D(t') \ . \qquad (5.2.23)$$

Up to time t_m when $C(t)$ passes through its maximum, $D(t)$ is given by its elastic form. Putting

$$D(t') = D(t') - D_e(t') + D_e(t') \ , \qquad (5.2.24)$$

we obtain that

$$k_e \int_{t_m}^{t} dt'\, l(t-t')[D(t')) - D_e(t')] = F(t-\theta_1(t))\, D_e(t)$$

$$\hspace{4cm} - k_e \int_{\theta_1(t)}^{t} dt'\, l(t-t')\, D_e(t') \ . \qquad (5.2.25)$$

Solving for $D(t)$, using (2.3.15), gives

$$D(t) = D_e(t) + \frac{1}{k_e} \int\limits_{t_m}^{t} dt' \, k(t-t') [F(t'-\theta_1(t')) D_e(t')$$

$$- k_e \int\limits_{\theta_1(t')}^{t'} dt'' \, l(t'-t'') D_e(t'')] \ . \tag{5.2.26}$$

Alternatively, from (5.2.24) we have

$$k_e \int\limits_{-\infty}^{t} dt' \, l(t-t') D(t') = k_e \int\limits_{t_m}^{t} dt' \, l(t-t') [D(t') - D_e(t')]$$

$$+ k_e \int\limits_{-\infty}^{t} dt' \, l(t-t') D_e(t') \ . \tag{5.2.27}$$

Substituting from (5.2.25) gives

$$k_e \int\limits_{-\infty}^{t} dt' \, l(t-t') D(t') = F(t-\theta_1(t)) D_e(t) + k_e \int\limits_{-\infty}^{\theta_1(t)} dt' \, l(t-t') D_e(t')$$

$$= \int\limits_{-\infty}^{\theta_1(t)} dt' \, j(t,t') D_e(t') \tag{5.2.28}$$

so that

$$D(t) = l_e \int\limits_{-\infty}^{t} dt' \, k(t-t') \int\limits_{-\infty}^{\theta_1(t')} dt'' \, j(t',t'') D_e(t'') \ . \tag{5.2.29}$$

We will confine ourselves to axisymmetric indentors, noting of course that fairly explicit results can also be given for an ellipsoidal indentor, by virtue of the classical Hertz solution. For a spherical indentor we may approximately take

$$S(r) = \frac{r^2}{2R} = cr^2 \ , \tag{5.2.30}$$

where R is the radius of the indentor. The relations between penetration, load and the radius of contact $a(t)$ then have the form [see Sneddon (1965)]:

$$D_e(t) = 2a^2(t) c \tag{5.2.31a}$$

$$W_e(t) = \frac{16}{3} l_e c a^3(t) \ , \tag{5.2.31b}$$

while $p_e(r,t)$ is given by

$$p_e(r,t) = p_0 [a^2(t) - r^2]^{1/2} \tag{5.2.32}$$

$$p_0 = \frac{8 \, l_e c}{\pi} = \frac{3}{2} \frac{W_e(t)}{\pi a^3(t)}$$

where $W_e(t)$ is the total load on the elastic medium characterized by l_e, that would cause a contact radius of $a(t)$. This quantity $a(t)$ links the elastic and viscoelastic solutions. One substitutes $p_e(r,t)$ and $D_e(t)$ in (5.2.22) and (5.2.29) to obtain $p(r,t)$ and $D(t)$.

For a more general indentor shape,

$$S(r) = cr^n, \quad n > 0 . \tag{5.2.33}$$

Equations (5.2.31) can be generalized according to formulae given by Galin (1961), [see also Sneedon (1965)]:

$$D_e(t) = A[a(t)]^n$$
$$A = 2^{n-2} \frac{nc[\Gamma(n/2)]^2}{\Gamma(n)} \tag{5.2.34}$$

and

$$W_e(t) = l_e A_1 [a(t)]^{n+1}$$
$$A_1 = \frac{4An}{n+1} . \tag{5.2.35}$$

The expression for the pressure is more complicated in the general case and will not be given. Equations (5.2.34) and (5.2.35) will find application in the next section.

5.2.1 The Steady-State Problem

We will discuss briefly the steady-state limit of the normal problem in three dimensions. This was analyzed in detail for the plane case in Sect. 3.11. The formulae developed in that section go over to three dimensions with minimal and obvious changes; essentially, these amount to substituting the three-dimensional elastic pressure distribution for the two dimensional form – for example (5.2.32) with $l_e = l_0$, replacing (3.11.46).

However, the depth of penetration was not discussed, since it is indeterminate in the plane problem. We will indicate the formulae determining this quantity and draw certain conclusions. As in Sect. 3.11, we focus on a period $[\Delta_1, \Delta_2]$ around the time t_0 when the contact area is a minimum. The values assigned to Δ_1, Δ_2, t_0 at the end of Sect. 3.11 for a load given by (3.11.58) also apply here. During the decreasing phase, before t_0, $D(t)$ is determined by (5.2.17), which becomes [see (2.4.18)]:

$$\int_{t_1(t)-\Delta}^{t} dt' \, \Pi_u^{(p)}(t, t') D(t') = \Pi_u(t) D_e(t)$$
$$\Pi_u^{(p)}(t, t') = \sum_{k=0}^{\infty} T_{2k}(t, t' - k\Delta) , \tag{5.2.36}$$

where $\Pi_u(t)$ is given by (3.11.5). During the increasing phase $[t_0, \Delta_2]$, $D(t)$ is determined by (5.2.20), now becoming [see (2.4.13)]:

$$D(t) = \int_{t-\Delta}^{t_1(t)} dt' \, \Gamma_u^{(p)}(t, t') D(t') + D_e(t)[1 - \Gamma_u(t)]$$
$$\Gamma_u^{(p)}(t, t') = \sum_{k=0}^{\infty} N_{2k+1}(t, t' - k\Delta) \tag{5.2.37}$$

where $\Gamma_u(t)$ is given by (3.11.8).

In the case of the standard linear model, $\Pi_u(t)$ is given by (3.11.17), while

$$\Pi_u^{(p)}(t, t') = l_0 \delta(t - t') + \frac{l_1}{1 - E_c(t)} e^{-\alpha(t - t')} . \tag{5.2.38}$$

Also, $\Gamma_u(t)$ is given by (3.11.22) and

$$\Gamma_u^{(p)}(t, t') = \frac{k_1 l_0}{1 - E(t)} \exp\{-\beta[t - t_1(t)] - \alpha[t_1(t) - t']\} . \tag{5.2.39}$$

Let us now agree that $t > t_0$, and write out the two equations more explicitly. Equation (5.2.37) becomes, on transforming the integral to $[t, t_1(t) + \varDelta]$:

$$D(t) = F_1(t) \int_t^{t_1(t) + \varDelta} dt' \, e^{\alpha t'} D(t') + H_1(t) D_e(t)$$

$$\tag{5.2.40}$$

$$F_1(t) = \frac{r\alpha E(t) e^{-\alpha t}}{1 - E(t)}$$

where $H_1(t)$ is given by (3.11.26) and r by (3.11.25). The notation has been altered here for reasons of clarity. Also, (5.2.36) becomes:

$$D(t_1(t)) = F_2(t) \int_t^{t_1(t) + \varDelta} dt' \, e^{\alpha t'} D(t') + H_2(t) D_e(t)$$

$$\tag{5.2.41}$$

$$F_2(t) = F_1(t) e^{\beta(t - t_1(t))} ,$$

where $H_2(t)$ is given by (3.11.31). Combining these two equations gives a result analogous to (3.11.32), namely

$$D(t_1(t)) = \sigma(t) D(t) + [1 - \sigma(t)] D_e(t)$$

$$\tag{5.2.42}$$

$$\sigma(t) = e^{\beta[t - t_1(t)]} .$$

We now transform (5.2.40) to an equation over the region $[t_0, \varDelta_2]$, using (5.2.42). This is a step analogous to (3.11.34). The result is

$$D(t) = F_1(t) \int_t^{\varDelta_2} dt' \, B_1(t') D(t') + Q(t) , \quad \text{where} \tag{5.2.43}$$

$$B_1(t') = e^{\alpha t'} - \dot{t}_1(t') \sigma(t') e^{\alpha[t_1(t') + \varDelta]}$$

$$\tag{5.2.44}$$

$$= e^{\alpha t'} \left(1 - \frac{\dot{t}_1(t')}{E(t')}\right) \quad \text{and}$$

$$Q(t) = H_1(t) D_e(t) - F_1(t) \int_t^{\varDelta_2} dt' \, \dot{t}_1(t')[1 - \sigma(t')] e^{\alpha[t_1(t') + \varDelta]} D_e(t') . \tag{5.2.45}$$

We differentiate (5.2.43) with respect to t, as in (3.11.37), to obtain

$$\frac{d}{dt}\left(\frac{D(t)}{F_1(t)}\right) = -B_1(t) D(t) + \frac{d}{dt}\left(\frac{Q(t)}{F_1(t)}\right) \quad \text{or} \tag{5.2.46}$$

$$\dot{D}(t) = -\beta D(t) + h(t) \quad \text{where} \tag{5.2.47}$$

$$h(t) = H_1(t)\dot{D}_e(t) + \beta D_e(t) , \tag{5.2.48}$$

which is similar in form to $b(t)$, given by (3.11.48). In order to solve (5.2.47), we need to know $D(t)$ at some particular time. From (5.2.43) and (5.2.45) it follows that

$$D(\Delta_2) = D_e(\Delta_2) , \tag{5.2.49}$$

since $H_1(\Delta_2)$ is unity. In other words, the elastic and viscoelastic indentations coincide when the contact area is maximum. The solution to (5.2.47) has the form

$$D(t) = -\int_t^{\Delta_2} dt' \, e^{-\beta(t-t')} h(t') + D_e(\Delta_2) e^{-\beta(t-\Delta_2)} . \tag{5.2.50}$$

All quantities on the right are known. The quantity $D_e(t)$ is related by (5.2.31) to $W_e(t)$ which is itself determined as outlined in Sect. 3.11. Note that, in line with the developments of that section, we replace l_e by l_0 in (5.2.31b). This is for $t \in [t_0, \Delta_2]$. The solution for $t \in [\Delta_1, t_0]$ is given by (5.2.42). Equation (5.2.50) can be rewritten in the form

$$D(t) = -\int_t^{\Delta_2} dt' \, H(t, t') D_e(t') + H_1(t) D_e(t) , \tag{5.2.51}$$

where

$$H(t, t') = \beta e^{-\beta(t-t')} - \frac{d}{dt'} [e^{-\beta(t-t')} H_1(t')] . \tag{5.2.52}$$

Note that we deduce from (5.2.40), in a manner similar to the derivation of (3.11.55), that

$$D_e(t_0) = \frac{1}{1-r} D(t_0) - \frac{r\alpha e^{-\alpha(t_0+\Delta)}}{(1-r)(1-e^{-\alpha\Delta})} \int_{t_0}^{t_0+\Delta} dt' \, e^{\alpha t'} D(t')$$

$$= \frac{1}{l_0(1-r)} \int_{-\infty}^{t_0} l(t_0-t') D(t') \, dt' . \tag{5.2.53}$$

Equations (5.2.49, 53) are special cases of results shown to be valid for a periodic loading of general viscoelastic bodies by Graham and Golden (1988).

A numerical example is given on Fig. 5.1 for a spherical indentor as described by (5.2.31) and (5.2.32). The overall level of indentation increases with r. The minimum point is shifted in a positive direction as r increases. This is a manifestation of the phase lag between stress and strain in viscoelastic materials. The parameter α/ω has little effect on the magnitude of the indentation, particularly at small values of r. It affects the position of the minimum and alters the values of Δ_1, Δ_2, as does the value of r. This is clear anyway from (3.11.60). Near the minimum, indentation decreases with increasing α/ω. At the maxima, this is reversed. Further data are presented in Golden and Graham (1987b) and Graham and Golden (1987).

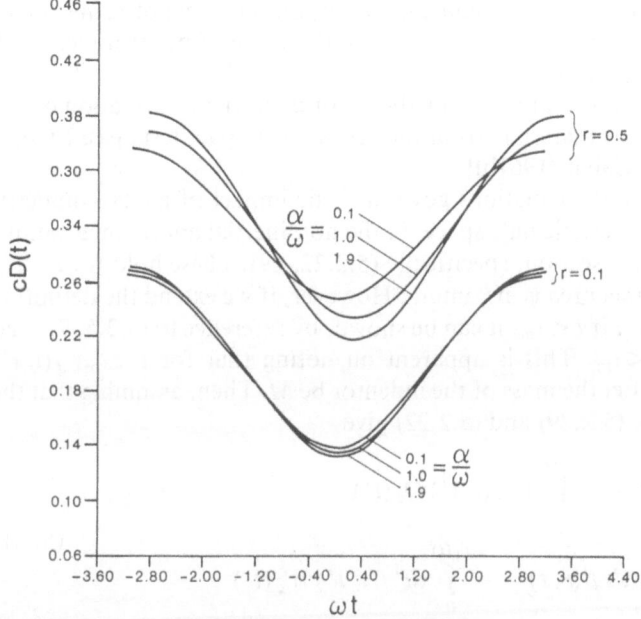

Fig. 5.1. The dimensionless indentation $cD(t)$ plotted over the time interval $[\Delta_1, \Delta_2]$ for $d = 2.0$, $c^2 K k_0 = 0.01$ and the parameters r and α/ω as shown

5.3 Impact Problems

A practical application of the general theory developed in the previous section is to the problem of a rigid indentor impacting once on a viscoelastic half-space and then losing contact with it. This corresponds to the case where $C(t)$ passes through a single maximum. Hunter (1960) first gave an approximate solution to this problem, deducing expressions for the total time of contact and the coefficient of restitution for a rigid sphere on a half-space composed of a Maxwell material or more generally, in the short contact time approximation, discussed later. He also considered the problem of a flat circular punch impacting on a viscoelastic half-space [Hunter (1966)], which is much simpler because the contact area does not vary. Calvit (1967a) has given a numerical procedure, based on the use of Taylor expansions, for solving the rigid sphere on the viscoelastic half-space problem. Earlier, a heuristic treatment of the problem was given by Pao (1955). An alternative numerical procedure was evolved by Graham (1978). Sabin and Graham (1983) propose a more elaborate and efficient numerical method.

All this work is based on the non-inertial approximation. However the practical utility of this approximation for the corresponding elastic problem has been demonstrated by Lord Rayleigh [Strutt (1906)], Hunter (1957) and Tsai (1968, 1971). If anything, the theory should be more realistic in the viscoelastic case. On the other hand, Aboudi (1979) who has developed a completely numerical theory, including inertial effects, might cast doubt on the value of using such a theory, but even this should not take away from the value of better understanding the implications of that theory.

These theories have applications to methods for deducing information on the mechanical properties of materials from the results of impact tests [see Lifshitz and Kolsky (1964) an Calvit (1967b)].

We now write down the equations governing the impact of an axisymmetric rigid indentor on a viscoelastic half-space, in the non-inertial approximation, using the results of the last section, specifically (5.2.22, 29). These hold for $t > t_m$, the time when the contact area is maximum. However, if we extend the definition of $\theta_1(t)$ so that $\theta_1(t) = t$ if $t < t_m$, it can be shown, by reference to (5.2.5, 7), that they also hold for $t < t_m$. This is apparent on noting that for $t < t_m$, $j(t, t')$ reduces to $k_e l(t - t')$. Let the mass of the indentor be M. Then, assuming that the impact begins at $t = 0$, (5.2.29) and (5.2.22) give

$$k_e \int_0^t dt' \, l(t - t') D(t') = \int_0^{\theta_1(t)} dt' \, j(t, t') D_e(t')$$

(5.3.1)

$$M \frac{d^2}{dt^2} D(t) = - \int_{C(t)} ds \, p(r, t) = - \int_0^{\theta_1(t)} dt' \, j(t, t') W_e(t') \ .$$

Assuming that the punch shape is given by (5.2.33), relations (5.3.1) give, by virtue of (5.2.34, 35):

$$\int_0^t dt' \, l(t - t') D(t') = A \int_0^{\theta_1(t)} dt' \, l_1(t, t') a^n(t')$$

(5.3.2a)

$$\frac{d^2}{dt^2} D(t) = - \frac{A_1}{M} \int_0^{\theta_1(t)} dt' \, l_1(t, t') a^{n+1}(t')$$

(5.3.2b)

where A is defined by (5.2.34), and A_1 is given by (5.2.35). Also

$$l_1(t, t') = l_e \, j(t, t') \ ,$$

(5.3.3)

the quantity $j(t, t')$ being defined by (5.2.22). If $t < t_m$, then

$$l_1(t, t') = l(t - t') \ .$$

(5.3.4)

Equations (5.3.2) are the fundamental dynamical equations governing the impact problem, to which must be added initial conditions. We have

$$D(0) = a(0) = 0 \ , \quad \dot{D}(0) = V \ ,$$

(5.3.5)

where V is the relative speed of the indentor and the half-space, just before impact, and the dot indicates time differentiation.

In general, these equations are susceptible only to numerical treatment, which has been given by Graham (1978), Sabin and Graham (1983). However, if viscoelastic effects are small, they can be regarded as a perturbation of the exact elastic theory, and some approximate formulae may be given. One way in which viscoelastic effects may be small is when the total collision time is short compared with the decay times of the material, so that we can approximate the viscoelastic functions by their Taylor expansions, keeping only terms linear in time and constant terms. It is noteworthy that in this approximation, the half-space acts like a Maxwell material with long decay time (Sect. 1.6). This approach has been explored by Hunter (1960) and Graham (1978).

However, viscoelastic effects may be intrinsically small compared with elastic effects, as assumed in Sect. 3.7. In this case, there is no assumption, necessarily, that the decay times are long.

The treatment here will merely assume that the time-dependent part of the viscoelastic functions is small compared with the time-independent part, which covers both possibilities mentioned above. We let $l(t), k(t)$ be given by the form (3.7.1) but where the dimensional time variable is used:

$$l(t) = l_0 [\delta(t) + \varepsilon(t)] , \quad k(t) = k_0 [\delta(t) - \varepsilon(t)] , \tag{5.3.6}$$

where $\varepsilon(t)$ is small and where l_0, k_0 depend on the instantaneous moduli. Subsequent calculations will be carried out to first order in $\varepsilon(t)$. The function $F(t - \theta(t))$, given by (5.2.21b) has the form

$$F(t - \theta_1(t)) = \frac{L(t - \theta_1(t))}{l_e} \tag{5.3.7}$$

by virtue of (5.2.2p), where

$$L(t) = l_0 [1 + \varepsilon_1(t)] , \tag{5.3.8}$$

the quantity $\varepsilon_1(t)$ being defined by

$$\dot{\varepsilon}_1(t) = \varepsilon(t) , \quad \varepsilon_1(0) = 0 . \tag{5.3.9}$$

We henceforth in this section choose $l_e = l_0$, the instantaneous value. First, consider the elastic limit [Deresiewicz (1968), Graham (1973)]. Equations (5.3.2) reduce to

$$D_0(t) = A a_0^n(t) , \quad \ddot{D}_0(t) = - \frac{A_1}{M} l_0 a_0^{n+1}(t) . \tag{5.3.10}$$

Substituting for $a_0(t)$ in the second equation in terms of $D_0(t)$ from the first equation gives

$$\ddot{D}_0(t) = - B_1 l_0 D_0^{\beta-1}(t)$$

$$\beta = \frac{2n+1}{n} \tag{5.3.11}$$

$$B_1 = \frac{A_1}{M A^{\beta-1}} .$$

We now proceed to obtain the first integral of this equation, essentially the energy integral of particle mechanics. Multiplying across by $\dot{D}_0(t)$ and integrating, we find that

$$\dot{D}_0^2(t) = V^2 - B_2 D_0^\beta(t) , \quad \text{where} \tag{5.3.12}$$

$$B_2 = \frac{2 B_1 l_0}{\beta} . \tag{5.3.13}$$

It follows immediately that the maximum indentation is given by

$$D_{0m} = \left(\frac{V^2}{B_2}\right)^{1/\beta} ,$$

(5.3.14)

and the maximum radius has the form

$$a_{0m} = \left(\frac{V^2}{B_2}\right)^{1/(2n+1)} \left(\frac{1}{A}\right)^{1/n} .$$

(5.3.15)

We can rewrite (5.3.12) in dimensionless form, defining first of all

$$u = \frac{D_0(t)}{D_{0m}} ,$$

(5.3.16)

in terms of which we have

$$\dot{u}^2 = E_0^2 h^2(u)$$

(5.3.17)

where

$$E_0 = \frac{V}{D_{0m}} = V^{1/(2n+1)} B_2^{1/\beta} , \quad h(u) = (1 - u^\beta)^{1/2} .$$

(5.3.18)

The solution of (5.3.17) is

$$t = \frac{1}{E_0} \int_0^u \frac{dv}{h(v)} = \frac{1}{\beta E_0} B_x[1/\beta, 1/2] ,$$

(5.3.19)

where

$$x = u^\beta$$

(5.3.20)

and $B_x[p, q]$ is the incomplete Beta function (Table A1.4). The time at which maximum indentation is achieved is given in terms of the complete Beta function by

$$t_{0m} = \frac{1}{E_0} \int_0^1 \frac{dv}{h(v)} = \frac{1}{\beta E_0} B[1/\beta, 1/2] .$$

(5.3.21)

Equation (5.3.19) applies to the first phase, when indentation is increasing. For the second, or ejection phase, the solution is

$$t - t_{0m} = -\frac{1}{E_0} \int_1^u \frac{dv}{h(v)} .$$

(5.3.22)

It may be shown without difficulty that the time when contact ends, or the total collision time T_0, is given by $2 t_{0m}$. In fact, there is complete symmetry between the two phases.

We now consider the viscoelastic correction.

5.3.1 The First Phase

During the period that the indentation is increasing, (5.3.2) becomes

$$D(t) = A a^n(t) , \qquad \ddot{D}(t) = -\frac{A_1}{M} \int_0^t dt' \, l(t-t') a^{n+1}(t') . \tag{5.3.23}$$

As in the elastic case, substitution from the first equation in the second gives

$$\ddot{D}(t) = -B_1 \int_0^t dt' \, l(t-t') D^{\beta-1}(t')$$

$$= -B_1 l_0 \left[D^{\beta-1}(t) + \int_0^t dt' \, \varepsilon(t-t') D_0^{\beta-1}(t') \right] . \tag{5.3.24}$$

The latter form is correct to first order in $\varepsilon(t)$. We multiply across by $\dot{D}(t)$ and integrate to obtain

$$\dot{D}^2(t) = V^2 - B_2 D^\beta(t) + g_1(t) , \qquad \text{where} \tag{5.3.25}$$

$$g_1(t) = -2B_1 l_0 \int_0^t dt' \, \dot{D}_0(t') \int_0^{t'} dt'' \, \varepsilon(t'-t'') D_0^{\beta-1}(t'') \tag{5.3.26}$$

and the other quantities are defined by (5.3.11) and (5.3.13). This is the energy integral. Note that $g_1(t)$ can be expressed in terms of u defined by (5.3.16):

$$g_1(t) = g_2(u) = \frac{2V^2}{E_0} \int_0^u du_1 \int_0^{u_1} du_2 \, \varepsilon(u_1, u_2) \frac{d}{du_2} h(u_2) \tag{5.3.27}$$

where

$$\varepsilon(u_1, u_2) \equiv \varepsilon(t' - t'') \tag{5.3.28}$$

with t', u_1 and t'', u_2 related by (5.3.19). In deriving (5.3.27), one transforms the inner integral from t to u by means of (5.3.19). Noting that

$$g_1(t_{0m}) \equiv g_2(1) , \tag{5.3.29}$$

we can write the maximum indentation in the form

$$D_m = D_{0m} (1 + g_2(1)/\beta V^2) \tag{5.3.30}$$

to first order, by virtue of (5.3.25). Furthermore, (5.3.25) may be written in the form

$$\dot{D}^2(t) = B_2 [D_m^\beta - D^\beta(t)] + g_1(t) - g_1(t_{0m}) . \tag{5.3.31}$$

It is convenient now to go over to dimensionless coordinates, defining, in this case

$$u = D(t)/D_m , \tag{5.3.32}$$

and writing (5.3.31) as

$$\dot{u}^2 = E^2 h^2(u) + g(u) , \tag{5.3.33}$$

where $h(u)$ is given by (5.3.18) and

$$g(u) = \frac{g_2(u) - g_2(1)}{D_{0m}^2} \qquad (5.3.34\,a)$$

$$E^2 = B_2 D_m^{\beta-2} = E_0^2 \left(1 + \frac{\beta-2}{\beta} \frac{g_2(1)}{V^2}\right) . \qquad (5.3.34\,b)$$

Note that u in $g(u)$ is defined by (5.3.16) rather than (5.3.32). However, to first order, this does not matter. We can treat them as being identical. Equation (5.3.33) will now be solved to first order. Let us divide across by $E^2 h^2(u)$, take square roots and integrate to obtain

$$\frac{1}{E} \int_0^u \frac{dv}{h(v)} - \frac{1}{2E_0^3} \int_0^u dv \frac{g(v)}{h^3(v)} = t . \qquad (5.3.35)$$

Note that the time integration in the second integral has been replaced by an integration over v, with the aid of (5.3.19). In particular, this gives immediately the time of maximum indentation to be

$$t_m = \frac{1}{E} \int_0^1 \frac{du}{h(u)} - \frac{1}{2E_0^3} \int_0^1 du \frac{g(u)}{h^3(u)} . \qquad (5.3.36)$$

On expressing E in terms of E_0 by means of (5.3.34) and recalling (5.3.21), we deduce that

$$t_m = t_{0m} \left[1 - \frac{\beta-2}{2\beta V^2} g_2(1)\right] - \frac{1}{2E_0^3} \int_0^1 du \frac{g(u)}{h^3(u)} . \qquad (5.3.37)$$

5.3.2 The Second Phase

For the ejection phase, (5.3.2a) becomes, in the light of (5.2.22), (5.3.3, 7, 8) and (5.3.10):

$$A \, a''(\theta_1(t)) = D(t) - \varepsilon_1(t - \theta_1(t)) D_0(t) + \int_{\theta_1(t)}^t dt' \, \varepsilon(t - t') D_0(t') . \qquad (5.3.38)$$

We see that (5.3.23) and (5.3.38) give

$$D(\theta_1(t)) = D(t) + m_1(t)$$

$$m_1(t) = -\varepsilon_1(t - \theta_1(t)) D_0(t) + \int_{\theta_1(t)}^t dt' \, \varepsilon(t - t') D_0(t') . \qquad (5.3.39)$$

Equation (5.3.2b) may be written as

$$\ddot{D}(t) = -B_1 \int_0^{\theta_1(t)} dt' \, l_1(t, t') D^{\beta-1}(t')$$

$$\qquad\qquad\qquad\qquad\qquad\qquad\qquad\qquad\qquad (5.3.40)$$

$$\qquad\quad = -B_1 l_0 D^{\beta-1}(t) + m_2(t)$$

with the aid of (5.3.39), where B_1 is given by (5.3.11) and

$$m_2(t) = -B_1 l_0 \left\{ (2-\beta)\varepsilon_1(t-\theta_1(t))D_0^{\beta-1}(t) \right.$$

$$\left. + \int_0^{\theta_1(t)} dt'\, \varepsilon(t-t')D_0^{\beta-1}(t') + (\beta-1)D_0^{\beta-2}(t) \int_{\theta_1(t)}^t dt'\, \varepsilon(t-t')D_0(t') \right\}.$$

$$(5.3.41)$$

Note that $\theta_1(t)$ occurs only in the same terms as $\varepsilon(t)$ and $\varepsilon_1(t)$ so that, to first order, we can put

$$\theta_1(t) = 2t_{0m} - t . \tag{5.3.42}$$

Note also that

$$\int_0^{\theta_1(t)} dt'\, \varepsilon(t-t')D_0^{\beta-1}(t') = \int_t^{T_0} dt'\, \varepsilon(t'-\theta_1(t))D_0^{\beta-1}(t')$$

$$(5.3.43)$$

$$\int_{\theta_1(t)}^t dt'\, \varepsilon(t-t')D_0(t') = \int_{t_{0m}}^t dt'\, [\varepsilon(t-t') + \varepsilon(t'-\theta_1(t))]D_0(t')$$

where $T_0 = 2t_{0m}$, the duration time of the collision, in the elastic limit. Multiplying (5.3.40) across by $\dot{D}(t)$ and integrating from t_m gives

$$\dot{D}^2(t) = B_2 [D_m^\beta - D^\beta(t)] + m_3(t) , \quad \text{where} \tag{5.3.44}$$

$$m_3(t) = 2 \int_{t_{0m}}^t dt'\, \dot{D}_0(t')m_2(t') . \tag{5.3.45}$$

Equation (5.3.44) corresponds to (5.3.31). We now wish to go over to the variable u, as before. First, let us re-express $m_3(t)$ in terms of u. We have

$$m_3(t) = m_4(u) = V \left[-\beta(\beta-2) V \int_u^1 dv\, \varepsilon_2(v) v^{\beta-1} \right.$$

$$- 2D_{0m} \int_u^1 dv \int_0^v dv'\, \varepsilon_3(v, v') \frac{dh(v')}{dv'}$$

$$\left. + \beta(\beta-1) D_{0m} \int_u^1 dv\, v^{\beta-2} \int_v^1 dv'\, \frac{\varepsilon_4(v, v')v'}{h(v')} \right] , \tag{5.3.46}$$

where u, v, v' are related to the corresponding time variables t, t', t'' through (5.3.22) and

$$\varepsilon_2(v) = \varepsilon_1(t' - \theta_1(t')) = \varepsilon_1(2(t' - t_{0m}))$$

$$\varepsilon_3(v, v') = \varepsilon(t'' - \theta_1(t')) \tag{5.3.47}$$

$$\varepsilon_4(v, v') = \varepsilon(t' - t'') + \varepsilon(t'' - \theta_1(t')).$$

In dimensionless variables, (5.3.44) becomes

$$\dot{u}^2 = E^2 h^2(u) + m(u) \quad \text{where} \tag{5.3.48}$$

$$m(u) = \frac{m_4(u)}{D_{0m}^2} . \tag{5.3.49}$$

The solution of (5.3.48), obtained in the same manner as (5.3.35), is given by

$$t - t_m = -\frac{1}{E} \int_1^u \frac{dv}{h(v)} + \frac{1}{2E_0^3} \int_1^u \frac{dv \, m(v)}{h^3(v)} \ . \tag{5.3.50}$$

In this phase, the two parameters of interest are the total collision time and the coefficient of restitution. We will denote the former quantity by T. It is the time at which the contact region reduces to zero. Using (5.3.38) and remembering that $a(\theta_1(t)) = a(t)$, we see that, to first order,

$$D(T) = -\int_0^{T_0} dt' \, \varepsilon(t-t') D_0(t') \ , \tag{5.3.51}$$

where $T_0 = 2t_{0m}$ is the total elastic collision time. The quantity $D(T)$ is non-zero to first order. Thus, T is the time at which

$$u = u_1 \equiv -\frac{1}{D_m} \int_0^{T_0} dt' \, \varepsilon(t-t') D_0(t') \tag{5.3.52}$$

in (5.3.50). In contrast to the elastic case, this is not the time at which the indentation returns to zero. After the contact has ceased, the deformation reduces gradually to zero, since no forces are operating. This is an example of the phenomenon of creep. In principle, the return to the completely undeformed state may take a long time, though for many materials, the deformation will quickly become negligible. After time T, (5.3.2a, b) are no longer applicable. We have, from (5.3.50):

$$T = t_m + \frac{1}{E} \int_{u_1}^1 \frac{dv}{h(v)} - \frac{1}{2E_0^3} \int_{u_1}^1 du \, \frac{m(u)}{h^3(u)}$$

$$= T_1 - \frac{u_1}{E_0} \tag{5.3.53}$$

to first order, where

$$T_1 = t_m + \frac{1}{E} \int_0^1 \frac{dv}{h(v)} - \frac{1}{2E_0^3} \int_0^1 du \, \frac{m(u)}{h^3(u)}$$

$$= 2t_m + \frac{1}{2E_0^3} \int_0^1 du \, \frac{g(u) - m(u)}{h^3(u)} \tag{5.3.54}$$

$$= 2t_{0m} - \frac{(\beta - 2)}{\beta V^2} t_{0m} \, g_2(1) - \frac{1}{2E_0^3} \int_0^1 du \, \frac{g(u) + m(u)}{h^3(u)}$$

where (5.3.36, 37) have been used. The coefficient of restitution e is determined by observing that (5.3.48) gives, at $u = 0$ (strictly $u = u_1$ but, to first order, it makes no difference):

$$\dot{u}^2 = \frac{V_e^2}{D_m^2} = E^2 + m(0) \ , \tag{5.3.55}$$

where V_e is the ejection velocity. Therefore

$$e^2 = \frac{V_e^2}{V^2} = \frac{E^2 D_m^2}{V^2} + \frac{m(0)}{E_0^2}$$

$$= 1 + \frac{g_2(1)}{V^2} + \frac{m(0)}{E_0^2} \tag{5.3.56}$$

by virtue of (5.3.30, 34b). The quantity $(1 - e^2)$ is the fraction of energy lost in the collision. It is proportional to the integrals of the non-conservative terms of the equations of motion (5.3.24) and (5.3.40) multiplied by $\dot{D}_0(t)$; see (2.11.7).

5.3.3 The Short-Time Approximation

This approximation assumes that T is much less than the decay times of the viscoelastic medium. It is the assumption underlying the work of Hunter (1960) and Graham (1978). The viscoelastic functions are expanded in a Taylor expansion about $t = 0$, and only linear terms are retained. Therefore, from (5.3.9):

$$\varepsilon_1(t) = -\alpha t , \quad \text{where} \tag{5.3.57}$$

$$\alpha t \ll 1 , \tag{5.3.58}$$

so that

$$\varepsilon(t) = -\alpha . \tag{5.3.59}$$

In this approximation, the creep function has the form associated with a Maxwell's solid, as given by (1.6.15). Equation (5.3.27) becomes

$$g_2(u) = -\frac{2V^2}{E_0} \alpha \int_0^u dv \, [h(v) - 1] , \tag{5.3.60}$$

so that (5.3.30) gives

$$D_m = D_{0m} \left[1 + \frac{2\alpha}{\beta E_0} (1 - I_1) \right] , \quad \text{where} \tag{5.3.61}$$

$$I_1 = \int_0^1 du \, h(u) = \frac{1}{\beta} B(1/\beta, 3/2) . \tag{5.3.62}$$

Also, from (5.3.37) and (5.3.34a):

$$t_m = t_{0m} \left[1 - \frac{(\beta - 2)\alpha}{\beta E_0} (1 - I_1) \right] + \frac{\alpha}{E_0^2} (I_3 - I_4) \tag{5.3.63}$$

where, according to (5.3.21)

$$t_{0m} = \frac{I_2}{E_0} , \quad I_2 = \int_0^1 \frac{du}{h(u)} \quad \text{and} \tag{5.3.64}$$

$$I_3 = \int_0^1 \frac{du}{h^3(u)} (1 - u)$$

$$I_4 = \int_0^1 \frac{du}{h^3(u)} \int_u^1 dv \, h(v) . \tag{5.3.65}$$

The expressions for T and e are a little more complicated to derive. From (5.3.46, 47, 50, 57) and (5.3.59):

$$m(u) = \frac{2Va}{D_{0m}} \left\{ \beta(\beta - 2) \int_u^1 dv\, v^{\beta-1} \int_v^1 \frac{dv'}{h(v')} + \int_u^1 dv\, [h(v) - 1] \right.$$

$$\left. - \beta(\beta - 1) \int_u^1 dv\, v^{\beta-2} \int_v^1 dv'\, \frac{v'}{h(v')} \right\} . \tag{5.3.66}$$

Partial integration gives

$$m(u) = \frac{2Va}{D_{0m}} \left\{ -(\beta - 2) u^\beta \int_u^1 \frac{dv}{h(v)} + \beta u^{\beta-1} \int_u^1 \frac{dv\, v}{h(v)} \right.$$

$$\left. - 2 \int_u^1 dv\, \frac{v^\beta}{h(v)} + \int_u^1 dv\, [h(v) - 1] \right\} . \tag{5.3.67}$$

It follows from (5.3.56) that

$$e^2 = 1 + \frac{4\alpha}{E_0} (I_1 - I_2) = 1 - 2T_0\alpha \left(1 - \frac{I_1}{I_2} \right) . \tag{5.3.68}$$

Also, from (5.3.54, 34a, 60 and 5.3.66):

$$T_1 - 2t_{0m} = \frac{2t_{0m}^2 \alpha}{I_2^2} \left[-\frac{1}{2n+1} I_2(1 - I_1) + I_3 - 2I_4 + I_5 \right] \tag{5.3.69}$$

where

$$I_5 = \int_0^1 \frac{v\, dv}{h(v)} . \tag{5.3.70}$$

Equation (5.3.69) may be derived with the aid of the identity

$$\int_0^1 \frac{du}{h^3(u)} \left\{ (2 - \beta) u^\beta \int_u^1 \frac{dv}{h(v)} + \beta u^{\beta-1} \int_u^1 \frac{dv\, v}{h(v)} - 2 \int_u^1 \frac{dv}{h(v)} \right\} = -2I_5 , \tag{5.3.71}$$

which may be proved by combining parts of the first term on the left with the second and third, and carrying out a partial integration with the aid of the relation

$$\frac{d}{du} \frac{1}{h(u)} = \frac{1}{2} \frac{\beta u^{\beta-1}}{h^3(u)} . \tag{5.3.72}$$

In the short time approximation, u_1, defined by (5.3.52), has the form

$$u_1 = \frac{2\alpha I_5}{E_0} \tag{5.3.73}$$

so that, from (5.3.53, 64 and 5.3.69):

$$T = 2t_{0m} + \frac{2t_{0m}^2 \alpha}{I_2^2} \left[-\frac{1}{(2n+1)} I_2 (1 - I_1) + I_3 - 2I_4 \right] .$$

(5.3.74)

These formulae agree with those of Hunter (1960) in the special case he considered ($\beta = 5/2$) and with the more general results of Graham (1978). Values of I_2, $i = 1, 2, 3, 4$ for $n = 1, 2$ are given in Table 5.1.

Table 5.1

	I_1	I_2	I_3	I_4
$n = 1$	0.8413	1.402	0.7549	0.4633
$n = 2$	0.8176	1.472	0.8462	0.4759

The ratio I_1/I_2 for $n = 2$ is close to 5/9, so that

$$e^2 = 1 - \tfrac{8}{9} \alpha T_0 ,$$

(5.3.75)

where $T_0 = 2t_{0m}$, the total contact time in the elastic limit, may be deduced from (5.3.11, 13, 18) and (5.3.64) to be

$$T_0 = \frac{2I_2}{E_0} = 2.944 \, V^{-1/5} \left(\frac{32 \, l_0 R^{1/2}}{15 \, M} \right)^{-2/5} ,$$

(5.3.76)

where R is the radius of curvature of the indentor at the point of impact. We have used (5.2.34) to give that for $n = 2$, $A = 2c = 1/R$ and also (5.2.35). For $n = 2$, (5.3.74) and Table 5.1 give

$$T = T_0 (1 - 0.037 \, \alpha T_0) .$$

(5.3.77)

Equations (5.3.76) and (5.3.77) were first given by Hunter (1960).

Note that (5.3.77) gives that $T < T_0$. This may not always be the case, however. Experimental measurements and numerical solutions given by Calvit (1967), based on a constant loss tangent model together with calculations by Graham (1978) and Sabin and Graham (1983), indicate that, generally speaking, $T > T_0$.

5.4 Hysteretic Friction

We consider the moving load problem in this section to the extent of deriving an expression for the coefficient of hysteretic friction in the small viscoelasticity and small velocity approximations, respectively.

5.4.1 Small Viscoelasticity Approximation

For motion along the negative x direction, and neglecting friction, (2.11.13) becomes

$$f_H = \frac{1}{W} \int_{C(t)} ds \, p(r, t) \frac{\partial}{\partial x} u(r, t) \; , \tag{5.4.1}$$

where $u(r, t)$ is the displacement in the z direction, into the half-space. Solving (5.1.2) for $u_z(r, t)$ gives

$$f_H = \frac{1}{W} \int_{C(t)} ds \, p(r, t) \frac{\partial}{\partial x} \int_{-\infty}^{t} dt' \, k(t - t') \frac{1}{2\pi} \int_{C(t')} ds' \frac{p(r', t')}{|r' - r|} \; . \tag{5.4.2}$$

It is easy to show that the instantaneous (delta function) contribution to $k(t)$ gives zero in (5.4.2). Assuming that $k(t)$ is given by (5.3.6), we deduce that to first order:

$$f_H = -\frac{k_0}{2\pi W} \int_{C_0(t)} ds \, p_0(r, t) \frac{\partial}{\partial x} \int_{-\infty}^{t} dt' \, \varepsilon(t - t') \int_{C_0(t')} ds' \frac{p_0(r', t')}{|r' - r|} \; , \tag{5.4.3}$$

where $p_0(r, t)$, $C_0(t)$ are the pressure distribution and the contact region in the limiting instantaneous elastic problem. Using the instantaneous elastic limit of (5.1.2), we can write this in the form

$$f_H = -\frac{k_0}{W} \int_{C_0(t)} ds \, p_0(r, t) \frac{\partial}{\partial x} v_c(r, t)$$

$$v_c(r, t) = l_0 \int_{-\infty}^{t} dt' \, \varepsilon(t - t') u_0(r, t') \; , \tag{5.4.4}$$

where $l_0 = k_0^{-1}$. If the solution to the elastic problem is known, then f_H is given by quadrature.

Let us consider in more detail the case where the indentor is spherical and uniform motion at velocity V, under steady-state conditions, prevails. Choosing coordinates moving with the indentor allows us to write (5.4.4) in the form

$$f_H = -\frac{k_0}{W} \int_{C_0} ds \, p_0(r) \frac{\partial}{\partial x} v_c(r)$$

$$\frac{\partial}{\partial x} v_c(r) = l_0 \int_{-\infty}^{x} \frac{dx'}{V} \varepsilon\left(\frac{x - x'}{V}\right) u_0'(x', y) \; , \tag{5.4.5}$$

where C_0 is the (fixed) elastic contact area in these coordinates, and $u_0'(x', y)$, the x derivative of $u_e(x', y)$; the derivative can be transferred onto the displacement by means of partial integration, remembering that the upper limit of the integral is x^+ and that $\varepsilon(t)$ contains as a factor the Heaviside step function $H(t)$ where $H(0^-)$ is zero.

Problem 5.4.1: If the material has a unique Poisson's ratio and the shear creep function of the material has the step function form

$$J(t) = J_0 + J_1 H(t - t_0) \; , \qquad t_0 > 0 \; , \tag{5.4.1\,p}$$

where J_1/J_0 is small, then show that

$$f_H = \frac{J_1}{W J_0} \int\limits_{C_0(t)} ds \, p_0(r) u_0'(x - V t_0, y) \ . \tag{5.4.2p}$$

Under a spherical indentor of radius R, the contact region is circular with radius a_0, given by (see (5.2.31) with l_0 replacing l_e):

$$a_0 = \left(\frac{3}{8} W R k_0 \right)^{1/3} \tag{5.4.6}$$

under the same normal load W. The pressure $p_0(r)$ has the form (5.2.32), which we write as

$$p_0(r) = p_1 (1 - \varrho^2)^{1/2} \ , \quad \varrho^2 = \frac{r^2}{a_0^2} \ , \quad p_1 = \frac{3 W}{2 \pi a_0^2} \ . \tag{5.4.7}$$

The displacement will be written as (Lur'e (1964), pg. 277):

$$u_0(r) = \begin{cases} D_0 - c_1 \varrho^2 \ , & \varrho < 1 \\ \dfrac{D_0}{\pi} [(2 - \varrho^2) \cot^{-1}(\sqrt{\varrho^2 - 1}) + \sqrt{\varrho^2 - 1}] \ , & \varrho > 1 \ , \end{cases} \tag{5.4.8}$$

where D_0 is given by (5.2.31), and

$$c_1 = \frac{a_0^2}{2R} = \frac{D_0}{2} \ . \tag{5.4.9}$$

It follows that

$$u_0'(r) = \begin{cases} -\dfrac{x}{R} \ , & \varrho < 1 \\ -\dfrac{2}{\pi} \dfrac{x}{R} \left[\cot^{-1}(\sqrt{\varrho^2 - 1}) - \sqrt{\dfrac{\varrho^2 - 1}{\varrho^2}} \right] \ , & \varrho > 1 \ . \end{cases} \tag{5.4.10}$$

On substituting (5.4.7, 10) into (5.5.5), the quantity f_H may be calculated numerically. In the case of a spectrum model (1.6.24):

$$\varepsilon(t) = -\frac{1}{G(0)} \sum_{i=1}^{N} \alpha_i G_i e^{-\alpha_i t} H(t) \ , \quad \alpha_i = \frac{1}{\tau_i}$$

$$G(0) = G_0 + \sum_{i=1}^{N} G_i \ . \tag{5.4.11}$$

Consider a situation where $\partial v_c / \partial x$ is slowly varying over the contact region. For example, this will be the case at high and low velocities (in the latter case, $\varepsilon(x/V)$ approaches a delta function) if we assume that $u_0'(x, y)$ is slowly varying under the indentor [Golden (1982)]. We can then put

$$v_c'(x, y) = v_c'(0, 0) + \text{linear terms} + \dots \tag{5.4.12}$$

so that keeping only the first term of (5.4.12) in (5.5.5) gives

$$f_{\mathrm{H}} = -\int_0^\infty \frac{dx}{V}\, \varepsilon\left(\frac{x}{V}\right) u_0'(x, 0)$$

$$= \frac{a_0^2}{VR}\left\{\int_0^1 d\varrho\, \varepsilon\left(\frac{a_0\varrho}{V}\right)\varrho\right. \tag{5.4.13}$$

$$\left. + \frac{2}{\pi}\int_1^\infty d\varrho\, \varepsilon\left(\frac{a_0\varrho}{V}\right)\varrho\left[\cot^{-1}(\varrho^2-1)^{1/2} - \frac{(\varrho^2-1)^{1/2}}{\varrho^2}\right]\right\} .$$

This can be shown to have the standard humped shape, as a function of velocity, discussed in general terms in Sect. 3.8, at least for simple choices of viscoelastic functions. Golden (1982) gives a more elaborate derivation of this result, with different notational conventions. The formula in that reference contains a misprint, namely a missing derivative sign.

5.4.2 Small Velocity Approximation

For low velocities, an expression for f_{H} linear in V can be given, without restriction on the size or nature of viscoelastic effects [Golden (1978)]. Indeed, a complete solution of the problem is possible. We will discuss here the special case of a spherical indentor, though the results may be generalized without difficulty to an ellipsoidal indentor. For steady-state motion in the negative x direction, (5.1.2) may be written in the form

$$\int_0^\infty dt'\, l(t')u(x - Vt', y) = \frac{1}{2\pi}\int_C \frac{dx'\, dy'\, p(x', y')}{[(x-x')^2 + (y-y')^2]^{1/2}} , \tag{5.4.14}$$

using coordinates moving with the punch, where C is the (fixed) contact region. For small velocities, we approximate the left-hand side as:

$$\int_0^\infty dt'\, l(t')u(x - Vt', y) = l_e u(x, y) + Vl_{1e}\frac{\partial}{\partial x}u(x, y) , \tag{5.4.15}$$

where

$$l_e = \int_0^\infty l(t)\, dt = \hat{l}(0) = L(\infty) \tag{5.4.16}$$

in terms of the Fourier transform $\hat{l}(\omega)$, given by (3.1.15), $L(t)$ defined by (5.2.2p), and

$$l_{1e} = -\int_0^\infty dt\, t l(t) = -i\hat{l}'(0) . \tag{5.4.17}$$

For a spherical indentor

$$l_e u(x, y) + Vl_{1e}\frac{\partial}{\partial x}u(x, y) = l_e\left(D_0 - \frac{r^2}{2R}\right) - Vl_{1e}\frac{x}{R}$$

$$= l_e\left(D_1 - \frac{x_1^2 + y^2}{2R}\right) , \tag{5.4.18}$$

where

$$x_1 = x + \delta , \quad D_1 = D_0 + \frac{\delta^2}{2R} , \quad \delta = \frac{Vl_{1e}}{l_e} . \tag{5.4.19}$$

Changing the x' integration variable on the right-hand side of (5.4.14) to $x'' = x' + \delta$ gives

$$l_e D_1 - \frac{x_1^2 + y^2}{2R} = \frac{1}{2\pi} \int_{C_s} \frac{dx'' \, dy' \, p_1(x'', y')}{[(x'' - x_1)^2 + (y' - y)^2]^{1/2}} \tag{5.4.20}$$

$$p_1(x'', y') \equiv p(x', y') ,$$

where C_s is the contact area in terms of the new coordinates. Equation (5.4.20) is identical to the elastic form. Therefore, $p_1(x'', y')$ has the same form as the elastic pressure distribution, in terms of the new variable x'', namely that given by (5.2.32). Also, the contact region C_s is circular with radius $a(t)$ determined by the total load according to (5.2.31b), while D_1 is related to $a(t)$ by (5.2.31a). It follows from (5.4.19) that the maximum indentation D_0 is less than D_1 by $\delta^2/2R$. Thus, the indentor penetrates less into the medium when it is moving than it would under the same load and stationary. This is clear, on observing that, in the stationary limit, δ goes to zero, D_1 reduces to D_0 and so on, with the result that the relevant equation has the elastic form with modulus determined by l_e. Strictly, the viscoelastic correction to the indentation is negligible to first order in V. The centre of the contact region is shifted to $x = -\delta$, or by an amount $|\delta|$ in the direction of motion.

Let us now consider the coefficient of hysteretic friction, given by (5.4.1). This becomes

$$f_H = -\frac{1}{RW} \int_C dx' \, dy' \, p(x', y') x'$$

$$= -\frac{1}{RW} \int_{C_s} dx'' \, dy' \, p_1(x'', y')(x'' - \delta) , \tag{5.4.21}$$

which simplifies to

$$f_H = \frac{\delta}{R} , \tag{5.4.22}$$

since the elastic form $p_1(x'', y')$ is symmetric over C_s. If the proportionality assumption applies, then

$$\delta = -\frac{V}{G(\infty)} \int_0^\infty dt \, \mu(t) t , \tag{5.4.23}$$

where $\mu(t)$ is related to the shear viscoelastic function $G(t)$ by (1.2.33a). Writing $G(t)$ in the form

$$G(t) = G(\infty) + g(t) , \tag{5.4.24}$$

we obtain

$$\delta = \frac{V}{G(\infty)} \int_0^\infty dt\, g(t)$$

$$f_{\mathrm{H}} = \frac{V}{R G(\infty)} \int_0^\infty dt\, g(t) \ .$$

(5.4.25)

For the standard linear model, defined by (1.6.1),

$$\delta = \frac{G_1}{G_0} V\tau \ ; \quad f_{\mathrm{H}} = \frac{G_1}{G_0} \frac{V\tau}{R} \ .$$

(5.4.26)

This approximate solution, linear in V, describes the initial rise of f_{H} as a function of V, well below where it rises to a maximum value, in other words, well below the hysteretic peak.

5.5 Summary

Three-dimensional contact problem solutions are obtained and analyses of impact and hysteretic friction are made.

The development relies upon formula (5.1.2) relating normal surface displacement to a distribution of pressure acting on an arbitrarily shaped patch of the surface of a viscoelastic half-space. Shear stress distributions on the half-space boundary are neglected.

The problem of a rigid punch of curved profile indenting the half-space is considered for monotone increasing, monotone decreasing and generally varying contact area. Formulae are given for the depth of penetration of the indentor into the half-space and the pressure in terms of the history of the contact area and the viscoelastic properties of the material. Specific results are obtained for the case of steady-state loading. Results for the case when the contact area has a single maximum, together with Newton's law of motion, provide equations (5.3.2) that may be used to describe the impact of a rigid indentor on a viscoelastic half-space. These equations are analyzed in the case of small viscoelasticity and formulae for the coefficient of restitution (5.3.68, 75) and the total contact time (5.3.74, 77) are derived.

Sect. 5.4 contains approximate formulae for the coefficient of hysteretic friction. The approximations apply in the cases of small viscoelasticity (5.4.4, 13) and small velocity (5.4.22, 23, 26).

6. Thermoviscoelastic Boundary Value Problems

As pointed out in Sect. 1.7, the viscoelastic functions of many materials depend strongly on temperature. The simplest realistic way of incorporating this dependence is to assume that the material is thermorheologically simple (TRS) in the sense defined in Sect. 1.7. This implies a non-linear dependence on the temperature field which renders the solution of most problem categories very difficult, in particular those where the temperature field is not given a priori but must be determined as part of the solution. A way out of this is to adopt a fully linear theory, as developed for example by Christensen (1982), Chap. 3. The assumption behind such a theory is that the effects of temperature variation on the viscoelastic functions is sufficiently small that its product with the field variables can be neglected. In many cases, this is very restrictive on the allowed range of temperature variation. A fully linear theory will not be considered here. We remark however that such a theory is susceptible to treatment by the Correspondence Principle-based methods, discussed in Chap. 2.

Attention will be confined to problems where the temperature field is specified in advance. This at least removes the difficulties associated with trying to determine field variables which occur non-linearly in the equations. However, there remains a significant difficulty. The TRS assumption destroys time homogeneity so that integral transforms can no longer be used to establish correspondence with elastic solutions. As discussed in Sect. 1.7, the convolution form can be retrieved by going over to reduced variables, but at the expense of adding an extra term to the equations of equilibrium, which again destroys the possibility of equivalence. This extra term vanishes if the temperature field is independent of the space variables, and one retrieves the Correpondence Principles and methods of Chap. 2. This case where the temperature field is purely time-dependent is not particularly interesting, however. If the temperature field is purely space-dependent, on the other hand, then time homogeneity is restored and a correspondence will exist with space-inhomogeneous Elasticity Theory [Sternberg (1964)]. Such problems are generally difficult to solve.

Therefore, in order to solve problems involving TRS materials, new methods must be adopted. We note that such materials are aging materials in the technical sense that time homogeneity does not apply — though not necessarily in the more conventional sense of undergoing irreversible changes with time. We are therefore looking for a method that applies to aging materials. The method outlined in very general terms in Sect. 2.12 is clearly of this type, since at no point is time homogeneity assumed. The aim of the present chapter is to provide a simple illustration of this method. In Sect. 6.1, the problem of a sphere in a specified temperature field is considered.

Muki and Sternberg (1961) solved problems involving special types of TRS materials. This work was generalized by Sternberg and Gurtin (1964). An elastic response in dilatation was assumed. Sternberg (1964) discussed the topic in general terms. Edelstein (1969a) presented a method, essentially that described in Sect. 2.12, for handling TRS materials and applied it to be the problem of a hollow cylinder with TRS response in both shear and dilatation. He extended the analysis to the case where the inner boundary is ablating [Edelstein (1969b)]. F. Williams (1975a) presents a general formulation of this method, applicable to materials with very general constitutive relations. She applies it to the problem of a spherical shell and the problem of a floating ice plate [see also F. Williams (1975b, 1976)]. We present here a special case of the spherical shell problem, to illustrate the approach.

The Edelstein method applies to materials that are aging and with spatial in-homogeneity – due to the dependence of its mechanical properties on tempera-ture and other environmental variables. This is clear from F. Williams' (1975a) analysis. However, the final result is an integral equation which, in all but the simplest cases, must be solved numerically.

Graham and Williams (1972) give generalized Papkovich-Neuber and other representations of solutions for aging materials.

There is some work on TRS materials which does not take the temperature field as given, but seeks to solve for it. Hunter (1967) manages to obtain solutions for the temperature field in a semi-infinite rod subjected to forced oscillations. It is necessary however to assume that the loss tangent depends linearly on the temperature field. Lockett and Morland (1967) assume that temperature and me-chanical fields are decoupled, so that the temperature field may be solved for in-dependently and specified as input to the mechanical equations.

For a survey that includes Russian contributions to this topic, see Il'iushin and Pobedria (1970).

6.1 A Sphere in a Specified Temperature Field

The treatment of this problem is based on the more general considerations of F. Williams (1975a, b). We use spherical polar co-ordinates defined conven-tionally by

$$x = r \sin\theta \cos\phi, \quad y = r \sin\theta \sin\phi, \quad z = r \cos\theta. \tag{6.1.1}$$

The definition of the strain tensor in terms of displacements u_r, u_θ, u_ϕ and the form of the equilibrium equations for these coordinates are given by Sokolnikoff (1956), pp. 184, for example. The sphere is taken to be centred at the origin. All specified quantities, namely the boundary functions and the temperature field, are assumed to be spherically symmetric. Body forces are neglected. The dis-placements have the form

$$u_r = u(r, t), \quad u_\theta = u_\phi = 0, \tag{6.1.2}$$

giving

$$\varepsilon_{rr} = \frac{\partial u}{\partial r}, \qquad \varepsilon_{\theta\theta} = \varepsilon_{\phi\phi} = \frac{u}{r}, \tag{6.1.3}$$

while the off-diagonal strain tensor components are zero. It follows that this is also true for the stress tensor. The constitutive equations (1.9.6) together with the fact that $\varepsilon_{\theta\theta} = \varepsilon_{\phi\phi}$ give that $\sigma_{\theta\theta} = \sigma_{\phi\phi}$.

Attention will be focussed on determining the stresses generated by the temperature field. The appropriate equations to use are compatibility equations of the kind discussed at the end of Sect. 1.9. We cannot use (1.9.26) however because the present problem is inhomogenous in space and time. Let us nevertheless follow a similar procedure to that used in deriving (1.9.26). In this context, there is only one compatibility equation,

$$\frac{\partial \varepsilon_{\theta\theta}}{\partial r} + \frac{1}{r}(\varepsilon_{\theta\theta} - \varepsilon_{rr}) = 0, \tag{6.1.4}$$

which may be verified by substitution from (6.1.3). The inverted constitutive relations (1.9.24) are the relevant relations here, since they express strain in terms of stress. In the present context, one obtains

$$\varepsilon_{rr} = \frac{1}{3}\left(\gamma + \frac{1}{3}\psi\right) * \sigma_{rr} + \frac{2}{3}\left(\frac{1}{3}\psi - \frac{1}{2}\gamma\right) * \sigma_{\theta\theta} + \alpha\theta$$

$$\varepsilon_{\theta\theta} = \frac{1}{3}\left(\frac{1}{2}\gamma + \frac{2}{3}\psi\right) * \sigma_{\theta\theta} + \frac{1}{3}\left(\frac{1}{3}\psi - \frac{1}{2}\gamma\right) * \sigma_{rr} + \alpha\theta, \tag{6.1.5}$$

where $\theta(r, t)$ is the specified temperature field and α is the coefficient of thermal expansion. Combining (6.1.4, 5) gives

$$\frac{\partial}{\partial r}\left[\frac{1}{3}\left(\frac{1}{2}\gamma + \frac{2}{3}\psi\right) * \sigma_{\theta\theta} + \frac{1}{3}\left(\frac{1}{3}\psi - \frac{1}{2}\gamma\right) * \sigma_{rr} + \alpha\theta\right]$$

$$+ \frac{1}{r}\left[\frac{1}{2}\gamma * (\sigma_{\theta\theta} - \sigma_{rr})\right] = 0. \tag{6.1.6}$$

The equations of equilibrium reduce to a single relation,

$$\frac{\partial \sigma_{rr}}{\partial r} + \frac{2}{r}(\sigma_{rr} - \sigma_{\theta\theta}) = 0, \tag{6.1.7}$$

which gives $\sigma_{\theta\theta}$ in terms of σ_{rr}:

$$\sigma_{\theta\theta} = \frac{r}{2}\left(\frac{\partial \sigma_{rr}}{\partial r}\right) + \sigma_{rr}. \tag{6.1.8}$$

We use this relation to eliminate $\sigma_{\theta\theta}$ from (6.1.6). Let $\sigma_{rr}(r, t)$ be denoted by $R(r, t)$. Equation (6.1.6) becomes

$$\frac{\partial}{\partial r}\left[r\left(\frac{1}{2}\gamma + \frac{2}{3}\psi\right) * \frac{\partial}{\partial r}R + 2\psi * R\right] + \frac{3}{2}\gamma * \frac{\partial R}{\partial r} = rq \tag{6.1.9a}$$

$$q = -\frac{6\alpha}{r}\frac{\partial}{\partial r}\theta(r,t) . \tag{6.1.9b}$$

Let us assume that the dependence of the dilatation function $\psi(t)$ on the temperature field $\theta(r,t)$ may be neglected. It could still depend on the average temperature T_0. Also, consider the instantaneous portion of $\gamma(t)$, given by (1.2.33b) namely $J(0)\delta(t)$. We will denote $J(0)$ by h_0. For a TRS material, it is clearly independent of $\theta(r,t)$. We put

$$\phi(r,t,t') = \tfrac{1}{2}\gamma(r,t,t') + \tfrac{2}{3}\psi(t-t') , \tag{6.1.10}$$

which explicitly indicates the space and time inhomogeneity resulting from TRS behaviour. Now

$$\phi(r,t,t') = \phi_0\delta(t-t') + \phi_1(r,t,t')H(t-t')$$

$$\phi_0 = \tfrac{1}{2}h_0 + \tfrac{2}{3}\psi_0 \tag{6.1.11}$$

$$\phi_1(r,t,t') = -\frac{1}{2}\frac{d}{dt'}J(r,t,t') + \frac{2}{3}\dot{Q}(t-t') ,$$

by virtue of (1.2.23) and (1.9.13). Note that $\psi_0 = Q(0)$. Equation (6.1.9a) can therefore be written in the form

$$\frac{1}{r^4}\frac{\partial}{\partial r}\left[r^4\phi * \frac{\partial}{\partial r}R(r,t)\right] = q \quad \text{or} \tag{6.1.12}$$

$$\frac{\phi_0}{r^4}\frac{\partial}{\partial r}\left[r^4\frac{\partial}{\partial r}R(r,t)\right] + \int_0^t dt'\left\{\frac{1}{r^4}\frac{\partial}{\partial r}\left[\phi_1(r,t,t')r^4\frac{\partial}{\partial r}R(r,t')\right]\right\} = q(r,t) \tag{6.1.13}$$

if we assume that temperature variation and deformation of the medium begins at $t = 0$. We write this in the form [see (2.12.1)]:

$$DR = q \quad \text{where} \tag{6.1.14}$$

$$D = D_0 + D_1 , \tag{6.1.15}$$

the operator D_0 being the differential operator in the first term of (6.1.13), while D_1 is the integro-differential operator in the second term. Following the scheme outlined in Sect. 2.12, we write (6.1.14) in the form

$$D_0 R = f \tag{6.1.16a}$$

$$f = -D_1 R + q . \tag{6.1.16b}$$

This is an ordinary differential equation with Green's function

$$G(r,u) = \frac{1}{3\phi_0}\left(\frac{1}{b^3} - \frac{1}{r^3}\right) u^4, u \le r$$

$$= \frac{1}{3\phi_0}\left(\frac{1}{b^3} - \frac{1}{u^3}\right) u^4, u > r , \tag{6.1.17}$$

where b is the radius of the sphere. The boundary conditions will be imposed that the radial stress is zero on the surface. The solution to (6.1.16a) is therefore given by

$$R(r, t) = \int_0^b du\, G(r, u) f(u, t) .\tag{6.1.18}$$

This is a special form of (2.12.4). There are no surface terms because of the zero stress boundary condition.

This is the first stage. The second is to substitute (6.1.18) into (6.1.16b) to obtain an integral equation for $f(r, t)$ of the form

$$f(r, t) = -\frac{1}{\phi_0 r^4} \frac{\partial}{\partial r} \left[\int_0^t dt'\, \phi_1(r, t, t') \int_0^r du\, u^4 f(u, t') \right] + q(r, t) ,\tag{6.1.19}$$

where the properties of $G(r, u)$, given by (6.1.17), have been used. Iteration similar to that for a simple Volterra integral equation (Appendix 4) gives the solution of (6.1.19) formally, as an infinite series. A space integration can be carried out at each stage. We finally obtain

$$f(r, t) = q(r, t) + \sum_{n=1}^{\infty} K_n(r, t)$$

$$K_n(r, t) = \left(\frac{-1}{\phi_0} \right)^n \frac{1}{r^4} \frac{\partial}{\partial r} \int_0^t dt_1\, \phi_1(r, t, t_1) \int_0^{t_1} dt_2\, \phi_1(r, t_1, t_2) \dots$$

$$\dots \int_0^{t_{n-1}} dt_n\, \phi_1(r, t_{n-1}, t_n) \int_0^r q(u, t_n) u^4 du .\tag{6.1.20}$$

For certain simple materials, this series can be summed [F. Williams (1975a)]. We summarize here the method used for a Maxwell material. Let us take it that the dilatational response of the material is elastic, so that it makes no contribution to $\phi_1(r, t, t')$, which therefore consists only of the shear contribution $\gamma(t')$, related to the creep function $J(t)$. This has the form (1.6.15). Therefore, from (6.1.11) and (1.7.5) – which of course applies equally to $J(t, t_0)$ – we have

$$J(t, t') = J + \frac{\xi(t) - \xi(t')}{\eta_0} ,\tag{6.1.21}$$

giving

$$\phi_1(r, t, t') = \frac{b(r, t')}{2\eta_0} , \qquad b(r, t') = \frac{1}{a(T(r, t'))} ,\tag{6.1.22}$$

where $a(T)$ is the factor introduced in Sect. 1.7 and η_0 is the temperature-independent part of the viscosity coefficient η, after $a(T)$ has been removed. Substituting into (6.1.20) gives

$$K_n(r, t) = \left(\frac{-1}{2\phi_0 \eta_0} \right)^n \frac{1}{r^4} \frac{\partial}{\partial r} \left[\int_0^t dt_1\, b(r, t_1) \int_0^{t_1} dt_2\, b(r, t_2) \dots \right.$$

$$\left. \dots \int_0^{t_{n-1}} dt_n\, b(r, t_n) \int_0^r q(u, t_n) u^4 du \right].\tag{6.1.23}$$

Consider for a moment

$$g(u, t) = \sum_{n=0}^{\infty} K_n^{(1)}(u, t) \quad \text{where} \tag{6.1.24}$$

$$K_n^{(1)}(u, t) = (-\beta)^n \int_0^t dt_1\, b(r, t_1) \int_0^{t_1} dt_2\, b(r, t_2) \dots \int_0^{t_{n-1}} dt_n\, b(r, t_n)\, q_1(u, t_n)$$

$$\tag{6.1.25}$$

$$K_0^{(1)}(u, t) = q_1(u, t) \;;$$

$$\beta = (2\,\phi_0\,\eta_0)^{-1} \;; \quad q_1(u, t) = u^4 q(u, t) \;.$$

This is the Neumann series expansion (Appendix 4) for the Volterra integral equation:

$$g(u, t) = -\beta \int_0^t dt'\, b(r, t')\, g(u, t') + q_1(u, t) \;, \tag{6.1.26}$$

which, on differentiation, reduces to the first order differential equation

$$\dot{g}(u, t) = -\beta\, b(r, t)\, g(u, t) + \dot{q}_1(u, t) \tag{6.1.27}$$

with initial condition

$$g(u, 0) = q_1(u, 0) \;, \tag{6.1.28}$$

which we take to be zero. The solution of this differential equation has the form

$$g(u, t) = \int_0^t dt'\, e^{-\beta(\xi - \xi')}\, \dot{q}_1(u, t') \;, \quad \text{where} \tag{6.1.29}$$

$$\xi(u, t) = \int_0^t dt_1\, b(r, t_1) \;, \quad \xi' = \xi(u, t') \;. \tag{6.1.30}$$

Returning to (6.1.20, 23), we deduce that

$$f(r, t) = \frac{1}{r^4} \frac{\partial}{\partial r} \int_0^t e^{-\beta(\xi - \xi')} \int_0^r u^4\, \dot{q}(u, t')\, du\, dt' \tag{6.1.31}$$

where $q(u, t)$ is given by (6.1.9b). Finally we substitute into (6.1.18) to obtain, after partial integration and cancellation:

$$R(r, t) = \frac{6\alpha}{\phi_0} \int_r^b dv\, \frac{1}{v^4} \int_0^t dt'\, e^{-\beta[\xi(v, t) - \xi(v, t')]} \int_0^v du\, u^3 \frac{\partial^2 \theta(u, t')}{\partial u\, \partial t'} \;. \tag{6.1.32}$$

Now ϕ_0 may be given in terms of the instantaneous elastic parameters of the medium as

$$\phi_0 = \frac{3}{2G} \frac{1 - \nu}{1 + \nu} \;, \tag{6.1.33}$$

where G is the shear modulus and ν is Poisson's ratio. If, in the light of (1.6.15), η_0 is written as $G\tau_0$ where τ_0 is a time constant (at the standard temperature where $a(T)$ is unity), then

$$\beta = \frac{1+v}{3\,\tau_0\,(1-v)}\ .\tag{6.1.34}$$

Equation (6.1.32), which was derived by F. Williams (1975a) using essentially the method described here, was given earlier by Rongved (1954) and Muki and Sternberg (1961).

6.2 Summary

This chapter has been used to illustrate the method discussed in Chap. 2, which may be used to solve thermoviscoelastic boundary value problems involving temperature fields simultaneously varying with position and time. The method relies upon the solution of integral equations in terms of Neumann series expansions.

The stress distribution is determined throughout a sphere whose surface $r = b$ is stress-free and which is subject to a prescribed temperature field $\theta(r, t)$. The series solution may be summed for simple material types.

7. Plane Inertial Problems

We consider plane contact and crack problems in this chapter, without neglecting inertial effects. Such problems are typically far more difficult than the non-inertial problems discussed in Chaps. 3 and 4, and require different techniques for their solution. This is an area still in the development stage so that it will not be possible to achieve the kind of synthesis or unification which is desirable. We confine our attention to steady-state motion at uniform velocity V in the negative x direction. We begin by deriving boundary relationships between displacement and stress. These are applied to moving contact problems in the small viscoelasticity approximation, and to Mode III crack problems without any approximation.

References to the relevant literature will be noted at appropriate places. Problems of material response to applied load have been treated by Morland (1963), Chu (1965) and Martinček (1979).

7.1 Displacement-Traction Relationships on the Boundary

In order to deal with moving contact problems, we consider as before a half-plane occupying $y > 0$, consisting of a homogeneous isotropic linear viscoelastic material. In this section, it will be assumed that the stresses are known on the x-axis, and are zero at infinity. Consider the two-dimensional version of (1.9.18):

$$\hat{\sigma}_{ij}(r, \omega) = 2\,\hat{\mu}(\omega)\,\hat{\varepsilon}_{ij}(r, \omega) + \delta_{ij}\,\hat{\lambda}(\omega)\hat{\varepsilon}(r, \omega)$$

$$-\varrho\omega^2 \hat{u}_i(r, \omega) = \hat{\mu}(\omega)\,\nabla^2 \hat{u}_i(r, \omega) + (\hat{\mu}(\omega) + \hat{\lambda}\,(\omega))\,\frac{\partial}{\partial x_i}\,\nabla \cdot \hat{u}(r, \omega) \qquad (7.1.1)$$

$$i, j = 1, 2\ , \qquad r = (x_1, x_2) = (x, y)\ .$$

Let

$$\hat{u}_i(k, \omega) = \int_{-\infty}^{\infty} dx\,dy\,dt\,u_i(r, t)\,e^{-i(k \cdot r + \omega t)}\ , \qquad k = (k_x, k_y) \qquad (7.1.2)$$

where the displacements are understood to be zero in the lower half-plane. The transforms $\hat{\sigma}_{ij}(k, \omega)$ may be defined similarly. In terms of these, the space-transformed version of (7.1.1) becomes

$$\hat{\sigma}_{ij}(k, \omega) = i\hat{\mu}(\omega)\,[k_j\,\hat{u}_i(k, \omega) + k_i\,\hat{u}_j\,(k, \omega)] + \delta_{ij}\,\hat{\lambda}(\omega)\,ik \cdot \hat{u}(k, \omega) \qquad (7.1.3\,a)$$

$$\left[\left(k^2 - \frac{\omega^2}{c_T^2(\omega)}\right)\delta_{ij} + \frac{c_L^2(\omega) - c_T^2(\omega)}{c_T^2(\omega)}\,k_i k_j\right]\hat{u}_j(k, \omega) = 0\ , \qquad i = 1, 2 \qquad (7.1.3\,b)$$

where

$$c_T^2(\omega) = \frac{\hat{\mu}(\omega)}{\varrho} , \qquad c_L^2(\omega) = \frac{\hat{\lambda}(\omega) + 2\hat{\mu}(\omega)}{\varrho} . \qquad (7.1.4)$$

The quantities $c_T(\omega)$, $c_L(\omega)$ are generalizations of the transverse and longitudinal speeds of sound in the medium. Note that they are complex since c_T^2, c_L^2 are proportional to the complex moduli. The presence of an imaginary part leads to exponential decay of sound waves in the medium, as a consequence of energy dissipation.

If the second equation of (7.1.3) is to have a solution, the determinant of the matrix acting on $\hat{u}_j(k, \omega)$ must be zero. The gives an equation which has solutions

$$k^2 = \frac{\omega^2}{c_T^2} \quad \text{or} \quad \frac{\omega^2}{c_L^2} , \qquad (7.1.5)$$

which leads to

$$k_y = \pm i n_T, \ \pm i n_L$$

$$n_T = \left(k_x^2 - \frac{\omega^2}{c_T^2(\omega)}\right)^{1/2} , \qquad n_L = \left(k_x^2 - \frac{\omega^2}{c_L^2(\omega)}\right)^{1/2} . \qquad (7.1.6)$$

Only the positive signs are acceptable, since the others give exponentially increasing solutions, as will be seen in a moment. We therefore have

$$k_y = i n_T \quad \text{or} \quad i n_L \qquad (7.1.7)$$

where n_T, n_L are assumed to have a positive real part. Consider the partial Fourier transforms

$$\hat{u}_i(k_x, y, \omega) = \int_{-\infty}^{\infty} dx\, dt\, u_i(r, t)\, e^{-i(k_x x + \omega t)} = \frac{1}{2\pi} \int_{-\infty}^{\infty} dk_y\, e^{i k_y y}\, \hat{u}_i(k, \omega) .$$

We can write these as

$$\hat{u}_i(k_x, y, \omega) = a_i(k_x, \omega)\, e^{-n_T y} + b_i(k_x, \omega)\, e^{-n_L y} , \qquad i = 1, 2 \qquad (7.1.8)$$

in view of the constraint imposed by (7.1.7). This can be seen more precisely by going through the above argument using $\hat{u}_i(k_x, y, \omega)$ and replacing k_y by $-i\partial/\partial y$. The dynamical equation (7.1.3b) may now be solved to give

$$a_2(k_x, \omega) = i\frac{k_x}{n_T} a_1(k_x, \omega) , \qquad b_2(k_x, \omega) = i\frac{n_L}{k_x} b_1(k_x, \omega) . \qquad (7.1.9)$$

Let us define

$$\hat{s}(k_x, \omega) = - \int_{-\infty}^{\infty} dx\, dt\, \sigma_{12}(x, 0, t)\, e^{-i(k_x x + \omega t)}$$

$$\qquad (7.1.10)$$

$$\hat{p}(k_x, \omega) = - \int_{-\infty}^{\infty} dx\, dt\, \sigma_{22}(x, 0, t)\, e^{-i(k_x x + \omega t)}$$

in terms of the specified boundary quantities σ_{12}, σ_{22}. Equation (7.1.3a) together with (7.1.8, 9) give (replacing k_y by $-i\partial/\partial y$):

$$\hat{s}(k_x, \omega) = \frac{2\hat{\mu}(\omega)}{n_T}(n_2 a_1 + n_T n_L b_1)$$

$$\hat{p}(k_x, \omega) = \frac{2\hat{\mu}(\omega)i}{n_T}\left(k_x n_T a_1 + \frac{n_2 n_T b_1}{k_x}\right), \qquad n_2 = \left(k_x^2 - \frac{\omega^2}{2c_T^2}\right),$$

$$(7.1.11)$$

which, on solution, result in

$$a_1(k_x, \omega) = \frac{n_T}{2\hat{\mu}(\omega)D}\left[i\frac{n_2}{k_x}\hat{s}(k_x, \omega) - n_L \hat{p}(k_x, \omega)\right]$$

$$b_1(k_x, \omega) = \frac{1}{2\hat{\mu}(\omega)D}[-ik_x n_T \hat{s}(k_x, \omega) + n_2 \hat{p}(k_x, \omega)] \qquad (7.1.12)$$

$$D = \frac{i}{k_x}(n_2^2 - k_x^2 n_L n_T) .$$

The steady-state assumption means that all quantities, expressed in terms of space and time variables are functions of $x + Vt$ rather than x, t separately.

Problem 7.1.1: Show that if $u_i(r, t)$, given by (7.1.2), has this property, then $\omega = Vk_x$.

This must hold true for all quantities. In fact, all Fourier transformed quantities reduce according to $\hat{f}(k_x, \omega) \to 2\pi\hat{f}(k_x)\delta(\omega - k_x V)$. We ignore the delta function in what follows. Replacing ω by Vk_x and defining

$$\delta_T = \frac{V^2}{c_T^2} ; \quad \delta_L = \frac{V^2}{c_L^2}$$

$$B_T = (1 - \delta_T)^{1/2} = \frac{n_T}{|k_x|}$$

$$B_L = (1 - \delta_L)^{1/2} = \frac{n_L}{|k_x|} \qquad (7.1.13)$$

$$B_2 = 1 - \frac{1}{2}\delta_T = \frac{n_2}{k_x^2} ,$$

we find that (7.1.12) becomes

$$ik_x a_1(k_x) = \frac{B_T}{2\hat{\mu}\Gamma_1}[iB_2 \hat{s}(k_x)\text{sgn}(k_x) - B_L \hat{p}(k_x)]$$

$$ik_x b_1(k_x) = \frac{1}{2\hat{\mu}\Gamma_1}[-iB_T \hat{s}(k_x)\text{sgn}(k_x) + B_2 \hat{p}(k_x)] \qquad (7.1.14)$$

$$\Gamma_1 = B_2^2 - B_T B_L ; \quad \text{sgn}(k_x) = \frac{k_x}{|k_x|} .$$

From (7.1.8, 9) and (7.1.14), we can deduce the following relations between displacement and traction on the boundary $y = 0$:

$$2\hat{u}(k_x\,V)\,\Gamma_1\,\mathrm{i}k_x\,\hat{u}_x(k_x) = -\Gamma_2(k_x)\hat{p}(k_x) - \frac{\mathrm{i}}{2}B_\mathrm{T}\,\delta_\mathrm{T}\,\hat{s}(k_x)\,\mathrm{sgn}(k_x) \qquad (7.1.15\,\mathrm{a})$$

$$2\hat{u}(k_x\,V)\,\Gamma_1\,\mathrm{i}k_x\,\hat{u}_y(k_x) = -\frac{\mathrm{i}}{2}B_\mathrm{L}\,\delta_\mathrm{T}\,\hat{p}(k_x)\,\mathrm{sgn}(k_x) + \Gamma_2(k_x)\hat{s}(k_x) \qquad (7.1.15\,\mathrm{b})$$

$$\Gamma_2 = B_\mathrm{T}\,B_\mathrm{L} - B_2 \; . \qquad (7.1.15\,\mathrm{c})$$

These equations have the same form as the corresponding elastic relations with moduli replaced by complex moduli, as required by the Classical Correspondence Principle discussed in Sect. 2.1. Relation (7.1.15b), which will be used in Sect. 7.2, was written down by Golden (1979b) from the elastic results of Eason (1965). More general results are derived by Golden (1982b).

7.1.1 Antiplane Strain

Relationships analogous to (7.1.15) will also be given for the antiplane strain configuration, which prevails for Mode III or tearing mode crack problems (Sect. 4.1). In this case,

$$u_x = u_y = 0 \; ; \qquad u_z = u_z(x,y) \; . \qquad (7.1.16)$$

The only non-zero strains are

$$\varepsilon_{zx} = \frac{1}{2}\frac{\partial u_z}{\partial x} \; ; \qquad \varepsilon_{zy} = \frac{1}{2}\frac{\partial u_z}{\partial y} \; . \qquad (7.1.17)$$

Consequently, only the components σ_{zx}, σ_{zy} of the stress tensor survive. On taking time Fourier transforms, these are given by $2\hat{\mu}(\omega)$ multiplied by the corresponding strains, as follows from (1.9.18). The dynamical equation takes the form

$$-\varrho\omega^2\hat{u}(r,\omega) = \hat{\mu}(\omega)\nabla^2\hat{u}(r,\omega) \qquad (7.1.18)$$

on dropping the subscript z. Let us now go over to $\hat{u}(k,\omega)$, defined by (7.1.2). Then (7.1.18) gives that $k^2 = \omega^2/c_\mathrm{T}^2$, so that

$$k_y = \mathrm{i}n_\mathrm{T} \; , \qquad (7.1.19)$$

and instead of (7.1.8), we write

$$\hat{u}(k_x,y,\omega) = a(k_x,\omega)\mathrm{e}^{-n_\mathrm{T}y} \; . \qquad (7.1.20)$$

Let $\hat{t}(k_x,\omega)$, given by

$$\hat{t}(k_x,\omega) = \int_{-\infty}^{\infty} dx\,dt\,\sigma_{zy}(x,0,t)\mathrm{e}^{-\mathrm{i}(k_x x + \omega t)} \; , \qquad (7.1.21)$$

be specified on the surface. Then it is easy to show that $\hat{u}(k,0,\omega)$, $\hat{t}(k,\omega)$ are related by

$$n_\mathrm{T}(k_x,\omega)\hat{\mu}(\omega)\hat{u}(k_x,0,\omega) = -\hat{t}(k_x,\omega) \; . \qquad (7.1.22)$$

In the steady-state case, this becomes

$$(k_x^2)^{1/2} B_T (k_x V) \hat{u}(k_x V) \hat{u}(k_x) = -\hat{t}(k_x) \ . \tag{7.1.23}$$

For convenience, the subscript will be dropped on k_x. Since the current stress cannot depend upon displacements in the future, the inverse transform of $n_T(k, \omega) \hat{u}(\omega)$ with respect to ω must be causal. In order to ensure this, we must, in line with the discussion in Sect. A3.2, use a contour in the ω plane, when taking the inverse transform, which is below all singularities of $n_T(k, \omega) \hat{u}(\omega)$. Let the singularities occur at points $\omega_i = f_i(k)$. Then the contour must be below each $f_i(k)$ for any value of k. In the steady-state case, the singularities in $B_T(kV) \hat{u}(kV)$ are determined by the conditions $Vk = f_i(k)$. The contour in the k plane corresponding to the causal contour in ω will be below $f_i(k)/V$ for all k and therefore in particular for k at the singularity. Thus, the inverse transform of $B_T(kV) \hat{u}(kV)$, using this contour, will be causal. Recall that n_T must have a positive real part. We define $B_T(kV)$ to have this property also. Let

$$(k^2)^{1/2} = \text{sgn}(\text{Re}\{k\}) k \ , \tag{7.1.24}$$

where account has been taken of the fact that we will deal with complex values of k in Sect. 7.3.

Problem 7.1.2: Show that (7.1.24) ensures that $n_T(k, kV)$ will have a positive real part.
  Hint: the difficult step is showing that $\text{sgn}(\text{Re}\{k\}) \text{Im}\{B_T\}$ is non-negative. Use (1.6.38) to show that $\text{Im}\{\hat{\mu}(\omega)\} = Q \, \text{sgn}(\text{Re}\{\omega\})$ where Q is positive. Relate $\text{Im}\{\hat{\mu}(\omega)\}$ to $\text{Im}\{B_T\}$ by squaring B_T and remembering that its real part is positive; see Golden (1986 b).

Finally, it will be observed that the results of this section amount to a solution of the first boundary value problem, as defined in Sect. 1.9, for the inertial case, omitting the important final step of taking inverse transforms.

7.2 Contact Problems on a Slightly Viscoelastic Medium

Not a great deal of work has been done to date on inertial, viscoelastic contact problems, essentially because of the considerable difficulty of such problems. Hunter (1968) has solved the problem of a sphere, in motion, embedded in an adhering viscoelastic medium with a view to providing a method for the dynamical testing of mechanical properties of certain materials. Atkinson and Coleman (1977) have given certain results in what may be regarded as the small decay time approximation for a standard linear solid. Golden (1979 b) has considered an expansion to first order in $V^2/c_T^2(\omega)$ and attempted to solve the problem by the methods developed for the non-inertial case, discussed in Chap. 3. This approach is algebraically very complicated and has limited applicability and so will not be considered here. We content ourselves with a more tractable problem discussed also by Golden (1979 b), namely contact problems on a medium exhibiting viscoelastic effects that are small compared with elastic effects.

Only the case of lubricated motion will be considered. This is conveniently simple, for purposes of illustration. The frictional case is discussed by Golden (1979b). Following in the spirit of Sects. 3.7, 5.3, we put

$$\mu(t) = g_0 [\delta(t) + \varepsilon_1(t)] \ , \qquad \lambda(t) = n_0 [\delta(t) + \varepsilon_2(t)] \tag{7.2.1}$$

where $\varepsilon_1(t)$, $\varepsilon_2(t)$ are small. If these two functions $\varepsilon_1(t)$, $\varepsilon_2(t)$ are equal, the material has the proportionality property which implies a unique Poisson's ratio. This restriction will not be imposed here. It follows from (7.2.1) that

$$\hat{\mu}(\omega) = g_0 [1 + \hat{\varepsilon}_1(\omega)] \ , \qquad \hat{\lambda}(\omega) = n_0 [1 + \hat{\varepsilon}_2(\omega)] \ . \tag{7.2.2}$$

We now write down (7.1.15b) for the frictionless case, again dropping the subscript on k_x:

$$i k \hat{l}(kV) \hat{u}(k) = i \hat{p}(k) \operatorname{sgn}(k) \tag{7.2.3a}$$

$$\hat{l}(kV) = -\frac{4 \hat{\mu}(kV) \Gamma_1(kV)}{B_L(kV) \delta_T(kV)} \tag{7.2.3b}$$

and put

$$\hat{l}(\omega) = l_0 [1 + \hat{\varepsilon}(\omega)] \ , \qquad \hat{\varepsilon}(\omega) = d_1 \hat{\varepsilon}_1(\omega) + d_2 \hat{\varepsilon}_2(\omega) \ , \tag{7.2.4}$$

where

$$l_0 = \hat{l}(\omega) \big|_{\hat{\varepsilon}_1 = \hat{\varepsilon}_2 = 0} \quad \text{and} \tag{7.2.5}$$

$$l_0 d_i = \frac{\partial \hat{l}(\omega)}{\partial \hat{\varepsilon}_i} \bigg|_{\hat{\varepsilon}_1 = \hat{\varepsilon}_2 = 0} . \tag{7.2.6}$$

More explicitly, we find that

$$l_0 = -\frac{4 g_0 (A_2^2 - A_T A_L)}{A_L \delta_{0T}} \quad \text{where} \tag{7.2.7}$$

$$A_T = (1 - \delta_{0T})^{1/2} \ , \quad A_L = (1 - \delta_{0L})^{1/2} \ , \quad A_2 = 1 - \tfrac{1}{2} \delta_{0T} \ ,$$

$$\delta_{0T} = \frac{\varrho V^2}{g_0} \ , \quad \delta_{0L} = \frac{\varrho V^2}{n_0 + 2 g_0} \ . \tag{7.2.8}$$

Also:

$$d_1 = 2 - \frac{4 g_0}{l_0} \left[\frac{A_2}{A_L} \left(1 - \eta \frac{A_2}{A_L^2} \right) - \frac{1}{2 A_T} \right]$$

$$d_2 = \frac{2 \eta n_0 A_2^2}{l_0 A_L^3} \ , \quad \eta = \frac{g_0^2}{(n_0 + 2 g_0)^2} \ . \tag{7.2.9}$$

We therefore have an explicit expression for $\hat{l}(\omega)$, which immediately gives the form of

$$l(t) = l_0 [\delta(t) + d_1 \varepsilon_1(t) + d_2 \varepsilon_2(t)] \ . \tag{7.2.10}$$

The function $l(t)$ will be causal, in the sense defined in Sect. 1.10. This is clear on physical grounds. It also follows from the kind of argument formulated in

Sect. 1.10 if we restrict the discussion to the subsonic case. For subsonic velocities, $\tilde{l}(\omega)$ can be expanded as a Taylor expansion in the velocity, where each term is a rational function of the moduli, as required by the argument in Sect. 1.10. We will see in the next section that the situation becomes more complex above the subsonic region. Taking the inverse Fourier transform of (7.2.3 a) gives

$$\int_{-\infty}^{x} \frac{dx}{V} l\left(\frac{x-x'}{V}\right) u'(x') = \frac{1}{\pi} \int_{C} \frac{p(x')}{x'-x} , \tag{7.2.11}$$

where (A3.2.8) and the Convolution Theorem discussed in Sect. A3.1 have been used. This equation has the same form as in the non-inertial limit. In fact, it is the steady-state version of (3.2.10). It is equivalent to the Hilbert problem (3.3.2) in the frictionless case. Therefore all the results for small viscoelasticity from Sects. 3.7, 8 apply here, with the modification that

$$\varepsilon(x) = d_1 \varepsilon_1(x) + d_2 \varepsilon_2(x) \tag{7.2.12}$$

in dimensionless coordinates moving with the punch. These functions are defined by the prescription (3.5.20). In particular, the expression for the hysteretic friction coefficient (3.8.15) applies where $v_c(x)$ is given by (3.7.21). Also, if ε_1 and ε_2 are both given by the same exponential decay term, (3.8.16) applies but where ε_0 depends on the moduli and the velocity of motion through d_1, d_2.

7.3 Crack Problems

The simplest inertial viscoelastic crack problem is that of a semi-infinite crack under Mode III or tearing mode displacement. This problem was first considered by Willis (1967), who used the Wiener-Hopf method to obtain a solution, for steady-state crack extension. More recently, Walton (1982) has given an alternative method of solving this problem, by casting it in the form of a Hilbert problem (with varying coefficients). Willis (1967) applies a crack extension criterion akin to that discussed in Sect. 4.7, to determine the velocity of extension, the point being that, in contrast to the quasi-static case, velocity of propagation now enters explicitly into the criterion. Atkinson and List (1972) consider a transient version of the Mode III problem. The steady-state Mode III problem for a layer composed of a standard linear solid is discussed in a limiting case by Atkinson and Coleman (1977) and Atkinson and Popelar (1979) using an exact, partially numerical application of the Wiener-Hopf method.

The Mode III problem has received most attention essentially because it is the simplest case. Mode I and II problems are considerably more difficult, but have been considered in a limiting case by Atkinson (1979), and again using an exact, partly numerical application of the Wiener-Hopf method, by Popelar and Atkinson (1980). But the physical importance of the Mode III problem is cast into doubt by experimental observations of Knauss (1970 b) indicating that such cracks do not grow rectilinearly, at least in brittle materials.

This topic will not be treated in depth here. To attempt to do so would be inappropriate, given the current state of the art. Instead, using relatively simple mathematical techniques, akin to those developed in earlier chapters, we will reproduce some important results obtained in the literature. Central to the method is the notion of Causality, which is discussed in Sect. 1.10. The approach adopted in this section was given by Golden (1986b). We focus attention on the mode III problem, as a conveniently simple illustrative example. Only the steady-state limit is considered. The semi-infinite crack is assumed to be moving in a negative x direction with velocity $(-V)$. In coordinates moving with the crack tip, it is taken to occupy the positive x-axis.

Taking into account (7.1.24), let us write (7.1.23) in the form

$$\hat{f}(k) i k \hat{u}(k) = -i \hat{t}(k) \operatorname{sgn}(\operatorname{Re}\{k\}) \ , \qquad \hat{f}(k) = \hat{\mu}(kV) B_T(kV) \ . \tag{7.3.1}$$

We wish to take the inverse transform with respect to k of this relation and utilize the causal property of $\hat{f}(k)$ mentioned at the end of Sect. 7.1. However, before we do this, it is necessary to examine the analytic structure of $\hat{f}(k)$ in the complex k plane, in particular to see whether it has any singularities in the lower half-plane. We deduce from the discussion in Sect. A3.2, that if such singularities exist, it is necessary to take the inverse transform along contours off the real axis, below these singularities, to preserve the causal property.

Since the properties of $\hat{\mu}(\omega)$ have been discussed in some detail in Sect. 1.5, we focus our attention on $B_T(\omega) = [1 - V^2/c^2(\omega)]^{1/2}$ where ω is used to denote kV. New singularities in this quantity, over and above those resulting from the singularities in $c^2(\omega) \ [= \hat{\mu}(\omega)/\varrho]$, will occur if, for some ω,

$$\hat{\mu}(\omega) = \varrho V^2 \ . \tag{7.3.2}$$

Now $\hat{\mu}(\omega)$, as the Fourier transform of a real function, will be a real function of $z = i\omega$. It is real along the real z- (imaginary ω-) axis. Also, in general, it will be complex elsewhere. At least, this will be true for spectrum models [(1.6.38)] which we will use as our background model for $\hat{\mu}(\omega)$ in this discussion. Therefore, solutions to (7.3.2) must be sought along the imaginary ω-axis. For spectrum models, the singularities of $\hat{\mu}(\omega)$ will occur along the positive imaginary axis. These will be isolated poles if the spectrum of decay times is discrete and cuts if the spectrum is continuous. We expect that solutions of (7.3.2) may exist in the vicinity of these singularities and perhaps elsewhere on the positive imaginary ω-axis. If so, the cut resulting from the branch point in $B_T(\omega)$ can be taken out along the positive direction and the lower half-plane remains singularity free. The real point of interest is under what conditions a solution of (7.3.2) can occur in the lower half-plane. Along the negative imaginary axis, $\hat{\mu}(\omega)$ varies between $\hat{\mu}(0)$ and $\hat{\mu}(\infty)$. Equations (1.5.10, 11) and the structure of $G(t)$ give that

$$\hat{\mu}(0) = G(\infty) \leq G(0) = \hat{\mu}(\infty) \ . \tag{7.3.3}$$

We claim that $\hat{\mu}(\omega)$ is monotonically increasing along the negative imaginary ω-axis, again relying on the properties of spectrum models. Therefore, singularities will occur in the lower half-plane only if $\hat{\mu}(0) \leq \varrho V^2 \leq \hat{\mu}(\infty)$ or

$$c_0^2 \leq V^2 \leq c_1^2 , \quad \text{where} \tag{7.3.4}$$

$$c_0^2 = \frac{\hat{\mu}(0)}{\varrho} , \quad c_1^2 = \frac{\hat{\mu}(\infty)}{\varrho} . \tag{7.3.5}$$

These are the limiting velocities of sound for the medium. If $V < c_0$, which is the subsonic case, no singularity can occur in the lower-half plane. If (7.3.4) is true, then a branch point will occur at a point on the negative imaginary axis. In the complex k plane, we denote this as $-iq_0$ where q_0 is positive. The branch cut is taken in a positive direction along the imaginary axis.

Problem 7.3.1: Demonstrate the above results explicitly for the standard linear model, discussed in Sect. 1.6. Show in particular that

$$q_0 = \frac{\varrho V^2 - G_0}{V\tau(G_0 + G_1 - \varrho V^2)} . \tag{7.3.1p}$$

Let us now proceed to take the inverse transform of (7.3.1), given that $B_T(kV)$ has a cut from $k = -iq_0$ upwards. In the subsonic case we put q_0 to zero. It must be assumed that this analytic structure is reflected in $\hat{u}(k)$, $\hat{t}(k)$.

The displacement $u(x)$ is zero on the negative x-axis. Also, $t(x)$ is known on the positive x-axis. Therefore, we write (7.3.1) in the form

$$\hat{f}(k)ik\,\hat{u}^+(k) = -i\,\mathrm{sgn}(\mathrm{Re}\{k\})[\hat{t}^+(k) + \hat{t}^-(k)] , \tag{7.3.6}$$

in the notation of (A3.2.5), where $t^+(k)$ is known. The functions $\hat{u}^+(k)$, $\hat{t}^-(k)$ are to be determined. We must choose a contour below the branch point at $k = -iq_0$ when taking the inverse transform. This and other points required in the present context are discussed in Sect. A3.2. Firstly let us consider $\hat{u}^+(k)$. Such a contour will ensure that we obtain a displacement function which is zero for negative x, as desired, though we also expect that it will diverge as $\exp(q_0 x)$ for large positive x. Similarly, $\hat{t}^-(k)$ will be analytic for $\mathrm{Im}\{k\} \geq -q_0$, or put differently, $t^-(x)$ will converge as $\exp(q_0 x)$ at large negative x. This latter observation means that we cannot choose a contour lower than $-iq_0$ without danger of meeting a singularity in $\hat{t}^-(k)$. Therefore, we take the inverse transform along a contour parallel to the real axis, a distance q_0 below it. Noting (A3.2.1p) and the Faltung theorem (A3.1.14, 17), we deduce that (7.3.1) becomes

$$v(x) = -\frac{1}{\pi} \int_{-\infty}^{\infty} dx' \frac{e^{q_0(x-x')}t(x')}{x'-x} , \quad v(x) = \int_{-\infty}^{x} dx' f(x-x')u'(x') , \tag{7.3.7}$$

where the causal property $f(x) = 0$, $x < 0$ has been incorporated. An immediate result of this property, and the fact that the displacement is zero on the negative axis, is that $v(x)$ also has this property, namely

$$v(x) = 0 , \quad x < 0 . \tag{7.3.8}$$

Inverting the infinite Hilbert transform in (7.3.7) according to (A2.4.22) gives

$$\frac{1}{\pi} \int_0^{\infty} dx' \frac{e^{q_0(x-x')}v(x')}{x'-x} = t(x) . \tag{7.3.9}$$

Since $t(x)$ is given for $x > 0$, let us regard (7.3.9) as a relation on the positive x-axis. Reinverting what is now a semi-infinite Hilbert transform, we obtain, by virtue of (A2.4.24):

$$v(x) = -\frac{1}{\pi\sqrt{x}} \int_0^\infty dx' \frac{e^{q_0(x-x')}t(x')\sqrt{x'}}{x'-x} , \qquad x > 0 . \qquad (7.3.10)$$

This is an explicit expression for $v(x)$. If the inverse of $f(x)$ [in the sense of (1.2.35)] is known, then we can solve $u'(x)$. However, neither $f(x)$ nor its inverse are known explicitly in general. Note that $v(x)$ diverges as $\exp(q_0 x)$ at large positive x.

Explicit results can be given for the stress on the negative x-axis and the stress intensity factor. Consider (7.3.9) for $x < 0$. Let us substitute from (7.3.10) for $v(x)$ and use (A2.4.25) to deduce that

$$t(x) = -\frac{1}{\pi\sqrt{-x}} \int_0^\infty dx' \frac{e^{q_0(x-x')}t(x')\sqrt{x'}}{x'-x} , \qquad x < 0 . \qquad (7.3.11)$$

Note that it goes to zero as $\exp(q_0 x)|x|^{-3/2}$ at large negative x. The stress intensity factor is given by

$$K_3 = \lim_{x \to 0} (-2x)^{1/2}t(x) = -\frac{\sqrt{2}}{\pi} \int_0^\infty dx \frac{e^{-q_0 x}t(x)}{\sqrt{x}} . \qquad (7.3.12)$$

If the motion is subsonic, the branch point moves above the real k-axis and we can take the inverse transform along this axis, but not above it, since we may meet zeros or branch points of $\hat{\mu}(kV)$. Therefore, q_0 is put to zero and instead of (7.3.11, 12) we have

$$t(x) = -\frac{1}{\pi\sqrt{-x}} \int_0^\infty dx' \frac{t(x')\sqrt{x'}}{x'-x} , \qquad x < 0 , \qquad (7.3.13\,\mathrm{a})$$

$$K_3 = -\frac{\sqrt{2}}{\pi} \int_0^\infty dx \frac{t(x)}{\sqrt{x}} . \qquad (7.3.13\,\mathrm{b})$$

These are in fact the same as for the non-inertial problem. The results (7.3.11 – 13) for the stress ahead of the crack and the stress intensity factor have been given by Willis (1967), for a standard linear solid, and Walton (1982) for a general material, using quite different methods to the one outlined here.

7.4 Summary

Problems where inertial effects are not neglected are considered briefly in this chapter. Generally speaking, little progress in analytic terms is possible for such problems.

I. Displacement-Traction Relationships. Displacement-traction relationships, in terms of Fourier transformed quantities, on the surface of a half-space, under plane strain conditions, are derived in Sect. 7.1 and given by (7.1.15) for steady-state uniform motion. A similar relationship between the Fourier transform of the displacement and tearing stress is given by (7.1.23) for tearing mode fracture, along the line of the crack. These have the same form as the equivalent elastic relations, with moduli replaced by complex moduli, as required by the Classical Correspondence Principle.

II. Contact Problems on Slightly Viscoelastic Media. If only first order visco-elastic effects are retained, it is shown that the equations have the same formal structure as in the non-inertial case and the results of Sects. 3.7 and 3.8 apply with the modification expressed by (7.2.12).

III. Tearing Mode Crack Problem. This problem is solved, utilizing constraints imposed by Causality. The form of the tearing stress off the crack face on the line of fracture is given by (7.3.11) for the case where the velocity is in the range of speeds of sound of the medium [see (7.3.4)] and by (7.3.13a) for the subsonic case. The quantity q_0 is defined by (7.3.1p) for a standard linear solid. The stress intensity factor for the two cases is given by (7.3.12) and (7.3.13b), respectively.

Appendix I
Tables of Relevant Integrals and Other Formulae

In Tables A1.1 – 3 we gather together a number of integrals, most of which occur in the main text and the other appendices. These are Hilbert transforms and related integrals. For a more detailed table of Hilbert transforms, we refer to Erdélyi et al. (1954), though certain of the transforms listed here are not contained in that reference, but are taken from elsewhere in the literature. These are discussed in Sect. A2.4. For Hilbert transforms where x is in the region of integration, the principle value of the integral is understood.

In Table A1.4, standard formulae involving a number of special functions are listed, in some cases as a supplement to the discussion of Sect. A2.4.

In drawing up these tables, the standard references Gradshteyn and Ryzhik (1965), Abramowitz and Stegun (1970), and Korn and Korn (1968) proved useful, as did Gladwell (1980).

Table A1.1 Hilbert Transforms on $[-1, 1]$

$$h(x) = \frac{1}{\pi} \int_{-1}^{1} dx' \frac{g(x')}{x' - x}; \qquad \begin{array}{l} m(x) = (1 - x^2)^{1/2}, \quad |x| < 1 \\ n(x) = (x^2 - 1)^{1/2}, \quad |x| > 1 \end{array}$$

$T_n(x)$, $U_n(x)$: Chebyshev Polynomials, defined on Table A1.4.

$D(\alpha, x) = x I_0(\alpha) - I_1(\alpha)$

$I_0(\alpha)$, $I_1(\alpha)$: Bessel Functions defined on Table A1.4.

$g(x)$	$h(x)$	
$\dfrac{1}{m(x)}$	$0, \quad \|x\| < 1$ $-\dfrac{\operatorname{sgn}(x)}{n(x)}, \quad \|x\| > 1$	(A1.1.1)
$\dfrac{x}{m(x)}$	$1, \quad \|x\| < 1$ $1 - \dfrac{x \operatorname{sgn}(x)}{n(x)}, \quad \|x\| > 1$	(A1.1.2)
$\dfrac{T_n(x)}{m(x)}$ $n = 0, 1, 2, \ldots$	$U_{n-1}(x), \quad \|x\| < 1$ $U_{n-1}(x) - \dfrac{\operatorname{sgn}(x)}{n(x)} T_n(x), \quad \|x\| > 1$	(A1.1.3)

Table A1.1 (cont.)

$g(x)$	$h(x)$	
$m(x)$	$-x, \quad \|x\| < 1$ $-x + \mathrm{sgn}(x)\, n(x), \quad \|x\| > 1$	(A1.1.4)
$x\, m(x)$	$\frac{1}{2} - x^2, \quad \|x\| < 1$ $\frac{1}{2} - x^2 + \mathrm{sgn}(x)\, n(x)\, x, \quad \|x\| > 1$	(A1.1.5)
$U_{n-1}(x)\, m(x)$ $n = 1, 2, \ldots$	$-T_n(x), \quad \|x\| < 1$ $-T_n(x) + \mathrm{sgn}(x)\, n(x)\, U_{n-1}(x), \quad \|x\| > 1$	(A1.1.6)
$e^{-\alpha x}\, m(x)$	$\alpha m(x)\, e^{-\alpha x} \int\limits_{-1}^{x} dy\, \dfrac{e^{\alpha y} D(\alpha, y)}{m(y)} - D(\alpha, x), \quad \|x\| < 1$ $\alpha n(x)\, e^{-\alpha x} \int\limits_{-\infty}^{x} dy\, \dfrac{e^{\alpha y} D(\alpha, y)}{n(y)} - D(\alpha, x), \quad x < -1$	(A1.1.7)
$\dfrac{e^{-\alpha x}}{m(x)}$	$\dfrac{\alpha e^{-\alpha x}}{m(x)} \int\limits_{-1}^{x} dy\, \dfrac{e^{\alpha y} D(\alpha, y)}{m(y)}, \quad \|x\| < 1$ $-\dfrac{\alpha e^{-\alpha x}}{n(x)} \int\limits_{-\infty}^{x} dy\, \dfrac{e^{\alpha y} D(\alpha, y)}{n(y)}, \quad x < -1$	(A1.1.8)
$(1+x)^{1-\theta}(1-x)^{\theta}$ $0 < \theta < 1$	$\dfrac{2\theta - 1 - x}{\sin(\pi\theta)} + g(x) \cot(\pi\theta), \quad \|x\| < 1$ $\dfrac{1}{\sin(\pi\theta)} \left[(2\theta - 1 - x) + \mathrm{sgn}(x)\, \|x+1\|^{1-\theta} \|x-1\|^{\theta}\right], \quad \|x\| > 1$	(A1.1.9)
$\dfrac{1}{(1+x)^{1-\theta}(1-x)^{\theta}}$ $0 < \theta < 1$	$-\dfrac{\cot(\pi\theta)}{(1+x)^{1-\theta}(1-x)^{\theta}}, \quad \|x\| < 1$ $\dfrac{-\mathrm{sgn}\, x}{\sin(\pi\theta)\, \|x+1\|^{1-\theta}\|x-1\|^{\theta}}, \quad \|x\| > 1$	(A1.1.10)

Table A1.2 Hilbert Transforms on $[a, b]$

$$h(x) = \frac{1}{\pi}\int_a^b dx' \frac{g(x')}{x'-x}; \quad m(x) = [(x-a)(b-x)]^{1/2}$$
$$n(x) = |(x-a)(x-b)|^{1/2}$$

$$\varepsilon(x) = \begin{cases} -1, & x < a \\ +1, & x > b \end{cases}$$

$$E(\beta, x) = xK_0(\beta) - K_1(\beta)$$

$K_0(\beta), K_1(\beta)$: Bessel Functions $-$ see Table A1.1.4.

$g(x)$	$h(x)$	
$\dfrac{1}{m(x)}$	$0, \quad a < x < b$	(A1.2.1)
	$-\dfrac{\varepsilon(x)}{n(x)}, \quad \begin{matrix} x < a \\ x > b \end{matrix}$	
$m(x)$	$\left(\dfrac{b+a}{2} - x\right), \quad a < x < b$	(A1.2.2)
	$\left(\dfrac{b+a}{2} - x\right) + \varepsilon(x)\,n(x), \quad \begin{matrix} x < a \\ x > b \end{matrix}$	
$\dfrac{x}{m(x)}$	$1, \quad a < x < b$	(A1.2.3)
	$1 - \dfrac{x\varepsilon(x)}{n(x)}, \quad \begin{matrix} x < a \\ x > b \end{matrix}$	
$xm(x)$	$\dfrac{1}{2}\left(\dfrac{b-a}{2}\right)^2 + x\left(\dfrac{b+a}{2} - x\right), \quad a < x < b$	(A1.2.4)
	$\dfrac{1}{2}\left(\dfrac{b-a}{2}\right)^2 + x\left(\dfrac{b+a}{2} - x\right)$	
	$\qquad + \varepsilon(x)xn(x), \quad \begin{matrix} x < a \\ x > b \end{matrix}$	
$(x-a)^{1-\theta}(b-x)^{\theta}$	$\dfrac{1}{\sin(\pi\theta)}[a(1-\theta) + b\theta - x + g(x)\cos(\pi\theta)],$	
	$\qquad\qquad\qquad\qquad a < x < b$	(A1.2.5)
$0 < \theta < 1$	$\dfrac{1}{\sin(\pi\theta)}[a(1-\theta) + b\theta - x$	
	$\qquad + \varepsilon(x)|x-a|^{1-\theta}|x-b|^{\theta}], \quad \begin{matrix} x > b \\ x < a \end{matrix}$	

Table A1.2 (cont.)

$g(x)$	$h(x)$	
$\dfrac{1}{(x-a)^{1-\theta}(b-x)^{\theta}}$ $0<\theta<1$	$-g(x)\cot(\pi\theta),\quad a<x<b$ $\dfrac{-\varepsilon(x)}{\sin(\pi\theta)\,\lvert x-a\rvert^{1-\theta}\lvert x-b\rvert^{\theta}},\quad \begin{array}{l}x>b\\x<a\end{array}$	(A1.2.6)
$\pi n(x)e^{\beta x}$ $(a=-\infty,b=-1,\beta>0)$	$-m(x)\beta e^{\beta x}\displaystyle\int_x^1 dy\,\dfrac{e^{-\beta y}E(\beta,y)}{m(y)}$ $+E(\beta,x),\quad \lvert x\rvert<1$ $-n(x)\beta e^{\beta x}\displaystyle\int_x^{\infty} dy\,\dfrac{e^{-\beta y}E(\beta,y)}{n(y)}$ $+E(\beta,x),\quad x>1$	(A1.2.7)
$\dfrac{1}{m(x)},\quad c<x<b$ $0,\quad a<x<c$	$\dfrac{1}{\pi m(x)}\log_e\left\lvert\dfrac{\sqrt{(x-a)}\sqrt{(b-c)}+\sqrt{(b-x)}\sqrt{(c-a)}}{\sqrt{(x-a)}\sqrt{(b-c)}-\sqrt{(b-x)}\sqrt{(c-a)}}\right\rvert,$ $a<c<b,\quad a<x<b.$	(A1.2.8)

Note: Transforms of polynomials multiplying the functions $g(x)$ in (A1.2.5, 6) are given by Golden (1979).

Table A1.3 Miscellaneous Integrals Associated with Hilbert Transforms

$$I=\frac{1}{\pi}\int_a^b dx\,g(x)\quad m(x,\theta)=(x-a)^{1-\theta}(b-x)^{\theta}$$
$$0<\theta<1$$

$g(x)$	I	
$\left(\dfrac{b-x}{x-a}\right)^{1/2}$	$\dfrac{b-a}{2}$	(A1.3.1)
$\left(\dfrac{b-x}{x-a}\right)^{1/2}x$	$\dfrac{b-a}{8}(b+3a)$	(A1.3.2)
$\left(\dfrac{x-a}{b-x}\right)^{1/2}$	$\dfrac{b-a}{2}$	(A1.3.3)
$\left(\dfrac{x-a}{b-x}\right)^{1/2}x$	$\dfrac{b-a}{8}(a+3b)$	(A1.3.4)

Table A1.3 (cont.)

$g(x)$	I	
$m(x, \frac{1}{2})$	$\dfrac{(b-a)^2}{8}$	(A1.3.5)
$\dfrac{1}{m(x, \frac{1}{2})}$	1	(A1.3.6)
$x\,m(x, \frac{1}{2})$	$\dfrac{(b+a)(b-a)^2}{16}$	(A1.3.7)
$\left(\dfrac{b-x}{x-a}\right)^{\theta}$ $0 < \theta < 1$	$\dfrac{(b-a)\,\theta}{\sin(\pi\theta)}$	(A1.3.8)
$m(x, \theta)$ $0 < \theta < 1$	$\dfrac{(b-a)^2\,\theta(1-\theta)}{2\sin(\pi\theta)}$	(A1.3.9)
$\dfrac{1}{m(x, \theta)}$ $0 < \theta < 1$	$\dfrac{1}{\sin(\pi\theta)}$	(A1.3.10)
$\dfrac{x}{m(x, \theta)}$ $0 < \theta < 1$	$\dfrac{b\theta + a(1-\theta)}{\sin(\pi\theta)}$	(A1.3.11)
$\dfrac{x^2}{m(x, \theta)}$ $0 < \theta < 1$	$\dfrac{1}{\sin(\pi\theta)}\left[b^2\theta + a^2(1-\theta) - \dfrac{(b-a)^2}{2}\theta(1-\theta)\right]$	(A1.3.12)

Note: One may obtain expressions for integrals involving $(b-x)^{1-\theta}(x-a)^{\theta}$ by interchanging θ and $(1-\theta)$ in (A1.3.9 – 12).

Table A1.4 Other Miscellaneous Integrals and Relationships

$$m(x) = (1-x^2)^{1/2}; \quad n(x) = (x^2-1)^{1/2}$$

Gamma and Beta Functions $\Gamma(z), B(\alpha, \beta)$

$$\int_0^{\infty} dt\, e^{-t} t^{z-1} = \Gamma(z), \quad z > 0 \tag{A1.4.1}$$

Note: $z\Gamma(z) = \Gamma(z+1)$; $\Gamma(1/2) = \pi^{1/2}$; $\Gamma(1-z)\Gamma(z) = \dfrac{\pi}{\sin(\pi z)}$

$$\int_0^1 dt\, t^{\alpha-1}(1-t)^{\beta-1} = B(\alpha, \beta)$$

$$= B(\beta, \alpha) = \frac{\Gamma(\alpha)\Gamma(\beta)}{\Gamma(\alpha+\beta)} \qquad \alpha > 0, \quad \beta > 0 \ . \tag{A1.4.2}$$

The incomplete Beta function $B_x(\alpha, \beta)$ is obtained by replacing the upper integration limit in (2) by x. Note that

$$\int_a^b dt(t-a)^{\alpha-1}(b-t)^{\beta-1} = (b-a)^{\alpha+\beta-1} B(\alpha, \beta) \qquad \alpha > 0, \quad \beta > 0. \tag{A1.4.3}$$

Chebyshev Polynomials $T_n(x)$, $U_n(x)$

Definition: $T_n(x) = \cos(n\theta)$

$$U_n(x) = \frac{\sin[(n+1)\theta]}{\sin\theta} \tag{A1.4.4}$$

$$x = \cos\theta\, , \qquad n = 0, 1, 2, \ldots \ .$$

[*Note:* $U_{-1}(x)$ used on occasion; it is equal to zero.]

Orthogonality relations

$$\frac{1}{\pi}\int_{-1}^1 dx\, \frac{T_n(x)\,T_m(x)}{m(x)} = \begin{cases} 0 & m \neq n \\ \frac{1}{2} & m = n > 0 \\ 1 & m = n = 0 \end{cases} \tag{A1.4.5}$$

$$\frac{1}{\pi}\int_{-1}^1 dx\, m(x)\,U_n(x)\,U_m(x) = \begin{cases} 0 & m \neq n \\ \frac{1}{2} & m = n \geq 0 \ . \end{cases} \tag{A1.4.6}$$

Recurrence and other relations

Note: prime denotes differentiation.

$$T_{n+1}(x) - 2x\,T_n(x) + T_{n-1}(x) = 0\, , \qquad U_{n+1}(x) - 2x\,U_n(x) + U_{n-1}(x) = 0 \tag{A1.4.7}$$

$$U_{n-1}(x) = \frac{1}{n}\,T_n'(x) \tag{A1.4.8}$$

$$T_{n+1}(x) = -\frac{m(x)}{n+1}\frac{d}{dx}[m(x)\,U_n(x)]\ .$$

The following integral formulae hold for $z \notin [-1, 1]$ [Gladwell (1980)]:

$$\frac{1}{\pi}\int_{-1}^1 dx'\, \frac{m(x')\,U_{n-1}(x')}{x'-z} = -[z - (z^2-1)^{1/2}]^n\, , \qquad n = 1, 2, 3, \ldots \tag{A1.4.9}$$

$$= -T_n(z) + (z^2-1)^{1/2}\,U_{n-1}(z)$$

and

$$\frac{1}{\pi}\int_{-1}^1 dx'\, \frac{T_n(x')}{m(x')(x'-z)} = -\frac{[z - (z^2-1)^{1/2}]^n}{(z^2-1)^{1/2}}\, , \qquad n = 0, 1, 2, \ldots \tag{A1.4.10}$$

where the standard branch of $(z^2-1)^{1/2}$ is chosen, namely that with the property that $\lim\limits_{|z|\to\infty} (z^2-1)^{1/2}/z = 1$. From (A1.4.9, 10) and (A2.2.9), one can deduce the Hilbert transforms, involving $T_n(x)$ and $U_n(x)$, which are given in Table A1.1.

Examples

$$
\begin{aligned}
T_0(x) &= 1 & U_{-1}(x) &= 0 \\
T_1(x) &= x & U_0(x) &= 1 \\
T_2(x) &= 2x^2-1 & U_1(x) &= 2x \\
T_3(x) &= 4x^3-3x & U_2(x) &= 4x^2-1 \;.
\end{aligned}
\tag{A1.4.11}
$$

Modified Bessel Functions with Imaginary Argument

$$I_n(\alpha), \quad K_n(\alpha), \quad n = 0, 1, 2, \ldots$$

$$I_n(\alpha) = \frac{(\alpha/2)^n}{\pi^{1/2}\,\Gamma(n+(1/2))} \int\limits_{-1}^{1} dx\,[m(x)]^{2n-1} e^{\pm\alpha x} \tag{A1.4.12}$$

$$= \frac{1}{\pi} \int\limits_{0}^{\pi} d\theta\, e^{\alpha\cos\theta}\cos(n\theta) \tag{A1.4.13}$$

$$= \frac{1}{\pi} \int\limits_{-1}^{1} \frac{dx}{m(x)}\, e^{\alpha x}\, T_n(x) \tag{A1.4.14}$$

$$I_n(-\alpha) = (-1)^n I_n(\alpha) \tag{A1.4.15}$$

$$K_n(\alpha) = \frac{\Gamma(1/2)(\alpha/2)^n}{\Gamma(n+(1/2))} \int\limits_{-\infty}^{-1} dx\,[n(x)]^{2n-1} e^{\alpha y} \tag{A1.4.16}$$

$$= \int\limits_{0}^{\infty} d\theta\, e^{-\alpha\cosh\theta}\cosh(n\theta) \tag{A1.4.17}$$

$$= \int\limits_{-\infty}^{-1} \frac{dy}{n(y)}\, e^{\alpha y}\, T_n(-y) \;. \tag{A1.4.18}$$

Recurrence relations

Note: prime indicates differentiation.

$$I_{n-1}(\alpha) - I_{n+1}(\alpha) = \frac{2n}{\alpha} I_n(\alpha) \tag{A1.4.19}$$

$$I_{n-1}(\alpha) + I_{n+1}(\alpha) = 2I_n'(\alpha) \tag{A1.4.20}$$

$$I_0'(\alpha) = I_1(\alpha) \tag{A1.4.21}$$

$$K_{n-1}(\alpha) - K_{n+1}(\alpha) = -\frac{2n}{\alpha} K_n(\alpha) \tag{A1.4.22}$$

$$K_{n-1}(\alpha) + K_{n+1}(\alpha) = -2K_n'(\alpha) \tag{A1.4.23}$$

$$K_0'(\alpha) = -K_1(\alpha) \;. \tag{A1.4.24}$$

Wronskian relation

$$I_n(\alpha)K_{n+1}(\alpha) + I_{n+1}(\alpha)K_n(\alpha) = 1/\alpha \;. \tag{A1.4.25}$$

Appendix II
Boundary Value Problems for Analytic Functions

The main purpose of this appendix is to summarize aspects of the theory of the linear relationship between boundary values of analytic functions on opposite sides of curves of discontinuity in the region of definition of the function, which are relevant to the solution of plane problems in Chaps. 3, 4. This topic is discussed in Sects. A2.2 und A2.3, while various introductory concepts are introduced in Sect. A2.1.

The problem of the linear relationship may be stated as follows: given the equation connecting the boundary values of an analytic function on a line of discontinuity together with some condition on the function at infinity, determine the analytic function, either uniquely or very nearly uniquely. This has been termed the Hilbert problem by Muskhelishvili (1953, 1963) and the Riemann problem by Gakhov (1966). We adopt the former terminology here.

A closely related problem is: given a linear relationship between the real and imaginary parts of the function on the boundaries of a domain in which it is analytic, find the function throughout the domain. This is termed the Riemann-Hilbert problem by Galin (1961) and also here.

So, in the case of the Riemann-Hilbert problem, a function is analytic in a given domain bounded by a particular contour, and information is supplied about the limiting values of the function on this contour. In the case of the Hilbert problem, the region of analyticity contains open or closed contours on which the function is not defined. These are cuts in the region of analyticity across which the function is discontinuous. The supplied information is a relationship between the limiting values on either side of these cuts or internal boundaries.

Plane elastic, and in a partial sense, viscoelastic boundary value problems can often be phrased in terms of either a Hilbert or a Riemann-Hilbert problem. Galin (1961) for example, bases his whole approach on the latter method. On the other hand, Muskhelishvili (1963) manages to cast various boundary value problems in the form of Hilbert problems, thereby solving them. Gakhov (1966) discusses the connection between the two problems. The Hilbert approach is adopted here, on the grounds that the theory is a little easier, though this may be a matter of taste.

The Riemann-Hilbert problem is therefore not discussed. The treatment of the Hilbert problem is not very general, being confined in particular to the case of constant coefficients. The reader is referred to the standard texts mentioned above for more complete treatments.

A2.1 Some Properties of Analytic Functions

We mention briefly, for the sake of convenient reference, some properties of analytic functions which are required in various contexts. For more complete treatment of these topics, we refer to the many standard references on the topic, for example Ahlfors (1966). Some of the discussion is also based on a section of Morse and Feshbach (1953).

Of all the functions defined on the xy plane, there is a very special class, termed analytic functions, which have the property that they are functions only of the combination $z = x + iy$ and have a uniquely defined derivative with respect to z at each point in the region. This latter requirement is very restrictive in that it means that the derivative is independent of which of the infinite number of directions the limit is approached. If we write such a function $F(z)$ in the form

$$F(z) = F(x, y) = u(x, y) + iv(x, y) , \qquad (A2.1.1)$$

then the uniqueness of the limit gives that

$$\frac{dF}{dz} = \frac{\partial F}{\partial x} = \frac{\partial F}{i\partial y} , \qquad (A2.1.2)$$

giving the Cauchy-Riemann conditions

$$\frac{\partial u}{\partial x} = \frac{\partial v}{\partial y} , \quad \frac{\partial u}{\partial y} = -\frac{\partial v}{\partial x} . \qquad (A2.1.3)$$

These conditions are necessary consequences of the analyticity assumption. If the derivatives are continuous at a given point, it may also be shown that they are sufficient to ensure analyticity.

Note that the Cauchy-Riemann equations imply that, if the real part of a complex function is known, its imaginary part is determined to within a constant. This point is discussed later, from another viewpoint.

A2.1.1 The Principle Value of a Singular Integral

Cauchy's Integral Formula states that if $F(z)$ is analytic within and on a closed contour C, then

$$F(z) = \frac{1}{2\pi i} \oint_C dz' \frac{F(z')}{z' - z} \qquad (A2.1.4)$$

if the contour is taken anti-clockwise, which is the conventional positive direction, and is a manifestation of a more basic convention, namely that angles are presumed to increase in an anti-clockwise direction. If z is outside of the contour C, this integral gives zero. Now let z approach the contour from the inside, towards a point z_0. The integral is undefined. However, it can be assigned a particular finite value as a result of a special limiting process which will now be described. We distort the contour into a small semi-circle around z_0. In the limit as this semi-circle gets smaller, it can be shown that there is a finite contribution of $\frac{1}{2}F(z_0)$, as a result of the integration over it. We define the Cauchy principal value of the integral as the value obtained by means of this limiting process minus the contribution $\frac{1}{2}F(z_0)$ of the semi-circle. Therefore, for z on the contour:

$$\frac{1}{\pi i} \oint_C \frac{F(z')}{z'-z} dz' = F(z) \tag{A2.1.5}$$

if the integral is interpreted as a Cauchy principal value. Let $F(z)$ be analytic in the lower half-plane and let it be zero at infinity, at least in this half-plane. We take C in (A2.1.5) to be the real axis and the infinite semi-circle enclosing the lower half-plane. Taking this contour clockwise, we obtain from (A2.1.5), for x on the real axis,

$$\frac{1}{\pi i} \int_{-\infty}^{\infty} dx' \frac{F(x')}{x'-x} = -F(x) \ , \tag{A2.1.6}$$

where the integral is a Cauchy principal value. It follows that if

$$F(z) = u(x,y) + i v(x,y) \ , \qquad \text{then} \tag{A2.1.7}$$

$$\frac{1}{\pi} \int_{-\infty}^{\infty} dx' \frac{u(x',0)}{x'-x} = v(x,0)$$

$$\tag{A2.1.8}$$

$$\frac{1}{\pi} \int_{-\infty}^{\infty} dx' \frac{v(x',0)}{x'-x} = -u(x,0) \ .$$

These relations are saying that $u(x,0)$, $v(x,0)$ are infinite Hilbert transforms of each other. This topic, mainly oriented towards finite Hilbert transforms, is discussed, in more detail, in Sect. A2.4. Note that if $F(z)$ were analytic in the upper half-plane, the signs in (A2.1.8) would be the other way.

A2.1.2 Analytic Continuation

In different parts of the complex plane, an analytic function may have different representations, as power series for example. Given two representations, the question arises as to whether they represent the same complex function or distinct functions. If they represent the same function, they are said to constitute analytic continuations of each other. There are well-known theorems which specify under what conditions two functions are analytic continuations of each other. The statements of interest in the present context are the following. Let the two regions on which the representations hold have a common boundary which is a smooth curve L (this concept is defined more precisely in Sect. A2.2). Then, if both representations have limiting values on L which are continuous and equal to each other, the two representations are analytic continuations of each other. Also, if a function is analytic in a region and zero along any continuous arc in this region, then it is zero everywhere in the region. It follows trivially that if two analytic functions are equal on a continuous arc, they are equal over the whole region on which both are analytic. This latter statement is required in Sect. 1.6.

A2.1.3 Liouville's Theorem

A theorem that is fundamental to the developments of Sect. A2.3 is Liouville's theorem, which we state in somewhat generalized form. Let a function $F(z)$ be analytic at every finite point on the complex plane and let it behave as z^n as z tends to infinity. Then it can only be a polynomial of degree n. In particular, if it is zero at infinity, it is zero everywhere.

A2.1.4 Singularities

What makes analytic functions interesting are their singularities, or points where they are not analytic. In fact, the content of Liouville's theorem is that if they have no singularities, they are trivial. The simplest singularities are poles, that is to say, behaving at z_0 as $(z - z_0)^{-n}$ where n is a positive integer called the order of the pole. These are isolated singularities. A function whose only singularities are poles is known as a meromorphic function. There is the well-known result, the basis of the Calculus of Residues, which states that the integral of an analytic function around a closed contour is equal to $(2\pi i)$ times the sum of the residues inside the contour. The residue of a pole at a point z_0 is equal to $\lim_{z \to z_0} [(z - z_0) F(z)]$. "Poles of infinite order" are called essential singularities.

A2.1.5 Branch Points

If one follows an analytic function around a contour to the initial point and it does not return to the same value, then the function is multi-valued. This is associated with the presence of a branch point within the contour. A branch point is a type of singularity, distinct from a pole or essential singularity. It is not isolated since, as we shall see below, its effects are not localized at one point. The function $(z - a)^\gamma$ is, for non-integral values of γ, a multi-valued function which is of considerable interest in the present context. We will, therefore, outline its properties. In the standard polar representation, it becomes

$$(z - a)^\gamma = |z - a|^\gamma e^{i\gamma\theta} \tag{A2.1.9}$$

where θ is the argument of $(z - a)$. Let γ be a real quantity. If it is rational, then let us write it as p/q, where the integers p, q have no common factors. Then, if we circle the point a, say r times, where $r < q$, the function returns to different values each time. When $r = q$, the function returns to its original value. We say that $(z - a)^\gamma$ has a branch point at $z = a$ and has q distinct branches. If γ is irrational, it has an infinite number of branches. Branch points always occur in pairs. The function $(z - a)^\gamma$ also has a branch point at infinity where it behaves as z^γ. One joins the point a to infinity by some convenient line and agrees that the function undergoes a discontinuous jump crossing this line. For the function under discussion, when a is real, this line is conventionally chosen to be the x-axis from the point a to $(-\infty)$. This is, however, an arbitrary choice. It is perfectly possible to define this function using another line of discontinuity. It would not be the same function however.

The complex plane, excluding the line of discontinuity is sometimes referred to as the cut plane, and the line itself as a cut. A multi-valued function with, say, q distinct branches can be completely characterized by taking q separate cut complex planes and defining a single-valued branch on each.

Returning to the function $F(z) = (z-a)^\gamma$ with a cut along $(-\infty, a]$, let $F^\pm(x)$ be the limiting values of $F(z)$ from above and below the real axis. One can show that

$$F^-(x) = e^{-2\pi i \gamma} F^+(x) . \tag{A2.1.10}$$

Note that this applies also if γ is complex, in which case there is a real as well as an imaginary exponential factor. Extending this procedure, one can show that if

$$F(z) = (z-a)^\gamma (z-b)^{1-\gamma} , \tag{A2.1.11}$$

where a, b are real and $b > a$, then

$$F^-(x) = \Gamma F^+(x) , \quad \text{where} \tag{A2.1.12}$$

$$\Gamma = \begin{cases} 1 , & x \notin [a, b] \\ e^{2\pi i \gamma} , & x \in [a, b] . \end{cases} \tag{A2.1.13}$$

Therefore, this function is analytic except in the interval $[a, b]$. It will in general have many branches, which we can represent as follows. Let $\arg(z-a) = \theta_a$ and $\arg(z-b) = \theta_b$. Then

$$F(z) = |z-a|^\gamma |z-b|^{1-\gamma} \exp\{i[\theta_a \gamma + \theta_b(1-\gamma) + 2\pi m \gamma + 2\pi n(1-\gamma)]\} ,$$
$$m, n : \text{integers} . \tag{A2.1.14}$$

Consider the branch for which $m = n = 0$. At large z, this behaves as z with no multiplying factor. This particular branch will be adopted for use in Sect. A2.3.

The cut for $F(z)$ given by (A2.1.11) must join a and b. The natural choice, and the one chosen here, is to take the portion of the real axis $[a, b]$ as the cut. In the next two sections, and in Chaps. 3, 4, we deal continually with analytic functions possessing discontinuities across particular line segments in the complex plane. These functions are multi-valued, with cuts along these segments.

A2.2 Cauchy Integrals

We consider integrals of the following type:

$$F(z) = \frac{1}{2\pi i} \int_L \frac{f(u) \, du}{u-z} , \tag{A2.2.1}$$

where L is a sectionally smooth curve on the complex plane. By this we mean a finite number of non-intersecting smooth arcs and contours. The term contour is used to indicate a closed curve, while arc refers to a curve that is not closed, and therefore has end points. Smoothness indicates that a tangent exists at each point and its slope varies continuously. In other words, each arc or contour, if represented parametrically, has continuous first derivative with respect to its parameter. It is also assumed that L has finite total length. In the present context, we will generally assume that L consists of finite segments of the real axis. The restriction of L to finite length can in fact be relaxed [Muskhelishvili (1963)] if suitable convergence conditions prevail. This is relevant in Sects. 3.2 and 7.3. We choose a positive direction along L. This is generally taken to be anti-clockwise for contours, but for arcs there is no set convention. For L on the real axis, we take the positive direction as the positive x direction. The region of the complex plane to the left as one moves along L in the positive direction is denoted by S^+, and the region on the right by S^-. These are the upper and lower half-planes, respectively, for L along the x-axis.

The function $f(t)$, referred to as the density function, is assumed to be bounded everywhere, except possibly at a finite number of points $c_k, k = 1, 2, \ldots$. In the present context, these points can always be taken to be end points of the arcs included in L. At the endpoints, it may have integrable singular points with

$$|f(u)| \le \frac{C}{|u - c_k|^\alpha} \qquad 0 \le \alpha < 1 \ . \tag{A2.2.2}$$

Furthermore, it is assumed that $f(t)$ is Hölder continuous at each point of L where it is not singular. This very important property is defined as follows: for any two points u_1, u_2, let there exist positive, real constants A, μ such that

$$|f(u_1) - f(u_2)| \le A \, |u_1 - u_2|^\mu \ . \tag{A2.2.3}$$

It is easy to show that if $\mu > 1$, the derivative of $f(u)$ is zero, so that it is a constant. This case is not of great interest, so it is always assumed that $\mu \le 1$. For $\mu = 1$, the Hölder condition is termed the Lipschitz condition and is obeyed by any differentiable function, and others not in this class. For $\mu < 1$, the condition implies continuity in the ordinary sense. The case $\mu = 0$, which is excluded, is consistent with discontinuity. A function obeying this condition at a point, or on a line, will be described as obeying the $H(\mu)$ condition on that set, if μ is specified or otherwise just the H condition.

At large z, the function $F(z)$ behaves as

$$F(z) \sim \frac{A}{z} \ , \qquad 2\pi i A = - \int_L du \, f(u) \tag{A2.2.4}$$

if this integral is non-zero. If it is zero, $F(z)$ falls off as some higher power of z. Consider its limiting value as z approaches a point u on L at which $f(u)$ is non-singular, and which is not an end point of an arc. We write

$$F(z) = \frac{1}{2\pi i} \int_L dt \frac{f(t) - f(u)}{t - z} + \frac{f(u)}{2\pi i} \int_L dt \frac{1}{t - z} \ . \tag{A2.2.5}$$

The Hölder condition (A2.2.3) implies that the first term approaches a well-defined integral,

$$\frac{1}{2\pi i} \int_L dt \frac{f(t) - f(u)}{t - u} \ , \tag{A2.2.6}$$

as $z \to u$, because the behaviour at the singularity is integrable. This step illustrates the central importance of the Hölder property. The singular and indeterminate behaviour is contained in the second term. Let

$$\Psi(z) = \frac{1}{2\pi i} \int_L \frac{dt}{t-z} \tag{A2.2.7}$$

and denote by $\Psi^+(u)$, $\Psi^-(u)$ the limiting values of $\Psi(z)$ as it approaches u from S^+ and S^-, respectively. In each of these cases, we deform the contour into a small semi-circle around u and consider the limit as this semi-circle shrinks to zero. It is easy to show that

$$\Psi^+(u) = \frac{1}{2} + \frac{1}{2\pi i} \oint_L \frac{dt}{t-u} , \qquad \Psi^-(u) = -\frac{1}{2} + \frac{1}{2\pi i} \oint_L \frac{dt}{t-u} , \tag{A2.2.8}$$

where the integrals on the right are Cauchy principle values. The more general formulae

$$F^+(u) = \frac{1}{2} f(u) + \frac{1}{2\pi i} \oint_L \frac{f(t)\,dt}{t-u}$$

$$F^-(u) = -\frac{1}{2} f(u) + \frac{1}{2\pi i} \oint_L \frac{f(t)\,dt}{t-u} \tag{A2.2.9}$$

follow from (A2.2.5) and (A2.2.8). These are the well-known Plemelj formulae, which are of great importance in the context of this volume. Another form of these relations is

$$F^+(u) - F^-(u) = f(u)$$

$$F^+(u) + F^-(u) = \frac{1}{\pi i} \oint_L dt \frac{f(t)}{t-u} , \tag{A2.2.10}$$

which shows clearly that $F(z)$, defined by (A2.2.1), is discontinuous across L at all points where $f(u)$ is non-zero. This implies the existence of branch points along L, resulting in cuts in the complex plane.

As noted elsewhere, we do not generally indicate a principle value integral explicitly, but leave it as understood that if the integrand is singular in the manner of a Cauchy integral, the integral is to be taken as a Cauchy principle value.

For $z \in L$, $F(z)$ given by (A2.2.1) is analytic since it is differentiable. It should perhaps be noted that this property requires no strong assumptions on $f(u)$. In particular, no analyticity requirements are imposed.

Consider the Cauchy integral over a single arc $[a, b]$:

$$F(z) = \frac{1}{2\pi i} \int_a^b dt \frac{f(t)}{t-z} . \tag{A2.2.11}$$

It is of interest to determine the behaviour of $F(z)$ as z approaches the end points. Consider z close to a. Let $f(a)$ be finite. Then, using the same trick as in (A2.2.5), we obtain

$$F(z) = \frac{f(a)}{2\pi i} \log\left(\frac{b-z}{a-z}\right) + \frac{1}{2\pi i} \int_a^b dt \, \frac{f(t)-f(a)}{t-z}$$

$$= \frac{f(a)}{2\pi i} \log\left(\frac{1}{a-z}\right) + F_1(z) ,$$

(A2.2.12)

where $F_1(z)$ has a definite, non-singular limit at $z = a$. Similarly, near $z = b$,

$$F(z) = \frac{f(b)}{2\pi i} \log(b-z) + F_2(z) ,$$

(A2.2.13)

where $F_2(z)$ is non-singular. Therefore, if $f(a)$ or $f(b)$ is finite, there is a logarithmic singularity in $F(z)$ at that end point. If an end value of $f(u)$ is zero, then $F(z)$ approaches a definite, finite limit at that point.

If $f(u)$ has a singularity at an end point of the type given by (A2.2.2), then $F(z)$ has a singularity of the same type. This may be seen intuitively by considering the dominant term in the integral. Rigorous arguments and detailed expressions are given by Muskhelishvili (1953) and Gakhov (1966), for example.

A2.3 The Hilbert Problem with Constant Coefficient

Consider a sectionally smooth curve L in the complex plane, in the sense defined in Sect. A2.2. We introduce the concept of a sectionally analytic function $F(z)$, with respect to L. This is a function analytic on the complex plane, excluding L and possibly infinite points. It has definite limits as z approaches a point on L from either side (in general not equal), with the exception of certain isolated points which we take to be end points of individual arcs included in L. Near such an isolated point, c, a sectionally analytic function, must obey the condition

$$|F(z)| \leq \frac{A}{|z-c|^\alpha} , \quad 0 \leq \alpha < 1 .$$

(A2.3.1)

Such points will be termed exceptional points,

The Hilbert problem with constant coefficient may now be posed. The discussion is based on that of Muskhelishvili (1963). Given L, we seek a sectionally analytic function $F(z)$ such that its limits at a point u on L, from S^+, S^-, namely $F^+(u)$, $F^-(u)$ obey the relation

$$F^+(u) = \eta F^-(u) + f(u) ,$$

(A2.3.2)

where $f(u)$ is assumed to obey the H condition everwhere on L. For simplicity, we exclude the possibility of singularities of the type given by (A2.2.2), for the general problem though not when discussing a special case below. At large $|z|$ it is assumed that $F(z)$ behaves as $|z|^k$ where k is an integer. If k is positive, then clearly $F(z)$ is singular at infinity.

The possible exclusion of end points from (A2.3.2) is related to the different and generally singular behaviour of Cauchy integrals at the end points of arcs, which was pointed out at the end of the last section. We shall see that such integrals play a central role in the solution of the Hilbert problem.

Let us consider first the simple special case where $\eta = 1$. In this case,

$$F^+(u) - F^-(u) = f(u) \ , \tag{A2.3.3}$$

so that the discontinuity across L is $f(u)$. In this case, we allow the possibility of singularities, as given by (A2.2.2), in $f(u)$. Plemelj's formulae (A2.2.10) gives that a particular solution of the problem is

$$F_1(z) = \frac{1}{2\pi i} \int_L \frac{f(u)\,du}{u-z} \ . \tag{A2.3.4}$$

We note that if $F(z)$ is a general solution to the problem, then $F_1(z) - F(z)$ has no discontinuity across L and so is analytic everywhere, except at certain isolated exceptional points. However, $F(z)$ is required to behave according to (A2.3.1) at such points. From the final paragraphs of Sect. A2.2, it follows that $F_1(z)$ will also have singularities of, at most, degree less than unity. Therefore $F(z) - F_1(z)$ will have this property. But such singularities cannot be isolated. They are branch points. Therefore $F(z) - F_1(z)$ must be bounded and analytic even at these points. It follows from Liouville's theorem (Sect. A2.1) that it is a polynomial of, at most, degree k. Therefore, the general solution $F(z)$ has the form

$$F(z) = \frac{1}{2\pi i} \int_L \frac{f(u)\,du}{u-z} + P_k(z) \ , \tag{A2.3.5}$$

where $P_k(z)$ is an arbitrary polynomial of degree k or less. If $F(z)$ is required to go to zero as $1/z$ at infinity, the general solution is unique and given by

$$F(z) = \frac{1}{2\pi i} \int_L \frac{f(u)\,du}{u-z} \ . \tag{A2.3.6}$$

Note that if $F(z)$ is required to go to zero faster than z^{-1} at infinity, then in general no solution exists unless subsidiary conditions are obeyed by $f(u)$. Specifically, if $F(z) \sim z^{-l}$ at large $|z|$, then we must have

$$\int_L u^r f(u)\,du = 0 \ , \qquad r = 0, 1, 2, \ldots, l-2 \ . \tag{A2.3.7}$$

Demonstration of this is left as a problem [see the comment after (A2.2.4)]. Noting the final remark of the previous section, we see that if $f(u)$ behaves as (A2.2.2) at the end points, then $F(z)$ behaves as in (A2.3.1) at those points.

We consider now the more general problem where $\eta \neq 1$. First we solve the homogeneous problem, where $f(u)$ is zero. Let L consist of a combination of arcs $[a_i, b_i]$, $i = 1, 2, \ldots, n$. Then we see from (A2.1.12) that a particular solution is given by

$$X_0(z) = \prod_{i=1}^n (z - a_i)^{-\theta}(z - b_i)^{\theta - 1} \ , \tag{A2.3.8}$$

provided that

$$\eta = e^{2\pi i \theta} \qquad \text{or} \tag{A2.3.9}$$

$$\theta = \frac{1}{2\pi i} \log \eta \ . \tag{A2.3.10}$$

The function $X_0(z)$ is formed by taking products of inverses (for later convenience) of the form given by (A2.1.11). It is discontinuous only across the intervals $[a_i, b_i]$. We choose that branch of $X_0(z)$ such that $z^n X_0(z) \to 1$ at large z, as discussed in Sect. A2.1. This is a convention. Any other branch would do equally well. Let

$$\theta = \alpha + i\beta \tag{A2.3.11}$$

where

$$\beta = -\frac{\log |\eta|}{2\pi} \ , \qquad \alpha = \frac{\arg(\eta)}{2\pi} \ . \tag{A2.3.12}$$

The function $X_0(z)$ behaves as $|z - a_i|^{-\alpha}$, $|z - b_i|^{\alpha - 1}$ near a_i, b_i respectively. Therefore, in order to satisfy (A2.3.1), we take

$$0 < \arg(\eta) < 2\pi \ . \tag{A2.3.13}$$

We exclude for the moment the case $\alpha = 0$, corresponding to η real and positive. If $F(z)$ is a general solution of the homogeneous problem, then, on L:

$$F^+(u) = \eta F^-(u) = \frac{X_0^+(u)}{X_0^-(u)} F^-(u) \qquad \text{or} \tag{A2.3.14}$$

$$\frac{F^+(u)}{X_0^+(u)} - \frac{F^-(u)}{X_0^-(u)} = 0 \ . \tag{A2.3.15}$$

Then, an argument identical to that leading to (A2.3.5) gives that

$$F(z) = X_0(z) P_k(z) \ , \tag{A2.3.16}$$

where $P_k(z)$ is a polynomial of degree not greater than $(k + n)$. If $F(z)$ is to behave as $1/z$ at infinity, then $P_k(z)$ can be of degree $(n - 1)$ at most.

The function $X_0(z)$ is singular at every end point a_i, b_i, since the case $\alpha = 0$ is excluded. If we require that $F(z)$ is bounded at specified ends $c_1, c_2, \ldots, c_p$, then we write (A2.3.16) in the form

$$F(z) = X_p(z) P_{k-p}(z) \ , \qquad \text{where} \tag{A2.3.17}$$

$$X_p(z) = X_0(z) \prod_{i=1}^{p} (z - c_i) \ . \tag{A2.3.18}$$

The polynomial $P_{k-p}(z)$ is arbitrary, with degree not more than $k - p + n$. Of particular interest is the solution bounded at all ends, where

$$X_{2n}(z) = X_b(z) = \prod_{i=1}^{n} (z - a_i)^{1-\theta} (z - b_i)^{\theta} \ . \tag{A2.3.19}$$

Note that $X_b(z)$ actually vanishes at the end points. It behaves as z^n for large z. Also, solutions bounded at the ends b_i would be given by

$$X_1(z) = X_0(z) \prod_{i=1}^{n} (z - b_i) = \prod_{i=1}^{n} \left(\frac{z - b_i}{z - a_i} \right)^{\theta}.$$ (A2.3.20)

This solution is of interest in that it allows us to include the case $\alpha = 0$ which was excluded because $X_0(z)$ diverges as $|z - b_i|^{-1}$ at these ends, thus breaking constraint (A2.3.1).

We move on to the non-homogeneous problem for which $f(u)$ is non-zero, and now restricted to be free of singularities. Equation (A2.3.15) becomes, using any solution $X(z)$ of the homogeneous problem with the required singularity structure,

$$\frac{F^+(u)}{X^+(u)} - \frac{F^-(u)}{X^-(u)} = \frac{f(u)}{X^+(u)} ,$$ (A2.3.21)

which has the same form as (A2.3.3). It may be checked that $f(u)/X^+(u)$ has the required properties of Hölder continuity and behaviour given by (A2.2.2) at the end points c_i where $X(z)$ is bounded. Therefore, from (A2.3.5), the general solution of the non-homogeneous problem is

$$F(z) = \frac{X(z)}{2\pi i} \int_L \frac{f(u)\,du}{X^+(u)(u - z)} + X(z)P(z) ,$$ (A2.3.22)

where $P(z)$ is an arbitrary polynomial consistent with the imposed behaviour of $F(z)$ for large z. The choice of $X(z)$ is determined by the boundedness requirements on $F(z)$. For example, if $F(z)$ is required to be bounded everywhere, then we choose $X(z) = X_b(z)$, given by (A2.3.19). That $F(z)$ is in fact bounded at the end points follows from the last paragraph of Sect. A2.2. Unlike $X(z)$, it is not necessarily zero at such points.

Note that in the particular case where $\eta = -1$, the quantity θ is equal to 1/2, and the powers in the various functions $X(z)$ become square roots. This is an important special case.

In the special case where L is the entire real axis, $F(z)$ is given by a different, simpler expression. Let us define a function $G(z)$ analytic off the real axis, and defined as

$$G(z) = F(z) , \quad z \in S^+ ; \quad G(z) = \eta F(z) , \quad z \in S^- .$$ (A2.3.23)

Then (A2.3.2) becomes

$$G^+(u) - G^-(u) = f(u) ,$$ (A2.3.24)

with solution given according to (A2.3.5) with appropriate alterations. Let us assume that $F(z) \sim z^{-1}$ as z becomes large. Then, from (A2.3.6), we have

$$F(z) = \frac{1}{2\pi i} \int_{-\infty}^{\infty} du \frac{f(u)}{u - z} , \quad z \in S^+$$
$$= \frac{1}{2\pi i \eta} \int_{-\infty}^{\infty} du \frac{f(u)}{u - z} , \quad z \in S^- .$$ (A2.3.25)

It is perhaps not apparent how the fact that L is the entire real axis has been used in the derivation of (A2.3.25). If L consists of part of the real axis and (A2.3.2) applies on L, the unstated assumption is that, on the remainder of the axis, the discontinuity is zero. If L consists of part of the real axis and the integrals in (A2.3.25) are over only this portion, it is clear that $F(z)$ is not continuous on the remainder. Instead, we have

$$F^+(x) = \eta F^-(x) \; . \tag{A2.3.26}$$

Therefore (A2.3.25) can apply only where the regions $S^\pm$ have no point in common.

A2.4 The Hilbert Transform

Consider the equation

$$\frac{1}{\pi} \int_L dx' \frac{\phi(x')}{x'-x} = g(x) \; , \quad x \in L \qquad \text{(with the possible exception of end points),} \tag{A2.4.1}$$

where $g(x)$ is given and $\phi(x)$ is to be determined. In general, L may be any sectionally smooth (see Sect. A2.2) curve in the complex plane. However, in the present context, it will always be a portion or all of the real axis. We will assume that it is the union of real intervals $[a_i, b_i]$, $i = 1, 2, \ldots, n$, as in the previous section. The function $g(x)$ is taken to obey the H condition while we look for a solution $\phi(x)$ with the same property but possibly with singular points, as defined by (A2.2.2).

Equation (A2.4.1) is a singular integral equation for $\phi(x)$ in terms of $g(x)$. The principle value of the integral is understood. It will emerge that the solution of this integral equation is closely related to the solution of the Hilbert problem discussed in the previous section.

Another way of looking at (A2.4.1) is as a statement that $g(x)$ is the Hilbert transform of $\phi(x)$ for the integration region L. Solution of the integral equation is equivalent to finding the inverse Hilbert transform. If L is the entire real axis, the integral on the left of (A2.4.1) is an infinite Hilbert transform. In Sect. A2.1, it was shown that the real and imaginary parts of the boundary values of an analytic function are infinite Hilbert transforms of each other, as given by (A2.1.8).

The treatment of this topic given here is based on that of Muskhelishvili (1953).

Consider the sectionally analytic function

$$F(z) = \frac{1}{2\pi i} \int_L \frac{\phi(x')dx'}{x'-z} \; . \tag{A2.4.2}$$

The Plemelj formulae (A2.2.10) gives that

$$F^+(x) + F^-(x) = -ig(x) \; , \quad x \in L \; , \tag{A2.4.3}$$

so that, from (A2.3.22), we have

$$F(z) = -\frac{X(z)}{2\pi} \int_L \frac{g(x')\,dx'}{X^+(x')(x'-z)} + X(z)P(z) \; , \tag{A2.4.4}$$

where $X(z)$ is a solution of the homogeneous relation

$$X^+(x) + X^-(x) = 0 \tag{A2.4.5}$$

chosen so as to give $F(z)$ the desired singularity structure. Also, $P(z)$ is a polynomial, arbitrary except that it must be consistent with the large z behaviour of $F(z)$. From (A2.4.2), we see that $F(z) \sim 1/z$ unless $\int_L \phi(x)\,dx$ is zero.

The function $\phi(x)$ may be deduced from the other Plemelj formula:

$$F^+(x) - F^-(x) = \phi(x) \; , \qquad x \in L \; , \tag{A2.4.6}$$

giving

$$\phi(x) = -\frac{X^+(x)}{\pi} \int_L dx' \frac{g(x')}{X^+(x')(x'-x)} + 2X^+(x)P(x) \; , \qquad x \in L \; . \tag{A2.4.7}$$

If we regard (A2.4.1) as the definition of $g(x)$, $x \in L$, then we have that

$$g(x) = 2iF(x)$$

$$= -\frac{iX(x)}{\pi} \int_L \frac{g(x')\,dx'}{X^+(x')(x'-x)} + 2iX(x)P(x) \; , \qquad x \notin L \; . \tag{A2.4.8}$$

Muskhelishvili (1953) discusses the problem in some generality. Here, we content ourselves with giving explicit formulae for the case where L is a single interval $[a, b]$. Take the case where the solution at both ends may be singular. Then

$$X(z) = [(z-a)(z-b)]^{-1/2} \qquad \text{and} \tag{A2.4.9}$$

$$\phi(x) = -\frac{1}{\pi m(x)} \int_a^b dx' \frac{g(x')m(x')}{x'-x} - \frac{2iP_0}{m(x)} \; , \qquad x \in [a, b] \; , \tag{A2.4.10}$$

where

$$m(x) = [(x-a)(b-x)]^{1/2} = \frac{-i}{X^+(x)} \; . \tag{A2.4.11}$$

Note that P_0 must be constant to give the correct large z behaviour of $F(z)$.

Problem A2.4.1: Show that

$$\int_a^b \frac{dx'}{m(x')(x'-x)} = 0 \; , \qquad x \in [a, b] \; . \tag{A2.4.1p}$$

This follows from (A2.4.1), (A2.4.10) on putting $g(x) = 0$. The first part of (A1.2.1) is a statement of this result. Deduce that

$$2iP_0 = -\frac{1}{\pi} \int_a^b dx \, \phi(x) \; . \tag{A2.4.2p}$$

Let us also give an expression for $g(x)$, $x \notin L$, from the general formula (A2.4.8). We define

$$n(x) = |(x-a)(x-b)|^{1/2} , \tag{A2.4.12}$$

observing that

$$X(x) = \frac{\varepsilon(x)}{n(x)} , \qquad \text{where} \tag{A2.4.13}$$

$$\varepsilon(x) = \begin{cases} 1 \; ; & x > b \\ -1 , & x < a . \end{cases} \tag{A2.4.14}$$

Equation (A2.4.13) is a consequence of the convention, noted in the previous section, that $zX(z) \to 1$ at large z. Substituting into (A2.4.8), we obtain

$$g(x) = \frac{\varepsilon(x)}{\pi n(x)} \int_a^b dx' \frac{g(x')m(x')}{x'-x} + \frac{2i\varepsilon(x)}{n(x)} P_0 , \qquad x \notin L . \tag{A2.4.15}$$

But we could also deduce the form of $g(x)$ outside of $[a,b]$ by combining (A2.4.1) and (A2.4.10) directly. Comparing the two results gives

$$\frac{1}{\pi} \int_a^b \frac{dx'}{(x'-x)(x''-x')m(x')}$$

$$= -\frac{\varepsilon(x)}{n(x)} \frac{1}{x''-x} , \qquad x \notin [a,b] , \qquad x'' \in [a,b] \tag{A2.4.16}$$

and

$$\frac{1}{\pi} \int_a^b \frac{dx'}{m(x')(x'-x)} = -\frac{\varepsilon(x)}{n(x)} , \qquad x \notin [a,b] . \tag{A2.4.17}$$

On decomposing the integrand in (A2.4.16) into partial fractions and using (A2.4.1p), we see that (A2.4.16) follows from (A2.4.17). Note that (A2.4.17) is the second relation of (A.1.2.1).

If $\phi(x)$ is required to be bounded at both ends, then

$$\phi(x) = -\frac{m(x)}{\pi} \int_a^b dx' \frac{g(x')}{m(x')(x'-x)} . \tag{A2.4.18}$$

This follows from the observation after (A2.3.22). We have the further requirement that

$$\int_a^b dx \frac{g(x)}{m(x)} = 0 , \tag{A2.4.19}$$

if $\phi(x)$ is to vanish as $1/x$. It also follows from (A2.4.18) that

$$\int_a^b dx \, \phi(x) = -\int_a^b dx \frac{g(x)x}{m(x)} \tag{A2.4.20}$$

by virtue of (A2.4.19) and (A1.2.2).

Problem A2.4.2: Deduce (A2.4.18 – 20) from (A2.4.10) and (A2.4.2p).

Consider (A2.4.1) when L is the entire real axis. Then

$$\frac{1}{\pi} \int_{-\infty}^{\infty} dx' \frac{\phi(x')}{x'-x} = g(x) \ . \tag{A2.4.21}$$

It follows from (A2.4.10) that the solution is

$$\phi(x) = -\frac{1}{\pi} \int_{-\infty}^{\infty} dx' \frac{g(x')}{x'-x} \ . \tag{A2.4.22}$$

This is the inversion formula for an infinite Hilbert transform, first introduced in a special context in Sect. A2.1. Consider also the semi-infinite Hilbert transform where we let $L = [0, \infty]$. Equation (A2.4.1) becomes

$$\frac{1}{\pi} \int_{0}^{\infty} dx' \frac{\phi(x')}{x'-x} = g(x) \ , \tag{A2.4.23}$$

while (A2.4.10) gives

$$\phi(x) = -\frac{1}{\pi\sqrt{x}} \int_{0}^{\infty} dx' \frac{g(x')\sqrt{x'}}{x'-x} \ , \qquad x > 0 \ . \tag{A2.4.24}$$

Equation (A2.4.16) becomes, in this case,

$$\frac{1}{\pi} \int_{0}^{\infty} \frac{dx'}{(x'-x)(x''-x')\sqrt{x'}} = \frac{1}{\sqrt{-x}} \frac{1}{x''-x} \ , \qquad x < 0 \ . \tag{A2.4.25}$$

These results are required in Sect. 7.3.

We now evaluate certain types of Hilbert transforms which are required in the main text. The interval $[a,b]$ will be assumed to have been transformed into $[-1,1]$. Consider

$$Z(x) = \frac{1}{\pi} \int_{-1}^{1} dx' \frac{q(x')}{(x'-x)m(x')} \ , \qquad m(x) = (1-x^2)^{1/2} \ , \tag{A2.4.26}$$

where $q(x)$ is a polynomial of degree n. We expand $q(x)$ in terms of Chebyshev polynomials. These polynomials $T_n(x)$, $U_n(x)$, $n = 0, 1, 2, \ldots$, are defined by the relations (A1.4.4). We put

$$q(x) = \sum_{r=0}^{n} a_r T_r(x) \ , \tag{A2.4.27}$$

where the a_r may be evaluated with the aid of the orthogonality relations (A1.4.5), though if n is small, and in the present context it always is, there is no need for these relations; the values of the a_r are obvious. It follows from (A1.1.3) that

$$Z(x) = \begin{cases} r(x) \ , & |x| < 1 \\ r(x) - \dfrac{\text{sgn}(x)}{n(x)} q(x) \ , & |x| > 1 \end{cases} \tag{A2.4.28}$$

where

$$r(x) = \sum_{r=1}^{n} a_r U_{r-1}(x) \ , \qquad n(x) = (x^2 - 1)^{1/2} \ . \tag{A2.4.29}$$

Also required (Sect. 3.6) is the integral

$$\Gamma(\beta, \alpha) = \frac{1}{\pi} \int_{-\infty}^{-1} dy \int_{-1}^{1} dx \, \frac{n(y) \exp(\beta y - \alpha x)}{(x - y) m(x)} \ . \tag{A2.4.30}$$

We need the modified Bessel functions with imaginary argument $I_n(\alpha)$, $K_n(\beta)$, $n = 0, 1, 2, \ldots$ to evaluate $\Gamma(\beta, \alpha)$. Some properties of these are given on Table A1.4. Consider the following integral, also used in Sect. 3.6,

$$J(\beta, x) = \int_{-\infty}^{-1} dy \, \frac{n(y) e^{\beta y}}{y - x} \ , \tag{A2.4.31}$$

which we first consider for $x > 1$, though finally we are interested in $|x| < 1$. It may be shown that

$$\frac{d}{dx} J(\beta, x) = \beta J(\beta, x) + \int_{-\infty}^{-1} dy \, \frac{y e^{\beta y}}{(y - x) n(y)} \tag{A2.4.32}$$

by transferring the derivative from x to y and partially integrating. Define

$$U(x) = \frac{J(\beta, x)}{n(x)} \ . \tag{A2.4.33}$$

With the aid of (A.2.4.32), (A1.4.18) and (A1.4.11), it may be shown that

$$\frac{d}{dx} U(x) - \beta U(x) = -(x^2 - 1)^{-3/2} \int_{-\infty}^{-1} dy \, \frac{e^{\beta y}(1 + xy)}{n(y)}$$

$$= -\frac{K_0(\beta) - x K_1(\beta)}{(x^2 - 1)^{3/2}} = \frac{d}{dx} g(x) \ , \tag{A2.4.34}$$

where

$$g(x) = \frac{E(\beta, x)}{n(x)} \ , \qquad E(\beta, x) = x K_0(\beta) - K_1(\beta) \ . \tag{A2.4.35}$$

The function $U(x)$ vanishes at infinity. Therefore, we can give the solution of this differential equation as

$$J(\beta, x) = -n(x) e^{\beta x} \int_{x}^{\infty} dx' \, e^{-\beta x'} g'(x')$$

$$= -n(x) e^{\beta x} \beta \int_{x}^{\infty} dx' \, e^{-\beta x'} g(x') + E(\beta, x) \ . \tag{A2.4.36}$$

This integral can be continued into $[-1, 1]$, since it is analytic also over this region. Replacing $n(x)/n(x')$ by $m(x)/m(x')$ when $|x| < 1$ and $|x'| < 1$ so that all quantities are real, we obtain

$$J(\beta, x) = -m(x) \beta e^{\beta x} \int_{x}^{1} dx' \, \frac{e^{-\beta x'}}{m(x')} E(\beta, x') + E(\beta, x) \ , \qquad |x| < 1 \ , \tag{A2.4.37}$$

since it can be shown that the integral over $[1, \infty)$ vanishes, with the aid of (A1.4.18). It follows that

$$\Gamma(\beta, \alpha) = \frac{1}{\pi} \int\limits_{-1}^{1} dx\, e^{-\alpha x} \left(\beta e^{\beta x} \int\limits_{x}^{1} dx'\, e^{-\beta x'} \frac{E(\beta, x')}{m(x')} - \frac{E(\beta, x)}{m(x)} \right) . \qquad \text{(A2.4.38)}$$

Interchanging the order of integration, we obtain

$$\Gamma(\beta, \alpha) = \frac{1}{\pi} \int\limits_{-1}^{1} dx'\, e^{-\beta x'} \frac{E(\beta, x')}{m(x')} \frac{\beta}{\beta - \alpha} [e^{(\beta - \alpha)x'} - e^{\alpha - \beta}] - \chi(\beta, \alpha) , \qquad \text{(A2.4.39)}$$

where, from (A1.4.14),

$$\chi(\beta, \alpha) = -[I_0(\alpha) K_1(\beta) + I_1(\alpha) K_0(\beta)] . \qquad \text{(A2.4.40)}$$

Note that, by virtue of (A1.4.25),

$$\chi(\alpha, \alpha) = -1/\alpha . \qquad \text{(A2.4.41)}$$

Carrying out the remaining integrations gives, with the aid of (A2.4.41),

$$\Gamma(\beta, \alpha) = \frac{\beta \chi(\beta, \alpha) + e^{\alpha - \beta}}{\beta - \alpha} - \chi(\beta, \alpha) = \frac{\alpha \chi(\beta, \alpha) + e^{\alpha - \beta}}{\beta - \alpha} . \qquad \text{(A2.4.42)}$$

We also require the integral

$$L(x) = \frac{1}{\pi} \int\limits_{-1}^{1} dx' \frac{e^{-\alpha x'}}{(x' - x) m(x')} , \qquad |x| < 1 . \qquad \text{(A2.4.43)}$$

Let us first consider the related integral

$$L_1(x) = \frac{1}{\pi} \int\limits_{-1}^{1} dx' \frac{e^{-\alpha x'} m(x')}{x' - x} , \qquad |x| < 1 . \qquad \text{(A2.4.44)}$$

This may be evaluated by means of the same strategy as was used to determine $J(\beta, x)$ [Golden (1977); in fact Morland (1967) was the first to evaluate this integral, using a different method]. The result is

$$L_1(x) = \alpha m(x) e^{-\alpha x} \int\limits_{-1}^{x} \frac{dx'}{m(x')} e^{\alpha x'} D(\alpha, x') - D(\alpha, x) , \qquad \text{(A2.4.45)}$$

where

$$D(\alpha, x) = x I_0(\alpha) - I_1(\alpha) . \qquad \text{(A2.4.46)}$$

Elementary manipulation gives that

$$L(x) = \frac{1}{1 - x^2} [L_1(x) + D(\alpha, x)] , \qquad \text{(A2.4.47)}$$

so that

$$L(x) = \frac{\alpha e^{-\alpha x}}{m(x)} \int\limits_{-1}^{x} \frac{dx'}{m(x')} e^{\alpha x'} D(\alpha, x') . \qquad \text{(A2.4.48)}$$

Appendix III
Fourier Transforms

The familiar concept of a Fourier transform is used in various places throughout the text. Here we summarize those properties which are relevant, mainly without proof. More detailed treatment of the topic may be found in many standard texts; notably Titchmarsh (1937), Sneddon (1951), (1972) and Morse and Feshbach (1953).

A3.1 Definition and Basic Properties

We define the Fourier transform (FT) of a function $f(t)$ as

$$\hat{f}(\omega) = \int_{-\infty}^{\infty} dt\, f(t)\, e^{-i\omega t} \; . \tag{A3.1.1}$$

The centrally important property of this transform is that an equally simple formula can be given for its inverse. In fact,

$$f(t) = \frac{1}{2\pi} \int_{-\infty}^{\infty} d\omega\, \hat{f}(\omega)\, e^{i\omega t} \; . \tag{A3.1.2}$$

This is the content of Fourier's Integral Theorem, namely that the function given by formula (A3.1.2) is in fact equal to the original function. A rigorous proof of this result requires some restrictions on the function $f(t)$, for example that $|f(t)|^2$ be integrable in the Lebesgue sense over the interval $(-\infty, \infty)$. However, we will now show that Fourier transforms are formally meaningful, and Fourier's Integral Theorem is formally true a more general class of functions. Consider the integral

$$I(\omega) = \frac{1}{2\pi} \int_{-\infty}^{\infty} dt\, e^{-i\omega t} \; , \tag{A3.1.3}$$

which, formally speaking, is the FT of the constant function $1/(2\pi)$. It is not convergent, but we will define it as a limiting process, as follows:

$$I(\omega) = \frac{1}{2\pi} \lim_{\varepsilon \to 0} \left(\int_{0}^{\infty} dt\, e^{-i(\omega - i\varepsilon)t} + \int_{-\infty}^{0} dt\, e^{-i(\omega + i\varepsilon)t} \right)$$

$$= \frac{1}{\pi} \lim_{\varepsilon \to 0} \frac{\varepsilon}{\omega^2 + \varepsilon^2} \; . \tag{A3.1.4}$$

We see that

$$I(\omega) = 0 , \quad \omega \neq 0 \tag{A3.1.5}$$

and if $\omega = 0$, it is infinite. Also, by direct integration and then taking the limit $\varepsilon \to 0$, one can show that

$$\int_{-\infty}^{\infty} d\omega\, I(\omega) = 1 = \int_{-a}^{b} d\omega\, I(\omega) , \tag{A3.1.6}$$

the second relation being a consequence of (A3.1.5), if a, b are positive. Therefore $I(\omega)$ is the singular delta function $\delta(\omega)$ and we have the fundamental formula

$$\frac{1}{2\pi} \int_{-\infty}^{\infty} dt\, e^{-i\omega t} = \delta(\omega) . \tag{A3.1.7}$$

It follows from (A3.1.5) and (A3.1.6) that for a function smooth in the vicinity of ω_0,

$$\int_{a}^{b} d\omega\, g(\omega)\, \delta(\omega - \omega_0) = g(\omega_0) , \quad \omega_0 \in [a, b] . \tag{A3.1.8}$$

Problem A3.1.1: Show that

$$\delta(at) = \frac{1}{|a|}\, \delta(t) . \tag{A3.1.1p}$$

If $f(x)$ has only one zero on the x-axis at x_0, show further that

$$\delta(f(x)) = \frac{1}{|f'(x_0)|}\, \delta(x - x_0) . \tag{A3.1.2p}$$

Problem A3.1.2: Use (A3.1.7) to give a formal proof of Fourier's Integral Theorem.

Problem A3.1.3: The Heaviside step function is defined by

$$H(t) = \begin{cases} 0 , & t < 0 \\ 1 , & t \geq 0 . \end{cases} \tag{A3.1.3p}$$

Using the same limiting procedure as in (A3.1.4), show that

$$\tilde{H}(\omega) = -\lim_{\varepsilon \to 0} \frac{i}{\omega - i\varepsilon} \tag{A3.1.4p}$$

and that Fourier's Integral Theorem applies in this case. To do this, one must consider the integral defining the transform as a contour integral closed over an infinite half-circle. The crucial point is to decide whether it must be the upper or the lower half-circle. This reasoning is presented in more detail in the next section.

Problem A3.1.4: Show that

$$\omega \int_0^\infty dt \sin(\omega t) = 1$$

$$\int_0^\infty dt \cos(\omega t) = \pi \delta(\omega) \tag{A3.1.5p}$$

where the integrals are defined in the limiting sense introduced above.

Problem A3.1.5: Show that

$$\lim_{\tau \to 0} \frac{1}{\tau} e^{-t/\tau} H(t) = \delta(t) \ . \tag{A3.1.6p}$$

If $f'(t)$ is the derivative of $f(t)$, then partial integration gives that

$$\hat{f}_1(\omega) = i\omega \hat{f}(\omega) \tag{A3.1.9}$$

where $\hat{f}_1(\omega)$ is the FT of $f'(t)$. A simple rigorous proof of this would require that $f(\pm\infty)$ vanish. However, formally, it applies to the case where $f(t) = 1$, since $\omega \delta(\omega) = 0$ everywhere. Also, we have, from (A3.1.4p) that

$$i\omega \hat{H}(\omega) = 1 \ , \tag{A3.1.10}$$

giving

$$H'(t) = \delta(t) \ . \tag{A3.1.11}$$

Similarly, if $\hat{f}_n(\omega)$ is the FT of the nth derivative of $f(t)$, then

$$\hat{f}_n(\omega) = (i\omega)^n \hat{f}(\omega) \ . \tag{A3.1.12}$$

If $f(t)$ is real, it follows that

$$\bar{\hat{f}}(\omega) = \hat{f}(-\omega) \ , \tag{A3.1.13}$$

where the bar denotes complex conjugation. The converse is also true. This is a result of some importance in Sect. 1.5. Furthermore, if $f(t)$ is a real, even function of t, $\hat{f}(\omega)$ is real; if $f(t)$ is a real odd function, $\hat{f}(\omega)$ is purely imaginary.

The Faltung or Convolution Theorem is a result of central importance for the Theory of Linear Viscoelasticity. It states that if

$$f(t) = \int_{-\infty}^\infty dt' \, g(t-t')h(t') = \int_{-\infty}^\infty dt' \, h(t-t')g(t') = [g*h](t) \tag{A3.1.14}$$

in a convenient and conventional shorthand, then

$$\hat{f}(\omega) = \hat{g}(\omega)\hat{h}(\omega) \ . \tag{A3.1.15}$$

This is readily shown in a formal sense by substituting the expressions for $g(t-t')$, $h(t')$ in terms of $\hat{g}(\omega)$, $\hat{h}(\omega)$.

A3.1.1 Causal Functions

Of particular importance in the main text are functions with the property that

$$g(t) = 0 , \quad t < 0 , \tag{A3.1.16}$$

which we shall term causal functions. In the next section, the properties of the transforms of such functions are discussed. Here we note that (A3.1.14) becomes, if $g(t)$ is causal,

$$f(t) = \int_{-\infty}^{t} dt' \, g(t-t')h(t') = \int_{0}^{\infty} dt' \, h(t-t') \, g(t') = [g*h](t) , \tag{A3.1.17}$$

which is a form that often occurs in the context of Linear Viscoelasticity. If $h(t)$ is also causal, then

$$f(t) = \int_{0}^{t} dt' \, h(t-t') g(t') . \tag{A3.1.18}$$

The important property (A3.1.15) applies equally well in these special circumstances.

In Sect. 1.5, the relation

$$\lim_{\omega \to \infty} i\omega \hat{f}(\omega) = f(0) \tag{A.3.1.19}$$

is required, where $f(t)$ is causal. This may be shown, at least formally, by writing out $\hat{f}(\omega)$ explicitly and changing the integration variable to $u = \omega t$.

A3.2 Analytic Properties of Fourier Integrals

We wish to discuss, in this section, the analytic properties of $\hat{f}(\omega)$ for complex values of ω. If the integral defining $\hat{f}(\omega)$ exists at a given ω, it will be analytic at that point. Conversely, singularities in $\hat{f}(\omega)$ are associated with lack of convergence in the defining integral. So we are also in fact discussing the convergence of the integral.

Let us first restrict ourselves to causal functions, which are mainly of interest. We have

$$\hat{f}(\omega) = \int_{0}^{\infty} dt \, e^{-i\omega t} f(t) . \tag{A3.2.1}$$

Let the integral exist for real ω. This will be true for example if $f(t)$ falls off as $t^{-\alpha}$, $\alpha > 1$. Then, for all ω in the lower half-plane, the integral is certainly convergent and $\hat{f}(\omega)$ is analytic. This is an immediate consequence of the fact that t is non-negative in (A3.2.1). Furthermore, along any radius from the origin to infinity in the lower half-plane, $\hat{f}(\omega)$ will decay exponentially to zero.

Consider now the converse of this statement. Let $\hat{f}(\omega)$ be analytic in the lower half-plane. Then we claim that $f(t)$, defined by (A3.1.2), is zero for $t < 0$. This is easily shown if we observe that for $t < 0$, we can close the path of integration in (A3.1.2) around the lower infinite half circle, which does not contribute because of the exponential decay factor in the integral. But this contour integral is zero, since $\hat{f}(\omega)$ has no singularities inside it. The result follows. Therefore, the causal property is equivalent to the property that $\hat{f}(\omega)$ is analytic for $\text{Im}\{\omega\} < 0$, provided that the integral in (A3.2.1) is convergent for real ω.

We now ask what happens if

$$f(t) \underset{t \to \infty}{\sim} e^{\alpha t} \tag{A3.2.2}$$

where $\alpha > 0$. It may be shown, by essentially the same argument, that $\hat{f}(\omega)$ is analytic for $\text{Im}\{\omega\} < -\alpha$, but not above that line. In other words, the region of analyticity of $\hat{f}(\omega)$ is reduced. The integral, as a definition of $\hat{f}(\omega)$, is meaningful only in this reduced region.

This result also applies when $\alpha < 0$, so that, if $f(t)$ decays exponentially, $\hat{f}(\omega)$ is analytic in part of the upper half-plane as well as the lower half-plane. Thus, its region of analyticity is expanded.

Consider now the converse of this statement. If $\hat{f}(\omega)$ is analytic for $\text{Im}\,\omega < -\alpha$, but not above this line, what can we say about $f(t)$? If we define $f(t)$ according to

$$f(t) = \frac{1}{2\pi} \int_L d\omega\, e^{i\omega t} \hat{f}(\omega) , \tag{A3.2.3}$$

where L is a line parallel to the real ω-axis just below $\text{Im}\{\omega\} = -\alpha$, then it is easy to show that $f(t)$ is causal by the same argument as used previously. If a line higher up is used, $f(t)$ will not be causal. Therefore, in the situation that $\hat{f}(\omega)$ has singularities in the lower half-plane, the definition of the inverse transform given by (A3.2.3) must be used if the causal property is to be preserved. It gives an $f(t)$ that diverges exponentially at large t, according to (A3.2.2).

Let us consider briefly the case where $f(t)$ is not causal. It is usually convenient to decompose it as follows:

$$f(t) = f^+(t) + f^-(t) , \quad \text{where} \tag{A3.2.4}$$

$$f^+(t) = \begin{cases} f(t) & t \ge 0 \\ 0 & t < 0 \end{cases}, \quad f^-(t) = \begin{cases} 0 & t \ge 0 , \\ f(t) & t < 0 , \end{cases} \tag{A3.2.5}$$

so that

$$\hat{f}(\omega) = \hat{f}^+(\omega) + \hat{f}^-(\omega) . \tag{A3.2.6}$$

All the statements made above apply to $f^+(t), \hat{f}^+(\omega)$. They also apply to $f^-(t)$, $\hat{f}^-(\omega)$ with simple but important modifications. If $\hat{f}^-(\omega)$ exists for real ω, it is analytic throughout the upper half-plane. If

$$f(t) \underset{t \to -\infty}{\sim} e^{-\alpha t} , \tag{A3.2.7}$$

then $f^-(\omega)$ is analytic down to $\text{Im}\{\omega\} = \alpha$.

Problem A3.2.1: Show that

$$\frac{1}{2\pi} \int_L d\omega\, \text{sgn}(\text{Re}\{\omega\})\, e^{i\omega t} = \frac{i e^{\alpha t}}{\pi t} , \tag{A3.2.1p}$$

where L is as in (A3.2.3) and $\operatorname{sgn}(\omega)$ is the signature function given in terms of the Heaviside function by

$$\operatorname{sgn}(\omega) = H(\omega) - H(-\omega) \ . \tag{A3.2.2p}$$

If L is the real axis, then $\alpha = 0$ in (A3.2.1p) and

$$\frac{1}{2\pi} \int_{-\infty}^{\infty} d\omega \, \operatorname{sgn}(\omega) e^{i\omega t} = \frac{i}{\pi t} \ . \tag{A3.2.8}$$

Appendix IV
Non-singular Integral Equations

In Sects. 1.2, 3.4 – 6, and 6.1, equations of the Volterra and Fredholm type play a part in the discussion. We sketch briefly some basic properties of these equations here, without proof. For detailed treatment of this topic, we refer to Rektorys (1969), Pogorzelski (1966), Tricomi (1957) and Lovitt (1950), for example.

A4.1 Fredholm Equations

An integral equation of the form

$$\phi(x) = f(x) + \lambda \int_a^b dy\, K(x, y)\, \phi(y) \tag{A4.1.1}$$

is a Fredholm equation if a, b are finite, if the real kernel $K(x, y)$ is defined for all $a \leq x \leq b$, $a \leq y \leq b$, and λ is an arbitrary complex number. It is necessary also to impose some integrability condition on the kernel. Somewhat different constraints are imposed by different authors; see for example Pogorzelski (1966), Rektorys (1969) and Tricomi (1957). The most convenient is probably that given by the latter authors, namely that $K(x, y)$ be square integrable, or

$$\int_a^b dx \int_a^b dy\, K^2(x, y) = N \tag{A4.1.2}$$

be finite. Strictly, this integral need exist only in a Lebesgue sense for most results to be valid, though this subtlety will be ignored in the present context. It is also assumed that $f(x)$ is square integrable over $[a, b]$. The homogeneous form of (A4.1.1) is where $f(x)$ is zero.

Equation (A4.1.1) is a Fredholm equation of the second kind. A Fredholm equation of the first kind has the form

$$\int_a^b dy\, K(x, y)\, \phi(y) = f(x) \;, \tag{A4.1.3}$$

the properties of which we discuss briefly below.

The basic result of Fredholm theory dealing with equations of the second kind, the so called Fredholm Alternative, may be stated as follows: for a given λ, either

- a unique square integrable solution exists, in which case the only solution of the homogeneous equation is the trivial solution $\phi(x) = 0$, $x \in [a, b]$;

or

- the homogeneous equation has at least one non-trivial solution and the in-homogeneous equation has more than one solution.

If the second option holds, λ is an eigenvalue of the equation and any solution of the homogeneous equation is an eigenvector. To any eigenvalue there are at most a finite number of linearly independent eigenvectors $\phi_r(x)$. Any solution of the inhomogeneous equation, in the second alternative, can be combined with any linear combination of the $\phi_r(x)$ to give another solution. This illustrates the lack of uniqueness of the solution in that case.

The inhomogeneous equation may be solved by iteration to obtain

$$\phi(x) = f(x) + \lambda \psi_1(x) + \lambda^2 \psi_2(x) + \ldots$$

$$\psi_1(x) = \int_a^b dy \, K(x, y) f(y) \, dy \tag{A4.1.4}$$

$$\psi_2(x) = \int_a^b dy \, K(x, y) \psi_1(y) \, dy = \int_a^b dy_1 \int_a^b dy_2 K(x, y_1) K(y_1, y_2) f(y_2) \; ,$$

and so on. This is the Neumann series. It converges if

$$|\lambda| < 1/N \tag{A4.1.5}$$

where N is given by (A4.1.2). On a more general domain, the solution can be expressed in terms of the Fredholm formulae, which will not be discussed.

It will be observed that the Fredholm Alternative corresponds very closely to well-known results concerning the solution of sets of linear algebraic equations.

Problem A4.1.1: Show that for a so-called degenerate or Pincherle-Goursat kernel, namely one which separates into a finite sum of products of the form

$$K(x, y) = \sum_{j=1}^{n} h_j(x) k_j(y) \; , \tag{A4.1.1p}$$

where $h_j(x)$, $k_j(y)$, $j = 1, 2, \ldots, n$ are two sets of linearly independent functions, the integral equation reduces to a finite set of linear, algebraic equations.

The theory of Fredholm equations of the first kind, given by (A4.1.3), differs significantly from that of the second kind. For example, if $K(x, y)$ has the product form (A4.1.1p), then a solution is not possible unless $f(x)$ is a linear sum of the functions $h_j(x)$. In Sect. 3.6 it was shown essentially that the equation of the second kind (3.6.7) reduces to an equation of the first kind, namely (3.6.24), with the above condition satisfied, for the case of a discrete spectrum. Pogorzelski (1966), for example, discusses equations of the first kind in some detail.

If the kernel in (A4.1.1) has the form $K(x - y)$, then it is possible to obtain a solution by means of integral transform techniques. In particular, by defining quantities as zero outside of $[a, b]$, we can write the kernel term as a convolution integral, as in (A3.1.14), so that taking Fourier transforms reduces the equation to an algebraic one. This method is equally powerful for equations of the first kind.

A4.2 Volterra Equations

These are special cases of the Fredholm equation where $K(x, y)$ is zero for $y > x$. Therefore we have

$$\phi(x) = f(x) + \lambda \int_a^x dy\, K(x, y)\, \phi(y) \tag{A4.2.1}$$

under the same square-integrability conditions as in the last section. The solution of this equation is, however, considerably simpler in many respects than that of the more general equation. The central result is that the equation of the second kind has a unique solution for all λ. Applying the Fredholm Alternative, outlined in the previous section, we deduce that the homogeneous version of (A4.2.1), that is, where $f(x) = 0$, has no non-trivial solution.

An iterative solution, essentially the same as (A4.1.4), is easily written down, and in this case is valid for all λ.

If the kernel $K(x, y)$ has the form $K(x-y)$, then, as remarked in the Fredholm case, taking Fourier transforms gives an equation of algebraic form, which is trivially solved. This is of course the reason why Fourier transform techniques are so powerful and important in discussing the constitutive equations of non-aging viscoelastic materials, such as (1.2.28).

References

Aboudi, J. (1979): The dynamic indentation and impact of a viscoelastic half-space by an axisymmetric rigid body. Comput. Meth. Appl. Mech. Eng. **20**, 135–150

Abramowitz, M., Stegun, I.A. (eds.) (1970): *Handbook of Mathematical Functions* (Dover, New York)

Ahlfors, L.V. (1966): *Complex Analysis* (McGraw-Hill, New York)

Alblas, J.B., Kuipers, M. (1970): The contact problem of a rigid cylinder rolling on a thin viscoelastic layer. Int. J. Eng. Sci. **8**, 363–380

Alfrey, T. Jr. (1948): *Mechanical Behavior of High Polymers* (Interscience, New York)

Arutyunyan, N.Kh. (1952): *Some Problems in the Theory of Creep* (Techteorizdat, Moscow) [in Russian], (Engl. transl.: Pergamon, Oxford (1966))

Atkinson, C. (1979): A note on some dynamic crack problems in linear viscoelasticity. Arch. Mech. Stos. **31**, 829–849

Atkinson, C., Coleman, C.J. (1977): On some steady-state moving boundary problems in the linear theory of viscoelasticity. J. Inst. Maths. Appl. **20**, 85–106

Atkinson, C., List, R.D. (1972): A moving crack problem in a viscoelastic solid. Int. J. Eng. Sci. **10**, 309–322

Atkinson, C., Popelar, C.H. (1979): Antiplane dynamic crack propagation in a viscoelastic layer. J. Mech. Phys. Solids **27**, 431–439

Barberan, J., Herrera, I. (1966): Uniqueness theorems and speed of propagation of signals in viscoelastic materials. Arch. Rat. Mech. Anal. **23**, 173–190

Barenblatt, G.I. (1962): The mathematical theory of equilibrium cracks in brittle fracture. Adv. Appl. Mech. **7**, 55–129

Battiato, G., Ronca, G., Varga, C. (1977): "Moving Loads on a Viscoelastic Double Layer: Prediction of Recoverable and Permanent Deformations", in *Proceedings of the Fourth International Conference on Structural Design of Asphalt Pavements* (The University of Michigan, Ann Arbor, Michigan, USA) pp. 459–466

Battiato, G., Varga, G. (1982): "The AGIP Viscoelastic Method for Asphalt Pavement Design", in *Proceedings of the Fifth International Conference on the Structural Design of Asphalt Pavements* (The Study Centre for Road Construction, The Netherlands) pp. 59–66

Bazant, Z.P. (1975): Theory of creep and shrinkage in concrete structures: a précis of recent developments. Mech. Today **2**, 1–93

Bilby, B.A., Cottrell, A.H., Swinden, K.H. (1963): The spread of plastic yield from a notch. Proc. Roy. Soc. **A272**, 304–314

Blackburn, W.S. (1971): Steady crack growth in a linear viscoelastic material. Int. J. Fract. Mech. **7**, 354–356

Bland, D.R. (1960): *The Theory of Linear Viscoelasticity* (Pergamon, Oxford)

Boltzmann, L. (1874): Zur Theorie der elastischen Nachwirkungen. Sitzungsber. Kaiserlich. Akad. Wiss. (Wien) Math.-Naturwiss. Classe **70** (2), 275–306; (also Vol. 1, p. 167 of his Collected Papers (1909): *Wissenschaftliche Abhandlungen*, (ed. F. Hasenöhrl, Leipzig; reprinted by Chelsea, New York, 1968)

Boucher, M. (1975): "Signorini's Problem in Viscoelasticity", in *The Mechanics of the Contact Between Deformable Bodies*, ed. by A.D. de Pater, J.J. Kalker (Delft University Press) pp. 41–53

Bowie, O.L., Freese, C.E. (1976): On the "overlapping" problem in crack analysis. Eng. Fracture Mech. **8**, 373–379

Calvit, H.H. (1967a) Numerical solution of the problem of impact of a rigid sphere onto a linear viscoelastic half-space, and comparison with experiment. Int. J. Solids Struct. **3**, 951–966

Calvit, H.H. (1967b): Experiments on rebound of steel spheres from blocks of polymers. J. Mech. Phys. Solids **15**, 141–150

Cherepanov, G.P. (1979): *Mechanics of Brittle Fracture* (McGraw-Hill, New York)

Christensen, R.M. (1972): Restrictions upon viscoelastic relaxation functions and complex moduli. Trans. Soc. Rheol. **16**, 603–614

Christensen, R.M. (1979): A rate-dependent criterion for crack growth. Int. J. Fract. **15**, 3–21

Christensen, R.M. (1982): *Theory of Viscoelasticity: An Introduction*, 2nd ed. (Academic Press, New York)

Christensen, R.M., McCartney, L.N. (1983): Viscoelastic crack growth. Int. J. Fract. **23**. R11–R13

Christensen, R.M., Wu, E.M. (1981): A theory of crack growth in viscoelastic materials. Eng. Fract. Mech. **14**, 215–225

Chu, B.T. (1965): Response of various material media to high velocity loadings: I. Linear elastic and viscoelastic materials. J. Mech. Phys. Solids **13**, 165–187

Coleman, B.D., Noll, W. (1961): Foundations of linear viscoelasticity. Rev. Mod. Phys. **33**, 239–249

Comninou, M. (1976): Contact between viscoelastic bodies. J. Appl. Mech. **43**, 630–632

Comninou, M., Dundurs, J. (1979): On the frictional contact in crack analysis. Eng. Fracture Mech. **12**, 117–123

Crochet, M.J., Naghdi, P.M. (1969): A class of simple solids with fading memory. Int J. Eng. Sci. **7**, 1173–1198

Crochet, M.J., Naghdi, P.M. (1970): On "Thermorheologically Simple" Solids, in *Proc. IUTAM Symposium on Thermoelasticity* (Springer-Verlag, New York) pp. 59–86

Day, W.A. (1970): Some results on the least work needed to produce a given strain in a given time in a viscoelastic material and a uniqueness theorem for dynamic viscoelasticity. Q. J. Mech. Appl. Math. **23**, 469–479

Deresiewicz, H. (1968): A note on Hertz' theory of impact. Acta Mech. **6**, 110–112

Doi, M., Edwards, S.F. (1986): *The Theory of Polymer Dynamics* (Clarendon Press, Oxford)

Dugdale, D.S. (1960): Yielding of steel sheets containing slits. J. Mech. Phys. Solids **8**, 100–104

Dundurs, J. (1975): "Properties of Elastic Bodies in Contact", in *The Mechanics of the Contact between Deformable Bodies*, ed. by A.D. de Pater, J.J. Kalker (Delft University Press) pp. 54–66

Dundurs, J., Stippes, M. (1970): Role of elastic constants in certain contact problems. J. Appl. Mech. **37**, 965–970

Eason, G. (1965): The stresses produced in a semi-infinite solid by a moving surface force. Int. J. Eng. Sci. **2**, 581–609

Edelstein, W.S. (1969a): The cylinder problem in thermoviscoelasticity. Res. Natl. Bur. Standards **73B**, 31–40

Edelstein, W.S. (1969b): Ablation and thermal effects in a viscoelastic cylinder. Acta Mech. **8**, 174–182

Edelstein, W.S., Gurtin, M.E. (1964): Uniqueness theorems in the linear dynamic theory of anistropic viscoelastic solids. Arch. Rat. Mech. Anal. **17**, 47–60

Efimov, A.B. (1966a): Axisymmetric contact problem for linear viscoelastic bodies. Vest. Moskov. Univ., Ser. 1, Mat. Mek. **21**, 120–127

Efimov, A.B. (1966b): Some quasi-static problems for a viscoelastic half-space. Mekhanika Polimerov **2**, 392–402

Eirich, F.R. (1956): *Rheology Theory and Applications* (Academic, New York)

England, A.H. (1971): *Complex Variable Methods in Elasticity* (Wiley, London)

Erdélyi, A. (ed.) (1954): *Tables of Integral Transforms*, Vol. 2 (McGraw-Hill, New York)

Eringen, A.C. (1967): Linear theory of micropolar viscoelasticity. Int. J. Eng. Sci. **5**, 191–204

Ferry, J.D. (1970): *Viscoelastic Properties of Polymers*, 2nd ed. (Wiley, New York)

Fichera, G. (1965): *Lectures on Linear Elliptic Differential Systems and Eigenvalue Problems*, Lecture Notes in Mathematics (Springer-Verlag, Berlin, Heidelberg, New York)

Fichera, G. (1972): "Boundary Value Problems of Elasticity with Unilateral Constraints", in Encyclopedia of Physics, Vol VIa/2: *Mechanics of Solids II*, ed. by C. Truesdell (Springer-Verlag, Berlin, Heidelberg)

Flügge, W. (1967): *Viscoelasticity* (Blaisdell, Waltham)

Gakhov, F. D. (1966): *Boundary Value Problems* (Pergamon, Oxford)

Galin, L. A. (1961): *Contact Problems in the Theory of Elasticity*, ed. by I. N. Sneddon [English translation by H. Moss] (Department of Mathematics, North Carolina State University, Raleigh)

Gladwell, G. M. L. (1980): *Contact Problems in the Classical Theory of Elasticity* (Sijthoff and Noordhoff, Alphen aan den Rijn)

Goldberger, M., Watson, K. (1964): *Collision Theory* (Wiley, New York)

Golden, J. M. (1975): A molecular theory of adhesive rubber friction. J. Phys. A. **8**, 966–979

Golden, J. M. (1977): Hysteretic friction of a plane punch on a half-plane with arbitrary viscoelastic behaviour. Q. J. Mech. Appl. Math. **30**, 23–49

Golden, J. M. (1978): Hysteretic friction in the small velocity approximation. Wear **50**, 259–273

Golden, J. M. (1979a): The problem of a moving rigid punch on an unlubricated viscoelastic half-plane. Q. J. Mech. Appl. Math. **32**, 25–52

Golden, J. M. (1979b): The problem of a rigid punch moving on a viscoelastic half-plane with inertial effects approximately included. J. Austral. Math. Soc. **21**, (Series B), 198–229

Golden, J. M. (1980): Hyteresis and lubricated rubber friction. Wear **65**, 75–87

Golden, J. M. (1982a): Approximate analytic treatment of the problem of a moving ellipsoidal punch on a viscoelastic half-space. Q. J. Mech. Appl. Math. **35** 155–171

Golden, J. M. (1982b): The problem of a rigid punch sliding on an elastic or a viscoelastic layer. Acta Mech. **43**, 201–221

Golden, J. M. (1986a): Frictional viscoelastic contact problems. Q. J. Mech. Appl. Math. **39**, 125–137

Golden, J. M. (1986b): Causality and viscoelastic boundary value problems. Int. J. Eng. Sci. **24**, 1141–1149

Golden, J. M., Graham, G. A. C. (1984a): Crack in a viscoelastic field of pure bending. Int. J. Eng. Sci. **22**, 801–811

Golden, J. M., Graham, G. A. C. (1984b): Fatigue crack growth in viscoelastic materials. The Arabian J. Sci. Eng. **9**, 77–85

Golden, J. M., Graham, G. A. C. (1987a): The transient quasi-static plane viscoelastic moving load problem. Int. J. Eng. Sci. **25**, 65–84

Golden, J. M., Graham, G. A. C. (1987b): The Steady-state plane normal viscoelastic contact problem. Int. J. Eng. Sci. **25**, 277–291

Golden, J. M., Graham, G. A. C. (1987c): Energy Balance Criteria for Viscoelastic Fracture (submitted for publication)

Gradshteyn, I. S., Ryzhik, I. M. (1965): *Tables of Integrals, Series and Products* (Academic, New York)

Graham, G. A. C. (1965a): The contact problem in the linear theory of viscoelasticity. Int. J. Eng. Sci. **3**, 27–46

Graham, G. A. C. (1965b): On the use of stress functions for solving problems in linear viscoelasticity theory that involve moving boundaries. Proc. R. Soc. (Edin.) A **67**, 1–8

Graham, G. A. C. (1967): The contact problem in the linear theory of viscoelasticity where the time dependent contact area has any number of maxima and minima. Int. J. Eng. Sci. **5**, 495–514

Graham, G. A. C. (1968): The correspondence principle of linear viscoelasticity theory for mixed boundary value problems involving time-dependent boundary regions. Q. Appl. Math. **26**, 167–174

Graham, G. A. C. (1969): The solution of mixed boundary value problems that involve time-dependent boundary regions, for viscoelastic materials with one relaxation function. Acta Mech. **8**, 188–204

Graham, G. A. C. (1970): Two extending crack problems in linear viscoelasticity. Q. Appl. Math. **27**, 497–507

Graham, G. A. C. (1973): A contribution to Hertz' theory of elastic impact. Int. J. Eng. Sci. **11**, 409–413

Graham, G. A. C. (1974): Extension, torsion and flexure of aging viscoelastic beams that have two relaxation functions. Bull. Math. de la Soc. Sci. Math. de la R.S. de Roumanie **18**, 283–293

Graham, G. A. C. (1975): Quasi-static crack growth in linear viscoelastic bodies that are acted upon by alternating tensile and compressive loads. Proc. R. Irish Acad. **75A**, 263–269

Graham, G. A. C. (1976): Stresses and displacements in cracked linear viscoelastic bodies that are acted upon by alternating tensile and compressive loads. Int. J. Eng. Sci. **14**, 1135–1142

Graham, G. A. C. (1978a): On the extension of Hertz' theory of elastic impact to viscoelasticity. Utilitas Mathematica **13**, 117–137

Graham, G.A.C. (1978b): Fracture of linear aging viscoelastic material. Utilitas Mathematica **14**, 181–187

Graham, G.A.C. (1980): Viscoelastic contact problems with friction. Int. J. Eng. Sci. **18**, 191–196

Graham, G.A.C. (1982): Viscoelastic crack in a field of pure bending. Mech. Res. Commun. **9**, 219–226

Graham, G.A.C., Golden, J.M. (1987): The three-dimensional steady-state viscoelastic indentation problem. Int. J. Eng. Sci. **26**, 121–126

Graham, G.A.C., Golden, J.M. (1988): The generalized partial correspondence principle in linear viscoelasticity. Q. Appl. Math. (to appear)

Graham, G.A.C., Sabin, G.C.W. (1973): The correspondence principle of linear viscoelasticity for problems that involve time-dependent regions. Int. J. Eng. Sci. **11**, 123–140

Graham, G.A.C., Sabin, G.C.W. (1978): The opening and closing of a growing crack in a linear viscoelastic body that is subject to alternating tensile and compressive loads. Int. J. Fracture **14**, 639–649

Graham, G.A.C., Sabin, G.C.W. (1981): Steady state solutions for a cracked standard linear viscoelastic body. Mech. Res. Commun. **8**, 361–368

Graham, G.A.C., Sneddon, I.N. (1981): The axisymmetric viscoelastic contact problem with rotational friction. J. Elasticity **11**, 33–42

Graham, G.A.C., Williams, F.M. (1972): Boundary value problems for time-dependent regions in aging viscoelasticity. Utilitas Mathematica **2**, 291–303

Green, A.E., Zerna, W. (1968): *Theoretical Elasticity* (Oxford University Press)

Griffith, A.A. (1921): The phenomena of rupture and flow in solids. Philos. Trans. R. Soc. A, **221**, 163–198

Griffith, A.A. (1925): "The Theory of Rupture", in *Proc. 1st Int. Congress Appl. Mech.* Delft, 1924, pp. 55–63

Gross, G. (1953): *Mathematical Structures of the Theories of Viscoelasticity* (Hermann & Cie, Paris)

Gurtin, M.E. (1972): "The Linear Theory of Elasticity", in Encyclopedia of Physics, Vol. VIa/2: *Mechanics of Solids II*, ed. by C. Truesdell (Springer-Verlag, Berlin, Heidelberg)

Gurtin, M.E. (1979): Thermodynamics and the cohesive zone in fracture. J. Appl. Math. Phys. (ZAMP) **30**, 991–1003

Gurtin, M.E., Sternberg, E. (1962): On the linear theory of viscoelasticity. Arch. Rat. Mech. Anal. **2**, 291–356

Harvey, R.B. (1975): On the deformation of a viscoelastic cylinder, rolling without slipping. Q. J. Mech. Appl. Math. **28**, 1–24

Hunter, S.C. (1957): Energy absorbed by elastic waves produced by Hertzian impact. J. Mech. Phys. Solids **5**, 162–171

Hunter, S.C. (1960a): "Viscoelastic Waves", *Progress in Solid Mechanics 1*, ed. by R. Hill, I.N. Sneddon (North-Holland, Amsterdam)

Hunter, S.C. (1960b): The Hertz problem for a rigid spherical indentor and a viscoelastic half-space. J. Mech. Phys. Solids **8**, 219–234

Hunter, S.C. (1961): The rolling contact of a rigid cylinder with a viscoelastic half-space. J. Appl. Mech. **28**, 611–617

Hunter, S.C. (1966): "The Solution of Boundary Value Problems in Linear Viscoelasticity", in *Mechanics and Physics of Solid Propellants*, Proceedings of the Fourth Symposium on Naval Structural Mechanics, ed. by A.C. Eringen, H. Liebowitz, S.L. Koh, J.M. Crowley (Pergamon, Oxford) pp. 257–295

Hunter, S.C. (1967): The transient temperature distribution in a semi-infinite viscoelastic rod, subject to longitudinal oscillations. Int. J. Eng. Sci. **5**, 119–143

Hunter, S.C. (1968): "The motion of a rigid sphere embedded in an adhering elastic or viscoelastic medium", *Proceedings of the Edinburgh Mathematical Society 16* (Series II), Part I, pp. 55–69

Hunter, S.C. (1983): *Mechanics of Continuous Media*, 2nd edition (Wiley, New York)

Il'iushin, A.A., Pobedria, B.E. (1970): *The Foundations of the Mathematical Theory of Thermoviscoelasticity* [in Russian] (Nauka, Moscow)

Irwin, G.R. (1948): "Fracture Dynamics", in *Fracturing of Metals* (American Society of Metals, Cleveland) pp. 147–166

Irwin, G.R. (1960): Fracture Mechanics, *Structural Mechanics*: Proceedings of the First Symposium on Naval Structural Mechanics (1958), ed. by J.N. Goodier, N. Hoff (Pergamon, Oxford)

Kachanov. L.M. (1961): On the kinetics of crack growth. J. Appl. Math. Mech. **25**, 739–745
Kalker, J.J. (1975): "Aspects of Contact Mechanics", in *The Mechanics of the Contact Between Deformable Bodies*, ed. by A.D. de Pater, J.J. Kalker (Delft University Press) pp. 1–25
Kalker, J.J. (1977): A survey of the mechanics of contact between solid bodies. J. Appl. Math. Phys. (ZAMP) **57**, 13–17
Kanninen, M.F., Popelar, C.H. (1985): *Advanced Fracture Mechanics* (Oxford University Press)
Knauss, W.G. (1970a): Delayed failure – the Griffith problem for linearly viscoelastic materials. Int. J. Fracture Mech. **6**, 7–20
Knauss, W.G. (1970b): An observation of crack propagation in anti-plane shear. Int. J. Fracture Mech. **6**, 183–187
Knauss, W.G. (1973): The mechanics of polymer fracture. Appl. Mech. Rev. **26**, 1–17
Knauss, W.G. (1974): "On the Steady Propagation of a Crack in a Viscoelastic Sheet: Experiments and Analysis", in *Deformation and Fracture of High Polymers*, ed. by H.H. Kausch, J.A. Hassell, R.I. Jaffee (Plenum, New York) 501–541
Knauss, W.G., Dietmann, H. (1970): Crack propagation under variable load histories in linearly viscoelastic solids. Int. J. Eng. Sci. **8**, 643–656
Koeller, R.C. (1984): Application of fractional calculus to the theory of viscoelasticity. J. Appl. Mech. **51**, 299–307
Kolsky, H. (1956): The propagation of stress pulses in viscoelastic solids. Philos. Mag. Ser. 8, **1**, 693–710
Korn, G.A., Korn, T.M. (1968): *Mathematical Handbook for Scientists and Engineers* (McGraw-Hill, New York)
Kostrov, B.V., Nikitin, L.V. (1970): Some general problems of mechanics of brittle fracture, Arch. Mech. Stosow. **6**, 749–776
Kramers, H.A. (1927): La diffusion de la lumiere par les atomes, Atti Congr. Int. Fis., Como **2**, 545–557
Kronig, R.L. (1926): On the theory of dispersion of X-rays, J. Opt. Soc. Am. **12**, 547–557

Landau, L.D., Lifshitz, E.M. (1959): *Theory of Elasticity* (Pergamon, London)
Lardner, R.W. (1968): A dislocation model of fatigue crack growth in metals. Philos. Mag. **17**, 71–82
Lardner, R.W. (1974): *Mathematical Theory of Dislocations and Fracture* (University of Toronto Press, Toronto)
Lee, E.H. (1955): Stress analysis in viscoelastic bodies. Quart. Appl. Math. **13**, 183–190
Lee, E.H. (1960): "Viscoelastic Stress Analysis", in *Structural Mechanics*, Proceedings of the First Symposium on Naval Structure Mechanics, ed. by J.N. Goodier, N.J. Hoff (Pergamon, Oxford) pp. 456–482
Lee, E.H. (1966): "Some Recent Developments in Linear Viscoelastic Stress Analysis", in *Proceedings of the Eleventh International Congress of Applied Mechanics*, ed. by H. Gortler (Springer-Verlag, Berlin) pp. 396–402
Lee, E.H., Radok, J.R.M. (1960): The contact problem for viscoelastic bodies. J. Appl. Mech. **27**, 438–444
Leitman, M.J., Fisher, G.M.C. (1973): "The Linear Theory of Viscoelasticity", in Encyclopedia of Physics, Vol. VIa/3: *Mechanics of Solids III*, ed. by C. Truesdell (Springer-Verlag, Berlin, Heidelberg)
Liebowitz, H., (ed.) (1968): *Fracture: An Advanced Treatise*, Vol. II (Academic, New York)
Lifshitz, J.M., Kolsky, H. (1964): Some experiments on anelastic rebound. J. Mech. Phys. Solids **12**, 35–43
Ling, F.F. (1973): *Surface Mechanics* (Wiley, New York)
Lockett, F.J. (1972): *Nonlinear Viscoelastic Solids* (Academic, London)
Lockett, F,J., Morland, L.W. (1967): Thermal stresses in a viscoelastic thin-walled tube with temperature dependent properties. Int. J. Eng. Sci. **5**, 879–898
Love, A.E.H. (1934): *A Treatise on the Mathematical Theory of Elasticity*, 4th ed. (Cambridge University Press)
Lovitt, W.V. (1950): *Linear Integral Equations* (Dover, New York)

Lubliner, J., Sackman, J. L. (1967): On uniqueness in general linear viscoelasticity. Q. Appl. Math. **25**, 129–138

Lur'e, A. I. (1964): *Three-Dimensional Problems of the Theory of Elasticity*, ed. by J. R. M. Radok (Interscience, New York)

Majidzadeh, K., Buranarom, C., Karakouzian, M (1976): *Application of Fracture Mechanics for Improved Design of Bituminous Concrete*, Vol. 1, Report No. FHWA-RD-76-91, Federal Highway Administration, Office of Research and Development, Washington, D. C.

Margetson, J. (1971): Rolling contact of a smooth viscoelastic strip between rotating rigid cylinders. Int. J. Mech. Sci. **13**, 207–215

Margetson, J. (1972): Rolling contact of a rigid cylinder over a smooth elastic or viscoelastic layer. Acta Mech. **13**, 1–9

Martincek, G. (1979): Dynamic response of a viscoelastic half-space. Acta Tech. Čsav. No. 4, 420–438

Markovitz, H. (1977): Boltzmann and the beginnings of linear viscoelasticity. Trans. Soc. Rheol. **23**, 381–398

Maxwell, J. C. (1867): On the dynamical theory of gases. Philos. Trans. R. Soc. **157**, 49–88 (Scientific Papers (1890) **2**, pp. 26–78

McCartney, L. N. (1977): Crack propagation, resulting from a monotonic increasing applied stress in a linear viscoelastic material. Int. J. Fract. **13**, 641–654

McCartney, L. N. (1978): Crack propagation in linear viscoelastic solids: some new results. Int. J. Fract. **14**, 547–554

McCartney, L. N. (1979): Crack growth laws for a variety of viscoelastic solids using energy and COD fracture criteria. Int. J. Fract. **15**, 31–40

McCartney, L. N. (1987): Mechanics of matrix-cracking in brittle-matrix fibre-reinforced composites. Proc. R. Soc. Lond. **A 409**, 329–350

McHenry, D. (1943): A new aspect of creep in concrete and its application to design. Proc. American Society for Testing Materials **43**, 1069–1086

Meyer, O. E. (1874): Theorie der Elastischen Nachwirkung. Ann. Phys. Chem. (Pogendorff) **151**, 108–118

Michell, J. H. (1899): On the direct determination of stress in an elastic solid, with application to the Theory of Plates. Proc. Lond. Math. Soc. **31**, 100–124

Misicu, M. (1963, 1964): Theory of viscoelasticity with couple stresses and some reductions to two-dimensional problems I. Revue de Mécanique Appliquée, Acad. République Populaire Roumaine **8** (6), 921–952; also Pt. II of this work in the same journal, **9** (1), 3–35

Morland, L. W. (1962): A plane problem of rolling contact in linear viscoelasticity theory. J. Appl. Mech. **29**, 345–352

Morland, L. W. (1963): Dynamic stress analysis for a viscoelastic half-plane subject to moving surface tractions. Proc. London Math. Soc. **13**, 471–492

Morland, L. W. (1967): Exact solutions for rolling contact between viscoelastic cylinders. Q. J. Mech. Appl. Math. **20**, 73–106

Morland, L. W. (1968): Rolling contact between dissimilar viscoelastic cylinders. Q. Appl. Math. **25**, 363–376

Morland, L. W., Lee, E. H. (1960): Stress analysis for linear viscoelastic materials with temperature variation. Trans. Soc. Rheol. **4**, 233–263

Morse, P. M., Feshbach, H. (1953): *Methods of Theoretical Physics*, Vol. I, II (McGraw-Hill, New York)

Mueller, H. K., Knauss, W. G. (1971): Crack propagation in a linearly viscoelastic strip. J. Appl. Mech. **38**, 483–488

Muki, R., Sternberg, E. (1961): On transient thermal stresses in viscoelastic materials with temperature dependent properties. J. Appl. Mech. **28**, 193–207

Muskhelishvili, N. I. (1953): *Singular Integral Equations*, 2nd ed. (Noordhoff, Groningen) [English translation by J. R. M. Radok]

Muskhelishvili, N. I. (1963): *Some Basic Problems of the Mathematical Theory of Elasticity*, 4th ed. (Noordhoff, Groningen) [English translation by J. R. M. Radok]

Nachman, A., Walton, J. R. (1978): The sliding of a rigid indentor over a power law viscoelastic layer. J. Appl. Mech. **45**, 111–113

Nowacki, W. (1965): *Théorie du Fluage* (Editions Eyrolles, Paris) [French translation by I. Kozniewska and A. Brandt]

Nuismer, R. J. (1974): On the governing equation for quasi-static crack growth in linearly viscoelastic materials. J. Appl. Mech. **41**, 631 – 634

Nutting, P. G. (1921): A new general law of deformation. J. Franklin Inst. **191**, 679 – 685

Onat, E. T., Breuer, S. (1963): "On Uniqueness in Linear Viscoelasticity", in *Progress in Applied Mechanics*, (The Prager Anniversary Volume), ed. by D. C. Drucker (McMillan, New York) pp. 349 – 353

Orowan, E. (1950): *Fatigue and Fracture in Metals* (Massachusetts Institute of Technology Symposium) (John Wiley, New York)

Pao, Y. H. (1955): Extension of the Hertz theory of impact to the viscoelastic case. J. Appl. Phys. **26**, 1083 – 1088

Pipkin, A. C. (1972): *Lectures on Viscoelasticity Theory* (Springer, New York)

Pogorzelski, W. (1966): *Integral Equations and Their Applications* (Pergamon, Oxford)

Popelar, C. H., Atkinson, C. (1980): Dynamic crack propagation in a viscoelastic strip. J. Mech. Phys. Solids **28**, 79 – 93

Predeleanu, M. (1965): Stress analysis in bodies with time-dependent properties. Bull. Math. Soc. Sci. Math. R.S. de Roumanie **9**, 115 – 127

Prokopovici, E. Y. (1956): On the plane contact problem in viscoelasticity theory. Prikl. Mat. Mekh. **20**, 680 – 687 [in Russian]

Rabotnov, Y. N. (1948): Equilibrium of elastic medium with after-effect. Prikl. Mat. Mekh. **12**, 53 – 62

Rabotnov, Y. N. (1969): *Creep Problems in Structural Members* (North Holland, Amsterdam)

Rabotnov, Y. N. (1980): *Elements of Hereditary Solid Mechanics* (Mir Publishers, Moscow)

Rapp, N. V., Romas'ko, V. S (1972): Analytical form of the dynamic functions of the theory of linear viscoelasticity. Sov. Phys. Dokl. **16**, 784 – 786

Read, W. T. (1950): Stress analysis for compressible viscoelastic materials. J. Appl. Phys. **21**, 671 – 674

Rektorys, K. (ed.) (1969): *Survey of Applicable Mathematics* (M.I.T. Press, Cambridge, Massachusetts)

Rogers, T. G. (1965): "Viscoelastic Stress Analysis", in *Proceedings of the Princeton University Conference on Solid Mechanics* (Princeton, New Jersey) pp. 49 – 74

Rongved, L. (1954): Residual stress in glass spheres. Technical Report No. 16 Contract 266(09), (Department of Civil Engineering and Engineering Mechanics, Columbia University)

Rouse, P. E. Jr. (1953): A theory of the linear viscoelastic properties of dilute solutions of coiling polymers. J. Chem. Phys. **21**, 1272 – 1280

Sabin, G. C. W. (1975): *"Some Dynamic Mixed Boundary Value Problems in Linear Viscoelasticity"*; Ph. D. Thesis, University of Windsor, Ontario

Sabin, G. C. W. (1987): The impact of a rigid axisymmetric indentor on a viscoelastic half-space. Int. J. Eng. Sci. **25**, 235 – 251

Sabin, G. C. W., Graham, G. A. C. (1980): The normal aging viscoelastic contact problem. Int. J. Eng. Sci. **18**, 751 – 757

Sabin, G. C. W. (1983): Efficient numerical solution of the viscoelastic impact problem. Utilities Mathematica **23**, 323 – 346

Scaife, B. K. P. (1971): *Complex Permittivity* (The English Universities Press)

Schapery, R. A. (1974): Viscoelastic Behaviour and Analysis of Composite Materials, *Mechanics of Composite Materials*, Vol. 2, ed. by G. P. Sendeckyj (Academic, New York) pp. 85 – 168

Schapery, R. A. (1975): A theory of crack initiation and growth in viscoelastic media I, II and III: Int. J. Fract. **11**, 141 – 159, 369 – 388, 549 – 562

Schapery, R. A. (1978): A method for predicting crack growth in nonhomogeneous viscoelastic media. Int. J. Fract. **14**, 293 – 309

Schapery, R. A. (1979): "On the Analysis of Crack Initiation and Growth in Nonhomogeneous Viscoelastic Media", in *Fracture Mechanics*, Proceedings of the Symposium in Applied Mathematics of the A.M.S. and S.I.A.M., Vol. XII, ed. by R. Burridge (American Mathematical Society, Providence) pp. 137 – 152

Signorini, A. (1959): Questioni die elasticita non linearizzata a semilinearizzata. Rend. die Matem. e delle sue appl. **18**, 95 – 139

Sih, G. C., Liebowitz, H. (1968): "Mathematical Theories of Brittle Fracture", in *Fracture: An Advanced Treaties*, Vol. II, ed. by H. Liebowitz (Academic, New York) pp. 67 – 190

Sneddon, I. N. (1951): *Fourier Transforms* (McGraw-Hill, New York)

Sneddon, I. N. (1965): The relation between load and penetration in the axisymmetric Boussinesq problem for a punch of arbitrary profile. Int. J. Eng. Sci. **3**, 47 – 57

Sneddon, I. N. (1972): *The Use of Integral Transforms* (McGraw-Hill, New York)

Sneddon, I. N., Lowengrub, M. (1969): *Crack Problems in the Classical Theory of Elasticity* (Wiley, New York)

Sokolnikoff, I. S. (1956): *Mathematical Theory of Elasticity*, 2nd ed. (McGraw-Hill, New York)

Staverman, A. J., Schwarzl, F. (1956): "Linear Deformation Behaviour of High Polymers", in *Die Physik der Hochpolymeren*, Vol. IV, ed. by H. A. Stuart (Springer, Berlin)

Sternberg, E. (1964): "On the Analysis of Thermal Stresses in Viscoelastic Solids", in *High Temperature Structures and Materials*, Proceedings of the Third Symposium on Naval Structural Mechanics, ed. by A. M. Freudenthal, B. A. Boley, H. Liebowitz (Pergamon, Oxford) pp. 348 – 382

Sternberg, E., Al Khozaie, S. (1964): On Green's functions and Saint-Venant's Principle in the Linear Theory of Viscoelasticity. Arch. Rat. Mech. Anal. **15**, 112 – 146

Sternberg, E., Gurtin, M. E. (1963): "Uniqueness in the Theory of Thermorheologically Simple Ablating Viscoelastic Solids", in *Progress in Applied Mechanics*, (The Prager Anniversary Volume) ed. by D. C. Drucker (MacMillan, New York) pp. 373 – 384

Sternberg, E., Gurtin, M. E. (1964): " Further Study of Thermal Stresses in Viscoelastic Materials with Temperature Dependent Properties", in *Proc. IUTAM Symposium on Second Order Effects in Elasticity, Plasticity and Fluid Mechanics*, Haifa, pp. 51 – 76

Stouffer, D. C. (1972): A thermal hereditary theory of linear viscoelasticity. J. Appl. Math. Phys. (ZAMP) **23**, 845 – 851

Stouffer, D. C., Wineman, A. S. (1971): Linear viscoelastic materials with environmental-dependent properties. Int. J. Eng. Sci. **9** 193 – 212

Stouffer, D. C., Wineman, A. S. (1972): A constitutive representation for linear aging environmental-dependent viscoelastic materials. Acta Mech. **13**, 31 – 53

Strutt, J. W. (Lord Rayleigh) (1906): On the production of vibrations by forces of relatively long duration, with application to the theory of collisions. Philos. Mag. **11**, 283 – 291; (reprinted in *Scientific Papers*, Vol. 5, Dover, 1964, pp. 292 – 299)

Tabor, D. (1952): The mechanism of rolling friction. Philos. Mag. **7**, 1055 – 1059

Tao, L. N. (1966): The associated elastic problems in dynamic viscoelasticity. Q. Appl. Math. **21**, 215 – 222

Ting, T. C. T. (1966): The contact stresses between a rigid indentor and a viscoelastic half-space. J. Appl. Mech. **33**, 845 – 854

Ting, T. C. T. (1968): Contact problems in the linear theory of viscoelasticity. J. Appl. Mech. **35**, 248 – 254

Ting, T. C. T. (1969): "A Mixed Boundary Value Problem in Viscoelasticity with Time-Dependent Boundary Regions", in *Proceedings of the Eleventh Midwestern Mechanics Conference*, ed. by H. J. Weiss, D. F. Young, W. F. Riley, T. R. Rogge (Iowa University Press) pp. 591 – 598

Titchmarsh, E. C. (1937): *Introduction to the Theory of Fourier Integrals* (Oxford University Press, Oxford)

Torvik, P. J, Bagley, R. L. (1984): On the appearance of the fractional derivative in the behavior of real materials. J. Appl. Mech. **51**, 294 – 298

Tricomi, F. G. (1957): *Integral Equations* (Interscience, New York)

Tsai, Y. M. (1968): A note on surface waves produced by Hertzian impact. J. Mech. Phys. Solids **16**, 133 – 136

Tsai, Y. M. (1971): Dynamic contact-stresses produced by the impact of an axisymmetrical projectile on an elastic half-space. Int. J. Solids Struct. **7**, 543 – 558

Volterra, V. (1909): Sulla equazioni integro-differenziali della teoria dell'elasticità. R. Acc. dei Lincei. Rend. **18** (2), 295

Volterra, V. (1959): *Theory of Functionals and of Integral and Integro-Differential Equations* (Dover, New York)

Walton, J.R. (1982): On the steady-state propagation of an anti-plane shear crack in an infinite general linearly viscoelastic body. Q. Appl. Math. **40**, 37–52

Walton, J.R. (1984): The sliding with Coulomb friction of a rigid indentor over a power-law inhomogeneous linearly viscoelastic half-plane. J. Appl. Mech. **51**, 289–293

Walton, J.R., Nachman, A., Schapery, R.A. (1978): The sliding of a rigid indentor over a power-law viscoelastic half-space. Q. J. Mech. Appl. Math. **31**, 296–321

Weber, W. (1835): Über die Elastizität der Seidenfäden. Ann. Phys. Chem. (Poggendorff) **34**, 247–257

Weber, W. (1841): Über die Elastizität fester Körper. Ann. Phys. Chem. (Poggendroff) **54**, 1–18

Williams, F.M. (1975a): "The Deformation of Viscoelastic Materials with Environment-Dependent Properties", Ph. D. Thesis (Department of Mathematics, Simon Fraser University)

Williams, F.M. (1975b): The spherical problem for a linear viscoelastic material with environment-dependent properties. Int. J. Eng. Sci. **13**, 993–1002

Williams, F.M. (1976): Time dependent deflections of nonhomogeneous ice plates. Acta Mech. **25**, 29–44

Williams, M.L. (1965): Initiation and growth of viscoelastic fracture. Int. J. Fract. Mech. **1**, 292–310

Williams, M.L. (1967): Fatigue-fracture growth in linearly viscoelastic materials. J. Appl. Phys. **38**, 4476–4480

Williams, M.L., Landel, R.F., Ferry, J.D (1955): The temperature dependence of relaxation mechanisms in amorphous polymers and other glass-forming liquids. J. Am. Chem. Soc. **77**, 3701–3707

Willis, J.R. (1967a): A comparison of the failure criteria of Griffith and Barenblatt. J. Mech. Phys. Solids **15**, 151–162

Willis, J.R. (1967b): Crack propagation in viscoelastic media. J. Mech. Phys. Solids **15**, 229–240

Wnuk, M.P. (1971): Subcritical growth of fracture (inelastic fatigue). Int. J. Fract. Mech. **7**, 383–407

Wnuk, M.P. (1973a): Slow growth of cracks in a rate sensitive Tresca solid. Eng. Fract. Mech. **5**, 605–626

Wnuk, M.P. (1973b): Prior to failure extension of flaws under monotonic and pulsating loadings. Eng. Fract. Mech. **5**, 379–396

Zener, C. (1948): *Elasticity and Anelasticity of Metals* (University of Chicago Press, Chicago)

Subject Index